Student Solutions Manual

Phillip E. Bedient • Richard E. Bedient

Eighth Edition

Elementary Differential Equations

Earl D. Rainville
Late Professor of Mathematics
University of Michigan

Phillip E. Bedient
Professor Emeritus of Mathematics
Franklin and Marshall College

Richard E. Bedient
Professor of Mathematics
Hamilton College

PRENTICE HALL, UPPER SADDLE RIVER, NJ 07458

Assistant Editor: *Audra Walsh*
Production Editor: *Carole Suraci*
Special Projects Manager: *Barbara A. Murray*
Supplement Cover Manager: *Paul Gourhan*
Manufacturing Buyer: *Alan Fischer*

Printed in the United States of America

10 9 8 7 6 5 4 3 2

ISBN 0-13-592783-8

Prentice-Hall International (UK) Limited,London
Prentice-Hall of Australia Pty. Limited, Sydney
Prentice-Hall Canada Inc., Toronto
Prentice-Hall Hispanoamericana, S.A., Mexico
Prentice-Hall of India Private Limited, New Delhi
Prentice-Hall of Japan, Inc., Tokyo
Pearson Education Asia Pte. Ltd., Singapore
Editora Prentice-Hall do Brasil, Ltda., Rio de Janeiro

Contents

CONTENTS

Chapter 1

Definitions; Families of Curves

1.2   Definitions

All answers in this section are determined by inspection.

1. The equation is ordinary, linear in x, and of order 2.

3. The equation is ordinary, nonlinear, and of order 1.

5. The equation is ordinary, linear in y, and of order 3.

7. The equation is partial, linear in u, and of order 2.

9. The equation is ordinary, linear in x or y, and of order 2.

11. The equation is ordinary, linear in y, and of order 1.

13. The equation is ordinary, nonlinear, and of order 3.

15. The equation is ordinary, linear in y, and of order 2.

1.3   Families of Solutions

1. Rewriting the equation yields $y = \int x^3 + 2x\,dx + c$. Integrating, we have $y = \frac{1}{4}x^4 + x^2 + c$.

3. Rewriting the equation yields $y = 4\int \cos 6x\,dx + c$. Integrating, we have $y = \frac{2}{3}\sin 6x + c$.

5. Rewriting the equation yields $y = 2\int \dfrac{1}{x^2 + 2^2}\,dx + c$. Integrating, we have $y = \arctan(x/2) + c$.

7. Rewriting the equation yields $y = 3\int e^x\,dx$. Integrating, we have $y = 3e^x + c$. Substituting the initial conditions gives $6 = 3 + c$ or $c = 3$ so $y = 3e^x + 3$.

9. As in Example 1.2, $y = ce^{4x}$. Substituting the initial conditions gives $3 = ce^0 = c$ so $y = 3e^{4x}$.

11. Rewriting the equation yields $y = 4\int \sin 2x\,dx$. Integrating, we have $-2\cos 2x + c$. Substituting the initial conditions gives $2 = -2\cos \pi + c = 2 + c$ or $c = 0$ so $y = -2\cos 2x$.

Chapter 2

Equations of Order One

2.1   Separation of Variables

1. The variables may be separated to give

$$\int \frac{dr}{r} = -4 \int t \, dt.$$

Integrating both sides we have

$$\ln r = -2t^2 + c \text{ or } r = e^{-2t^2 + c} \text{ or } r = e^{-2t^2} e^c.$$

To simplify this we replacing the constant e^c by another constant. We could use a different name like k to get, $r = ke^{-2t^2}$, but the convention is to reuse the name c to obtain

$$r = ce^{-2t^2} \text{ or } r_0 = c \text{ or } r = r_0 \exp\left(-2t^2\right).$$

3. The variables may be separated to give

$$\int \frac{y}{1 + y^2} \, dy = \int \frac{dx}{x}.$$

Integrating both sides we have

$$\tfrac{1}{2} \ln\left(1 + y^2\right) = \ln x + c \text{ or } \ln\left(1 + y^2\right)^{1/2} = \ln x + c.$$

Exponentiating both sides yields

$$(1 + y^2)^{1/2} = e^{(\ln x + c)} \text{ or } (1 + y^2)^{1/2} = e^{\ln x} e^c.$$

As in Exercise 1 replacing e^c with c;

$$(1 + y^2)^{1/2} = cx \text{ or } y^2 = cx^2 - 1 \text{ or } y = \sqrt{cx^2 - 1}.$$

From the given conditions, $3 = \sqrt{4c - 1}$, $9 = 4c - 1$, or $c = \frac{10}{4}$. Thus

$$y = \tfrac{1}{2}\sqrt{10x^2 - 4}.$$

5. Separating the variables we have,

$$2 \int \frac{dx}{x} = 3 \int \frac{dy}{y}.$$

Integrating both sides yields, $2 \ln x + c = 3 \ln y$ or $y^3 = cx^2$. Finally, we substitute the initial conditions to obtain, $1 = 4c$, or $c = \frac{1}{4}$. Thus $y = (x/2)^{2/3}$.

7. Separating the variables we have,

$$\int e^{-y} \, dy = \int x e^{-x^2} \, dx.$$

Integrating both sides yields, $-e^{-y} = -\frac{1}{2} e^{-x^2} + c$. Finally, we substitute the initial conditions to obtain, $-1 = -\frac{1}{2} = c$ or $c = -\frac{1}{2}$. Thus $2e^{-y} = e^{-x^2} + 1$ or $\ln 2 - y = \ln \left(e^{-x^2} + 1 \right)$ or $y = \ln 2 - \ln \left[1 + \exp \left(-x^2 \right) \right]$.

9. Separating the variables we have,

$$2a^2 \int r^{-3} \, dr - \int r^{-1} \, dr = \int \sin \theta \, d\theta.$$

Integrating both sides yields, $-a^2 r^{-2} - \ln r = -\cos \theta + c$. Finally, we substitute the initial conditions to obtain, $-a^2 a^{-2} - \ln a = -\cos 0 + c$ or $-\ln a = c$. Thus $-a^2 r^{-2} - \ln r = -\cos \theta - \ln a$ or $r^2 (\ln r - \ln a) = r^2 \cos \theta - a^2$ or $r^2 \ln (r/a) = r^2 \cos \theta - a^2$.

11. Separating the variables we have,

$$\int \frac{dy}{y^2} = \int \frac{dx}{1-x}.$$

Integrating both sides yields, $-y^{-1} = -\ln (1-x) + c$. Replacing c by $\ln c$, $y[\ln |1-c| + \ln c] = 1$. Thus $y \ln |c(1-x)| = 1$.

13. Separating the variables we have,

$$\int \frac{dy}{y^3} = -\int \frac{x \, dx}{e^{x^2}}.$$

Integrating both sides yields, $-\frac{1}{2} y^{-2} = \frac{1}{2} e^{-x^2} + c$. Replacing $-2c$ by c leaves $e^{-x^2} + y^{-2} = c$.

15. Separating the variables we have,

$$\int \frac{m \, dx}{x} = \int \frac{n \, dy}{y}.$$

Integrating both sides yields, $m \ln x = n \ln y + c$ or $x^m = cy^n$.

17. Separating the variables we have,

$$\int \frac{dV}{V} = -\int \frac{dP}{P}.$$

Integrating both sides yields, $\ln V = -\ln P + c$ or $\ln (PV) = c$ or $PV = c$.

19. Separating the variables we have,

$$\int \frac{dr}{r} = \int \frac{b\sin\theta\,d\theta}{1 - b\cos\theta}.$$

Integrating both sides yields, $\ln|r| = \ln|1 - b\cos\theta| + c$ or $r = c(1 - b\cos\theta)$.

21. Separating the variables we have,

$$\int y\,dy = -\int \frac{x^2}{x-1}\,dx = -\int \left((x+1) + \frac{1}{x-1}\right)\,dx.$$

Integrating both sides yields, $\frac{1}{2}y^2 = -(x+1)^2 - \ln(x-1) + c$ or $(x+1)^2 + y^2 + 2\ln|c(x-1)| = 0$.

23. Separating the variables we have,

$$\int \frac{y+1}{y^3}\,dy = -\int xe^x\,dx.$$

Integrating by parts yields, $\int \left(\frac{1}{y^2} + \frac{1}{y^3}\right)\,dy = -xe^x + \int e^x\,dx.$

Thus $-\dfrac{1}{y} - \dfrac{1}{2y^2} = -xe^x + e^x + c$ or $e^x(x-1) = (2y+1)/(2y^2) + c$.

25. Separating the variables we have,

$$\int ye^{-y}\,dy = \int x^{-2}\,dx.$$

Integrating both sides yields, $-ye^{-y} + \int e^{-y}\,dy = -x^{-1} + c$. Thus $-ye^{-y} - e^{-y} = -x^{-1} + c$ or $x(y+1) = (1 + cx)e^y$.

27. Separating the variables we have,

$$\int \sec y\,dy = \int \cos^2 x\,dx.$$

Integrating both sides yields, $\ln|\sec y + \tan y| = \frac{1}{2}x + \frac{1}{4}\sin 2x + c$.
Thus $4\ln|\sec y + \tan y| = 2x + \sin 2x + c$.

29. Separating the variables we have,

$$\int \cos^2 x\,dx = \int t(1 + t^2)\,dt.$$

Integrating both sides yields, $\frac{1}{2}x + \frac{1}{4}\sin 2x = \frac{1}{4}(1 + t^2)^2 + c$. Thus $2x + \sin 2x = c + (1 + t^2)^2$.

31. Separating the variables we have,

$$\int \frac{1 + 3\alpha}{\alpha}\, d\alpha = -\int \frac{1 + \beta}{\beta}\, d\beta$$

or

$$\int \left(\frac{1}{\alpha} + 3\right) d\alpha = -\int \left(\frac{1}{\beta} + 1\right) d\beta.$$

Integrating both sides yields, $\ln \alpha + 3\alpha = -\ln \beta - \beta + c$ or $\ln \alpha\beta = -3\alpha - \beta + c$. Thus $c\alpha\beta = \exp(-3\alpha - \beta)$.

33. Separating the variables we have,

$$\int \frac{x\, dx}{\sqrt{a^2 - x^2}} = \int dy.$$

Integrating both sides yields, $-\sqrt{a^2 - x^2} = y + c$. Thus $y - c = -\sqrt{a^2 - x^2}$, the lower half of the circle $x^2 + (y - c)^2 = a^2$.

35. Separating the variables we have,

$$\int \frac{a^2\, dx}{a\sqrt{x^2 - a^2}} = \int dy.$$

Integrating both sides yields, $a \arcsec(x/a) = y + c$. Thus $x = a \sec \dfrac{y + c}{a}$.

37. Separating the variables we have,

$$\int \frac{x}{x^2 + 1}\, dx = \int \frac{y^2 + 1}{y + 1}\, dy = \int \left(y - 1 + \frac{2}{y + 1}\right) dy.$$

Integrating both sides yields, $\frac{1}{2}\ln(x^2 + 1) = \frac{1}{2}y^2 - y + 2\ln|y + 1| + c$.
Thus $\ln(x^2 + 1) = y^2 - 2y + 4\ln|c(y + 1)|$.

2.2 Homogeneous Functions

All functions are homogeneous except those of Exercises 2, 5, 6, and 19 by examination.

2.3 Equations with Homogeneous Coefficients

1. Substituting $y = xv$, we have $3(3x^2 + v^2x^2)\, dx - 2x^2v(v\, dx + x\, dv) = 0$ or $(9 + v^2)\, dx = 2xv\, dv$. We first separate the variables to obtain

$$\int \frac{dx}{x} = 2\int \frac{v\, dv}{9 + v^2}.$$

Integrating yields $\ln x = \ln(9 + v^2) + c$. Replacing c by $\ln c$ and substituting for v leaves $x = c[9 + (y/x)^2]$ or $x^3 = c(9x^2 + y^2)$.

3. Substituting $y = xv$, we have $2(2x^2 + v^2x^2)\,dx - x^2v(v\,dx + x\,dv) = 0$ or $(4 + v^2)\,dx = xv\,dv$. We first separate the variables to obtain

$$\int \frac{dx}{x} = \int \frac{v\,dv}{4 + v^2}.$$

Integrating yields $\ln x = \frac{1}{2}\ln(4 + v^2) + c$. Replacing c by $\ln c$ and substituting for v leaves $x^2 = c^2[4 + (y/x)^2]$ or $x^4 = c^2(4x^2 + y^2)$.

5. Substituting $y = xv$, we have $(4x^2 + 7vx^2 + 2v^2x^2)\,dx - x^2(v\,dx + x\,dv) = 0$, or $(4 + 6v + 2v^2)\,dx = x\,dv$. We first separate the variables to obtain

$$\int \frac{dx}{x} = \int \frac{dv}{4 + 6v + 2v^2} = -\frac{1}{2}\int\left(\frac{1}{2 + v} + \frac{1}{1 + v}\right)dv.$$

Integrating yields $\ln x = -\frac{1}{2}[\ln(2 + v) - \ln(1 + v)] + c$ or $\dfrac{1 + v}{2 + v} = c^2x^2$. Substituting for v leaves $c[1 + (y/x)] = x^2[2 + (y/x)]$ or $x^2(y + 2x) = c(y + x)$.

7. Substituting $y = xv$, we have $(x - xv)(4x + xv)\,dx + x(5x - xv)(v\,dx + x\,dv) = 0$ or $(-2v^2 + 2v + 4)\,dx = (-5x + xv)\,dv$. We first separate the variables to obtain

$$\int \frac{dx}{x} = \int \frac{v - 5}{-2v^2 + 2v + 4}dv = -\frac{1}{2}\int\left(\frac{-1}{v - 2} + \frac{2}{v + 1}\right)dv.$$

Integrating yields $\ln x = -\frac{1}{2}\ln(v - 2) - \ln(v + 1) + c$ or $\dfrac{c(v - 2)^{1/2}}{v + 1} = x$. Substituting for v leaves $c[(y/x) - 2]^{1/2} = x[(y/x) + 1]$ or $x(y + x)^2 = c^2(y - 2x)$.

9. Substituting $x = yv$, we have $(v^2y^2 + 2y^2v - 4y^2)(v\,dy + y\,dv) - (v^2y^2 - 8y^2v - 4y^2)\,dy = 0$ or $(v^3 + v^2 + 4v + 4)\,dy = -y(v^2 + 2v - 4)\,dv$. We first separate the variables to obtain

$$\int \frac{dy}{y} = -\int \frac{v^2 + 2v - 4}{v^3 + v^2 + 4v + 4}dv = \int\left(\frac{1}{v + 1} - \frac{2v}{v^2 + 4}\right)dv.$$

Integrating yields $\ln x = \ln(v + 1) - \ln(v^2 + 4) + c$, or $\dfrac{c(v + 1)}{v^2 + 4} = y$. Substituting for v leaves $y[(y/x)^2 + 4] = c[(y/x) + 1]$ or $x^2 + 4y^2 = c(x + y)$.

11. Substituting $y = xv$, we have $(x^2 + v^2x^2)\,dx + x^2v(v\,dx + x\,dv) = 0$ or $(1 + 2v^2)\,dx = -xv\,dv$. We first separate the variables to obtain

$$-\int \frac{dx}{x} = \int \frac{v}{1 + 2v^2}dv.$$

Integrating yields $-\ln x = \frac{1}{4}\ln(1 + v^2) + c$, or $(1 + 2v^2)^{1/4}x = c$. Substituting for v leaves $[1 + 2(y/x)^2]x^4 = c^4$ or $x^2(x^2 + 2y^2) = c^4$.

13. Substituting $x = vy$, we have $v^2(v\,dy + y\,dv) + yv(yv + v)\,dv = 0$ or $(y^2 + 2y)\,dv = -v\,dy$. We first separate the variables to obtain

$$-\int \frac{dv}{v} = \int \frac{dy}{y^2 + 2y} = \tfrac{1}{2}\int \left(\frac{1}{y} - \frac{1}{y+2}\right)dy.$$

Integrating yields $-\ln v = \tfrac{1}{2}[\ln y - \ln(y+2)] + c$ or $\left(\dfrac{y}{y+2}\right)^{1/2} = \dfrac{c}{v}$. Substituting for y leaves $c[(x/v) + 2] = (x/v)v^2$ or $xv^2 = c(x + 2v)$.

15. Substituting $y = xv$, we have $(x + xv\sin^2 v)\,dx - x\sin^2 v(v\,dx + x\,dv) = 0$ or $(x + xv\sin^2 v - xv\sin^2 v)\,dx = x^2\sin^2 v\,dv$. We first separate the variables to obtain

$$\int \frac{dx}{x} = \int \sin^2 v\,dv.$$

Integrating yields $\ln x = \tfrac{1}{2}v - \tfrac{1}{4}\sin 2v + c$. Replacing c by $\ln c$ and substituting for v leaves $\ln|x/c| = \tfrac{1}{2}(y/x) - \tfrac{1}{4}\sin[2(y/x)]$ or $4x\ln|x/c| - 2y + x\sin(2y/x) = 0$.

17. Substituting $y = xv$, we have $(x - xv\arctan v)\,dx + x\arctan v(v\,dx + x\,dv) = 0$ or $-(1 - v\arctan v + v\arctan v)\,dx = x\arctan v\,dv$. We first separate the variables to obtain

$$-\int \frac{dx}{x} = \int \arctan v\,dv.$$

Integrating yields $-\ln x + c = v\arctan v - \tfrac{1}{2}\ln(1 + v^2)$. Replacing c by $\ln c$ leaves $\ln\left(\dfrac{c^2(1 + v^2)}{x^2}\right) = 2v\arctan v$. Then substituting for v gives $\ln\left(\dfrac{c^2[1 + (y/x)^2]}{x^2}\right) = 2(y/x)\arctan(y/x)$ or $2y\arctan(y/x) = x\ln[c^2(x^2 + y^2)/x^4]$.

19. Substituting $t = sv$, we have $vs(s^2 + v^2s^2)\,ds - s(s^2 - v^2s^2)(v\,ds + s\,dv) = 0$ or $(v + v^3 - v + v^3)\,ds = s(1 - v^2)\,dv$. We first separate the variables to obtain

$$\int \frac{ds}{s} = \int \frac{1 - v^2}{2v^3}\,dv = \frac{1}{2}\int \left(v^{-3} - v^{-1}\right)dv.$$

Integrating yields $\ln|s| + c = \tfrac{1}{2}(-\tfrac{1}{2}v^{-2} - \ln|v|)$ or $\ln|csv^{1/2}| = -\tfrac{1}{4}v^2$. Substituting for v leaves $-2t^2\ln|cs^2(t/s)| = s^2$ or $s^2 = -2t^2\ln|cst|$.

21. Substituting $y = xv$, we have $(3x^2 - 2x^2v + 3x^2v^2)\,dx + 4x^2v(v\,dx + x\,dv) = 0$ or $(3 - 2v - v^2)\,dx = 4xv\,dv$.
    We first separate the variables to obtain

$$\int \frac{dx}{x} = \int \frac{4v}{3 - 2v - v^2}\,dv = \int \left(\frac{-3}{3 + v} + \frac{1}{1 - v}\right)dv.$$

Integrating yields $\ln x = -3\ln(3 + v) - \ln(1 - v) + c$.
Substituting for v leaves $x[3 + (y/x)]^3[1 - (y/x)] = c$ or $(y - x)(y + 3x)^3 = cx^3$.

23. Substituting $y = xv$, we have $(x - xv)\,dx + (3x + xv)(v\,dx + x\,dv) = 0$ or $(1 - v + 3v + v^2)\,dx = -(3 + v)x\,dv$. We first separate the variables to obtain

$$-\int \frac{dx}{x} = \int \frac{3 + v}{1 + 2v + v^2}\,dv = \int \left(\frac{1}{1 + v} + \frac{2}{(1 + v)^2}\right)\,dv.$$

Integrating yields $-\ln x = \ln(1 + v) - 2(1 + v)^{-1} + c$ or $\dfrac{2}{1 + v} = \ln[x(1 + v)] + c$. Substituting for v leaves $\dfrac{2}{1 + (y/x)} = \ln[x(1 + [y/x])] + c$ or $2x = (x + y)\ln(x + y) + c(x + y)$. The initial conditions give $6 = (1)\ln(1) + c = c$. Thus $2(2x + 3y) + (x + y)\ln(x + y) = 0$.

25. Substituting $y = xv$, we have $(xv + \sqrt{x^2 + x^2 v^2})\,dx - x(v\,dx + x\,dv) = 0$ or $(v + \sqrt{1 + v^2} - v)\,dx = x\,dv$. We first separate the variables to obtain

$$\int \frac{dx}{x} = \int \frac{dv}{\sqrt{1 + v^2}}.$$

Integrating yields $\ln x = \ln(v + \sqrt{1 + v^2}) + c$. Substituting for v leaves $x = c[(y/x) + \sqrt{1 + (y/x)^2}]$. The initial conditions give $3 = c(1 + \sqrt{3 + 1})$ or $c = 1$. Thus $x^2 = 2y + 1$.

27. Substituting $y = xv$, we have $(x^2 v^2 + 7x^2 v + 16x^2)\,dx + x^2(v\,dx + x\,dv) = 0$ or $(v^2 + 7v + 16 + v)\,dx = -x\,dv$. We first separate the variables to obtain

$$\int \frac{dx}{x} = \int \frac{dv}{v^2 + 8v + 16} = -\int \frac{dv}{(v + 4)^2}.$$

Integrating yields $\ln x = \dfrac{1}{v + 4} + c$. Substituting for v leaves $[(y/x) + 4]\ln x = 1 + c[(y/x) + 4]$. The initial conditions give $0 = 1 + 5c$ or $c = -(1/5)$. Thus $(y + 4x)\ln x = x - \frac{1}{5}(y + 4x)$ or $x - y = 5(y + 4x)\ln x$.

29. Substituting $x = yv$, we have $y^2 v(y\,dv + v\,dy) + 2(y^2 v^2 + 2y^2)\,dy = 0$ or $(v^2 + 2v^2 + 4)\,dy = -vy\,dv$. We first separate the variables to obtain

$$-\int \frac{dy}{y} = \int \frac{v}{3v^2 + 4}\,dv.$$

Integrating yields $-\ln y = \frac{1}{6}\ln(3v^2 + 4) + c$. Substituting for v leaves $c = \ln\left([3(x/y)^2 + 4]y^6\right)$. The initial conditions give $c = 4$. Thus $y^4(3x^2 + 4y^2) = 4$.

31. Substituting $y = xv$, we have $xv(9x - 2xv)\,dx - x(6x - xv)(v\,dx + x\,dv) = 0$ or $(9v - 2v^2 - 6v + v^2)\,dx = (6 - v)x\,dv$. We first separate the variables to obtain

$$\int \frac{dx}{x} = \int \frac{6 - v}{-v^2 + 3v}\,dv = \int \frac{v - 6}{v(v - 3)}\,dv = \int \left(\frac{2}{v} - \frac{1}{v - 3}\right)\,dv.$$

Integrating yields $\ln x = 2\ln v - \ln(v-3) + c$. Substituting for v leaves $c = \dfrac{x[(y/x)-3]}{(y/x)^2}$ or

$c = \dfrac{x^2(y-3x)}{y^2}$. The initial conditions give $c = [1(1-3)]/1 = -2$. Thus $-2y^2 = x^2(y-3x)$ or $3x^3 - x^2y - 2y^2 = 0$.

33. Substituting $y = xv$, we have $(16x + 5xv)\,dx + (3x + xv)(v\,dx + x\,dv) = 0$ or $(16 + 5v + 3v + v^2)\,dx = (-3-v)x\,dv$. We first separate the variables to obtain

$$-\int \frac{dx}{x} = \int \frac{v+3}{v^2+8v+16}\,dv = \int \left(\frac{1}{v+4} - \frac{1}{(v+4)^2}\right)\,dv.$$

Integrating yields $-\ln x = \ln(v+4) + \dfrac{1}{(v+4)} + c$. Substituting for v leaves

$-\ln(x[(y/x)+4]) = \dfrac{1}{(y/x)+4} + c$ or $-\ln(y+4) = \dfrac{x}{y+4x} + c$. The initial conditions give $-\ln(-3+4) = 1/(-3+4) + c$ or $c = -1$. Thus $-(y+4x)\ln(y+4x) = x - y - 4x$ or $y + 3x = (y+4x)\ln(y+4x)$.

35. Substituting $x = yv$, we have $2y^2v(y\,dv + v\,dy) - (3y^2v^2 - 2y^2)\,dy = 0$ or $(3v^2 - 2 - 2v^2)\,dy = 2vy\,dv$. We first separate the variables to obtain

$$\int \frac{dy}{y} = \int \frac{2v}{v^2-2}\,dv.$$

Integrating yields $\ln y = \ln(v^2-2) + c$. Substituting for v leaves $\ln\left(\dfrac{y}{(x/y)^2-2}\right) = c$ or

$c = \dfrac{y^3}{x^2-2y^2}$ or $c = 1/2$. Thus $x^2 = 2y^2(y+1)$.

2.4 Exact Equations

1. Let $\dfrac{\partial F}{\partial x} = x + y$. Integrating with respect to x we have $F = x^2/2 + xy + T(y)$. Differentiating with respect to y yields $\dfrac{\partial F}{\partial y} = x + T'(y) = x - y$ so $T(y) = -y^2/2$. Thus $F = x^2/2 + xy - y^2/2 = c$ or $x^2 + 2xy - y^2 = c$.

3. Let $\dfrac{\partial F}{\partial x} = 2xy - 3x^2$. Integrating with respect to x we have $F = x^2y - x^3 + T(y)$. Differentiating with respect to y yields $\dfrac{\partial F}{\partial y} = x^2 + T'(y) = x^2 + y$ so that $T(y) = y^2/2$.
Thus $F = x^2y - x^3 + y^2/2 = c$ or $x^2y - x^3 + \frac{1}{2}y^2 = c$.

5. Let $\dfrac{\partial F}{\partial x} = x - 2y$. Integrating with respect to x we have $F = x^2/2 - 2xy + T(y)$. Differentiating with respect to y yields $\dfrac{\partial F}{\partial y} = -2x + T'(y) = 2y - 2x$ so that $T(y) = y^2$.
Thus $F = x^2/2 - 2xy + y^2 = c$ or $x^2 + 2y^2 = 4xy + c$.

7. This equation has homogeneous coefficients. Substituting $y = xv$, we have $(x - 2xv)\,dx + 2(xv - x)(v\,dx + x\,dv) = 0$ or $(1 - 2v + 2v^2 - 2v)\,dx = (-2v + 2)x\,dv$. We first separate the variables to obtain

$$\int \frac{dx}{x} = \int \frac{(-2v + 2)\,dv}{2v^2 - 4v + 1} = -\int \frac{(2v - 2)\,dv}{(2v^2 - 4v + 1)}.$$

Integrating yields $\ln x = -\frac{1}{2}\ln(2v^2 - 4v + 1) + c$ or $x^2(2v^2 - 4v + 1) = c$.
Substituting for v leaves $x^2[2(y/x)^2 - 4(y/x) + 1] = c$ or $x^2 + 2y^2 = 4xy + c$.

9. Let $\dfrac{\partial F}{\partial x} = y^2 - 2xy + 6x$. Integrating with respect to x we have $F = xy^2 - x^2y + 3x^2 + T(y)$.
Differentiating with respect to y yields $\dfrac{\partial F}{\partial y} = 2xy - x^2 + T'(y) = -x^2 + 2xy - 2$ so that
$T(y) = -2y$. Thus $F = xy^2 - x^2y + 3x^2 - 2y = c$.

11. Let $\dfrac{\partial F}{\partial x} = \cos 2y - 3x^2y^2$. Integrating with respect to x we have $F = x\cos 2y - x^3y^2 + T(y)$.
Differentiating yields $\dfrac{\partial F}{\partial y} = -2x\sin 2y - 2x^3y + T'(y) = \cos 2y - 2x\sin 2y - 2x^3y$ so that
$T(y) = \frac{1}{2}\sin 2y$. Thus $F = x\cos 2y - x^3y^2 + \frac{1}{2}\sin 2y = c$ or $\frac{1}{2}\sin 2y + x\cos 2y - x^3y^2 = c$.

13. Let $\dfrac{\partial F}{\partial x} = 1 + y^2 + xy^2$. Integrating with respect to x we have $F = x + xy^2 + \frac{1}{2}x^2y^2 + T(y)$.
Differentiating with respect to y yields $\dfrac{\partial F}{\partial y} = 2xy + x^2y + T'(y) = x^2y + y + 2xy$ so that
$T(y) = y^2/2$. Thus $F = x + xy^2 + \frac{1}{2}x^2y^2 + \frac{1}{2}y^2 = c$ or $2x + y^2(1 + x)^2 = c$.

15. Let $\dfrac{\partial F}{\partial y} = x^2 - x\sec^2 y$. Integrating with respect to y we have $F = x^2y - x\tan y + Q(x)$.
Differentiating with respect to x yields $\dfrac{\partial F}{\partial x} = 2xy - \tan y + Q'(x) = 2xy - \tan y$ so that
$Q(x) = 0$. Thus $F = x^2y - x\tan y = c$.

17. Let $\dfrac{\partial F}{\partial r} = r + \sin\theta - \cos\theta$. Integrating we have $F = r^2/2 + r(\sin\theta - \cos\theta) + T(\theta)$. Differentiating
with respect to θ yields $\dfrac{\partial F}{\partial \theta} = r(\cos\theta + \sin\theta) + T'(\theta) = r(\cos\theta + \sin\theta)$ so that $T(\theta) = 0$. Thus
$F = r^2/2 + r(\sin\theta - \cos\theta) = c$ or $r^2 + 2r(\sin\theta - \cos\theta) = c$.

19. Let $\dfrac{\partial F}{\partial r} = \sin\theta - 2r\cos^2\theta$. Integrating with respect to x we have $F = r\sin\theta - r^2\cos^2\theta + T(\theta)$. Differentiating yields $\dfrac{\partial F}{\partial \theta} = r\cos\theta + 2r^2\cos\theta\sin\theta + T'(\theta) = r\cos\theta(2r\sin\theta + 1)$ so that
$T(\theta) = 0$. Thus $F = r\sin\theta - r^2\cos^2\theta = c$.

21. Let $\dfrac{\partial F}{\partial x} = 2xy$. Integrating with respect to x we have $F = x^2y + T(y)$. Differentiating with
respect to y yields $\dfrac{\partial F}{\partial y} = x^2 + T'(y) = y^2 + x^2$ so that $T(y) = y^3/3$. Thus $F = x^2y + y^3/3 = c$
or $y(3x^2 + y^2) = c$.

23. Let $\dfrac{\partial F}{\partial x} = xy^2 + y - x$. Integrating with respect to x we have $F = \frac{1}{2}x^2y^2 + xy - \frac{1}{2}x^2 + T(y)$. Differentiating with respect to y yields $\dfrac{\partial F}{\partial y} = x^2y + x + T'(y) = x^2y + x$ so that $T(y) = 0$. Thus $F = \frac{1}{2}x^2y^2 + xy - \frac{1}{2}x^2 = c$ or $x^2y^2 + 2xy - x^2 = c$.

25. Let $\dfrac{\partial F}{\partial y} = y^2 + x^2(1-xy)^{-2}$. Integrating we have $F = y^3/3 + x(1-xy)^{-1} + Q(x)$. Differentiating with respect to x yields $\dfrac{\partial F}{\partial x} = xy(1-xy)^{-2} + (1-xy)^{-1} + Q'(x) = (1-xy)^{-2}$ so that $Q(x) = 0$. Thus $F = \frac{1}{3}y^3 + x(1-xy)^{-1} = c$. Substituting the initial conditions gives $\frac{1}{3} - 2 = c$ or $c = -\frac{5}{3}$. Thus $\frac{1}{3}y^3 + x(1-xy)^{-1} = -\frac{5}{3}$ or $xy^4 - y^3 + 5xy - 3x = 5$.

27. Let $\dfrac{\partial F}{\partial y} = x^2 + 3y^2 + \exp\left(-x^2\right)$. Integrating we have $F = yx^2 + y^3 + y\exp\left(-x^2\right) + Q(x)$. Differentiating with respect to x,
$$\dfrac{\partial F}{\partial x} = 2xy - 2xy\exp\left(-x^2\right) + Q'(x) = 6x^2 + 2xy - 2xy\exp\left(-x^2\right)$$ so that $Q(x) = 2x^3$. Thus $F = x^2y + y^3 + y\exp\left(-x^2\right) + 2x^3 = c$.

2.6 The General Solution of a Linear Equation

1. The equation in standard form is $\dfrac{dy}{dx} - (3/x)y = x^4$. Thus the integrating factor is

$$\exp\left(-3\int \frac{dx}{x}\right) = \exp\left(-3\ln x\right) = \frac{1}{x^3}.$$

Multiplying by the integrating factor we have $\dfrac{d}{dx}\left(\dfrac{y}{x^3}\right) = x$. Integrating and simplifying yields $y/x^3 = x^2/2 + c$ or $2y = x^5 + cx^3$.

3. The equation in standard form is $\dfrac{dx}{dy} + [4/(y+1)]x = y/(y+1)$. Thus the integrating factor is

$$\exp\left(4\int \frac{dy}{y+1}\right) = \exp\left[4\ln\left(y+1\right)\right] = (y+1)^4.$$

Multiplying by the integrating factor we have $\dfrac{d}{dy}[(y+1)^4 x] = y(y+1)^3$. Integrating and simplifying yields $20x(y+1)^4 = (4y-1)(y+1)^4 + c$ or $20x = 4y - 1 + c(y+1)^{-4}$.

5. The equation in standard form is $\dfrac{dx}{du} + [(1-3u)/u]x = 3$ Thus the integrating factor is

$$\exp\left(\int \frac{(1-3u)}{u}du\right) = \exp\left(\ln u - 3u\right) = u\exp\left(-3u\right).$$

Multiplying by the integrating factor we have $\dfrac{d}{du}(ue^{-3u}x) = 3ue^{-3u}$. Integrating and simplifying yields $ue^{-3u}x = -ue^{-3u} - \frac{1}{3}e^{-3u} + c$ or $xu = ce^{3u} - u - \frac{1}{3}$.

7. The equation in standard form is $\dfrac{dy}{dx} - (\cot x)y = \csc x$. Thus the integrating factor is

$$\exp\left(-\int \cot x\, dx\right) = \exp\left[-\ln(\sin x)\right] = \frac{1}{\sin x}.$$

Multiplying by the integrating factor we have $\dfrac{d}{dx}\left(\dfrac{y}{\sin x}\right) = \csc^2 x$. Integrating and simplifying yields $y/\sin x = -\cot x + c$ or $y = c\sin x - \cos x$.

9. The equation in standard form is $\dfrac{dy}{dx} + (1/\cos x)y = \cos x$. Thus the integrating factor is

$$\exp\left(\int \sec x\, dx\right) = \exp\left[\ln(\sec x + \tan x)\right] = \sec x + \tan x.$$

Multiplying by the integrating factor we have $\dfrac{d}{dx}[y(\sec x + tanx)] = 1 + \sin x$. Integrating and simplifying yields $y(\sec x + \tan x) = c + x - \cos x$.

11. The equation in standard form is $\dfrac{dy}{dx} + (\cot x + 1/x)y = 1$. Thus the integrating factor is

$$\exp\left(\int \left(\cot x + \frac{1}{x}\right) dx\right) = \exp\left[\ln(\sin x) + \ln x\right] = x\sin x.$$

Multiplying by the integrating factor we have $\dfrac{d}{dx}(yx\sin x) = x\sin x$. Integrating and simplifying yields $xy\sin x = c + \sin x - x\cos x$.

13. The equation in standard form is $\dfrac{dy}{dx} - [2x/(1+x^2)]y = (x^2+x^4)/(1+x^2)$. Thus the integrating factor is

$$\exp\left(-\int \frac{2x}{1+x^2} dx\right) = \exp\left[-\ln(1+x^2)\right] = (1+x^2)^{-1}.$$

Multiplying by the integrating factor we have $\dfrac{d}{dx}[y(1+x^2)^{-1}] = \dfrac{x^2}{1+x^2}$. Integrating and simplifying yields $y(1+x^2)^{-1} = x - \arctan x + c$ or $y = (1+x^2)(c + x - \arctan x)$.

15. The equation in standard form is $\dfrac{dy}{dx} - (m_2)y = c_1 e^{m_1 x}$. Thus the integrating factor is

$$\exp\left(\int -m_2\, dx\right) = \exp(-m_2 x).$$

Multiplying by the integrating factor we have $\dfrac{d}{dx}(e^{-m_2 x}y) = c_1 e^{(m_1-m_2)x}$. Integrating and simplifying yields $y = c_3 e^{m_1 x} + c_2 e^{m_2 x}$, where $c_3 = c_1/(m_1 - m_2)$.

17. The equation in standard form is $\dfrac{dy}{dx} + [2/(x(x^2+1)]y = (x^2+1)^2/x$. Thus the integrating factor is

$$\exp\left(\int \frac{2\,dx}{x(x^2+1)}\right) = \exp\left[2\ln x - \ln\left(x^2+1\right)\right] = \frac{x^2}{x^2+1}.$$

Multiplying by the integrating factor we have $\dfrac{d}{dx}\left(y\dfrac{x^2}{x^2+1}\right) = x^3 + x$. Integrating and simplifying yields $x^2y/(x^2+1) = \frac{1}{4}(x^2+1)^2 + c$ or $x^2y = \frac{1}{4}(x^2+1)^3 + c(x^2+1)$.

19. The equation in standard form is $\dfrac{dy}{dx} + [2/(x^2-1)]y = 1$. Thus the integrating factor is

$$\exp\left(\int \frac{2}{x^2-1}\,dx\right) = \exp\left(\ln\left(\frac{x-1}{x+1}\right)\right) = \frac{x-1}{x+1}.$$

Multiplying by the integrating factor we have $\dfrac{d}{dx}\left(y\dfrac{x-1}{x+1}\right) = \dfrac{x-1}{x+1}$. Integrating and simplifying yields $y\dfrac{x-1}{x+1} = x - 2\ln|x+1| + c$ or $(x-1)y = (x+1)(c + x - 2\ln|x+1|)$.

21. The equation in standard form is $\dfrac{dy}{dx} - (3\tan x)y = 1$. Thus the integrating factor is

$$\exp\left(-3\int \tan x\,dx\right) = \exp\left[-3\ln\left(\sec x\right)\right] = \cos^3 x.$$

Multiplying by the integrating factor we have $\dfrac{d}{dx}[y(\cos^3 x)] = \cos^3 x$. Integrating and simplifying yields $y\cos^3 x = \frac{1}{3}(2 + \cos^2 x)\sin x + c$ or $3y\cos^3 x = c + 3\sin x - \sin^3 x$.

23. The equation in standard form is $\dfrac{dy}{dx} - [6x/(x^2+a^2)]y = 2x(x^2+a^2)$. Thus the integrating factor is

$$\exp\left(-6\int \frac{x}{x^2+a^2}\,dx\right) = \exp\left[-3\ln\left(x^2+a^2\right)\right] = (x^2+a^2)^{-3}.$$

Multiplying by the integrating factor we have $\dfrac{d}{dx}[y(x^2+a^2)^{-3}] = \dfrac{2x}{(x^2+a^2)^2}$. Integrating and simplifying yields $y(x^2+a^2)^{-3} = -(x^2+a^2)^{-1} + c$ or $y = (x^2+a^2)^2[c(x^2+a^2) - 1]$.

25. If $n = 0$, the equation is separable so $\displaystyle\int dy = b\int \frac{x}{x+a}\,dx$ or $y = bx + c - ab\ln|x+a|$.

If $n = -1$, the equation in standard form is $\dfrac{dy}{dx} - y/(x+a) = bx/(x+a)$. Thus the integrating factor is

$$\exp\left(-\int \frac{dx}{x+a}\right) = \exp\left[-\ln\left(x+a\right)\right] = (x+a)^{-1}.$$

Multiplying by the integrating factor we have $\dfrac{d}{dx}[y(x+a)^{-1}] = \dfrac{bx}{(x+a)^2}$. Integrating and simplifying yields $y(x+a)^{-1} = b\ln|x+a|+ab/(x+a)+c$ or $y = ab+c(x+a)+b(x+a)\ln|x+a|$.

27. The equation in standard form is $\dfrac{dy}{dx} - [1/(2x+3)]y = (2x+3)^{-1/2}$. Thus the integrating factor is

$$\exp\left(-\int \frac{dx}{2x+3}\right) = \exp\left[-\tfrac{1}{2}\ln(2x+3)\right] = (2x+3)^{-1/2}.$$

Multiplying by the integrating factor we have $\dfrac{d}{dx}[y(2x+3)^{-1/2}] = (2x+3)^{-1}$. Integrating yields $y(2x+3)^{-1/2} = \tfrac{1}{2}\ln(2x+3) + c$. Substituting initial conditions gives $0 = \tfrac{1}{2}\ln(-2+3)+c$ or $c = 0$ so $2y = (2x+3)^{1/2}\ln(2x+3)$.

29. The equation in standard form is $\dfrac{di}{dt} + (R/L)i = E/L$. Thus the integrating factor is

$$\exp\left((R/L)\int dx\right) = \exp(Rt/L).$$

Multiplying by the integrating factor we have $\dfrac{d}{dx}[i\exp(Rt/L)] = (E/L)\exp(Rt/L)$. Integrating yields $i\exp(Rt/L) = (E/L)(L/R)\exp(Rt/L) + c$. Substituting initial conditions gives $0 = E/R + c$ or $c = -(E/R)$ so $i = \dfrac{E}{R}\left[1 - \exp\left(-\dfrac{Rt}{L}\right)\right]$.

31. The equation in standard form is $\dfrac{dy}{dx} + 2y = 4x$. Thus the integrating factor is

$$\exp\left(\int 2\,dx\right) = \exp 2x.$$

Multiplying by the integrating factor we have $\dfrac{d}{dx}(ye^{2x}) = 4xe^{2x}$. Integrating yields $ye^{2x} = 2xe^{2x} - e^{2x} + c$. Substituting initial conditions gives $-1 = -1 + c$ or $c = 0$ so $y = 2x - 1$.

33. The equation in standard form is $\dfrac{ds}{dt} + [(2t^3)/(1+t^2)]s = 6t(1+t^2)$. Thus the integrating factor is

$$\exp\left(2\int \frac{t^3}{1+t^2}dt\right) = \exp\left[t^2 - \ln(t^2+1)\right] = \frac{e^{t^2}}{t^2+1}.$$

Multiplying by the integrating factor we have $\dfrac{d}{dx}\left(s\dfrac{e^{t^2}}{t^2+1}\right) = 6te^{t^2}$. Integrating yields

$s\left(\dfrac{e^{t^2}}{t^2+1}\right) = 3e^{t^2} + c$. Substituting the initial conditions gives $2 = 3 + c$ or $c = -1$ so $s = (1+t^2)[3 - \exp(-t^2)]$.

Miscellaneous Exercises

1. The equation is separable.
   Separating the variables we have,

$$\int e^y \, dy = \int e^{2x} \, dx.$$

Integrating both sides yields, $e^y = \frac{1}{2}e^{2x} + c$. Thus $2e^y = e^{2x} + c$.

3. The equation is linear in y.
   The equation in standard form is $\dfrac{dy}{dx} + [3/(x+1)]y = 4/(x+1)^2$. Thus the integrating factor is

$$\exp\left(3 \int \frac{dx}{x+1}\right) = \exp\left[3\ln\left(x+1\right)\right] = (x+1)^3.$$

Multiplying by the integrating factor we have $\dfrac{d}{dx}[y(x+1)^3] = 4(x+1)$. Integrating and simplifying yields $y(x+1)^3 = 2(x+1)^2 + c$ or $y = 2(x+1)^{-1} + c(x+1)^{-3}$.

5. The equation is homogeneous.
   Substituting $y = xv$, we have $(v^2x^2)\,dx + (-2x^2 - 3x^2v)(v\,dx + x\,dv) = 0$ or $(v^2 - 2v - 3v^2)\,dx = (2+3v)x\,dv$. We first separate the variables to obtain

$$-\int \frac{dx}{x} = \frac{1}{2}\int \frac{3v+2}{v^2+v}\,dv = \frac{1}{2}\int \left(\frac{2}{v} + \frac{1}{v+1}\right)\,dv.$$

Integrating yields $-\ln x = \ln v + \frac{1}{2}\ln\left(v+1\right) + c$ or $x^2v^2(v+1) = c$. Substituting for v leaves $x^2(y/x)^2[(y/x)+1] = c$ or $y^2(x+y) = cx$.

7. The equation is linear in y.
   The equation in standard form is $\dfrac{dy}{dx} + (2x)y = x^3$. Thus the integrating factor is

$$\exp\left(\int 2x\,dx\right) = \exp\left(x^2\right).$$

Multiplying by the integrating factor we have $\dfrac{d}{dx}[y(e^{x^2})] = e^{x^2}x^3$. Integrating and simplifying yields $y(e^{x^2}) = \frac{1}{2}(x^2e^{x^2} - e^{x^2}) + c$, $\quad y = \frac{1}{2}(x^2 - 1) + ce^{-x^2}$.
   Substituting initial conditions gives $2 = ce^{-1}$ or $c = 2e$ so $2y = x^2 - 1 + 4\exp\left(1 - x^2\right)$.

9. The equation is homogeneous.
   Substituting $y = xv$, we have $xv(x + 3xv)\,dx + x^2(v\,dx + x\,dv) = 0$ or $(v + 3v^2 + v)\,dx = -x\,dv$. We first separate the variables to obtain

$$-\int \frac{dx}{x} = \int \frac{dv}{3v^2+2v} = \frac{1}{2}\int \left(\frac{1}{v} - \frac{3}{3v+2}\right)\,dv.$$

Integrating yields $-\ln x = \frac{1}{2}[\ln v - \ln\left(3v+2\right)] + c$ or $x^2v = c(3v+2)$. Substituting for v leaves $x^2(y/x) = c[3(y/x)+2]$ or $x^2y = c(2x+3y)$.

11. The equation is separable.
    Separating the variables we have,

    $$\int \frac{dy}{y^4} = \int \frac{x^4}{1+x^2}\,dx.$$

    Integrating both sides yields, $-\frac{1}{3}y^{-3} = \frac{1}{3}x^3 - x + \arctan x + c$.
    Thus $x^3y^3 + 1 = y^3(c + 3x - 3\arctan x)$.

13. The equation is separable.
    Separating the variables we have,

    $$\int \frac{dx}{\cos x} = \int \cos^2 t\,dx.$$

    Integrating both sides yields, $\ln|\sec x + \tan x| = \frac{1}{2}t + \frac{1}{4}\sin 2t + c$.
    Thus $4\ln|\sec x + \tan x| = 2t + \sin 2t + c$.

15. The equation is homogeneous.
    Substituting $y = xv$, we have $(x^2v - x^2v^2)\,dx - (x^2v + x^2)(v\,dx + x\,dv) = 0$ or
    $(v - v^2 - v^2 - v)\,dx = x(v + 1)\,dv$. We first separate the variables to obtain

    $$-\int \frac{dx}{x} = \int \frac{v+1}{v^2}\,dv = \frac{1}{2}\int \left(\frac{1}{v} + \frac{1}{2v^2}\right)\,dv.$$

    Integrating yields $-\ln x = \frac{1}{2}(\ln v - v^{-1}) + c$ or $1/v = \ln cx^2v$. Substituting for v leaves
    $x/y = \ln|cx^2(y/x)|$ or $x = y\ln|cxy|$.

17. The equation is exact.
    Let $\dfrac{\partial F}{\partial x} = x + 2y$. Integrating with respect to x we have $F = x^2/2 + 2xy + T(y)$. Differentiating
    with respect to y yields $\dfrac{\partial F}{\partial y} = 2x + T'(y) = 2x + y$ so $T(y) = y^2/2$. Thus $x^2 + 4xy + y^2 = c$.

19. The equation is exact.
    Let $\dfrac{\partial F}{\partial x} = x^3 + y^3$. Integrating with respect to x we have $F = x^4/4 + y^3x + T(y)$. Differentiating
    with respect to y yields $\dfrac{\partial F}{\partial y} = 3y^2x + T'(y) = 3y^2x + ky^3$ so $T(y) = ky^4/4$.
    Thus $ky^4 + 4xy^3 + x^4 = c$.

21. The equation is exact.
    Let $\dfrac{\partial F}{\partial x} = 3y + 2xy^3$. Integrating with respect to x we have $F = 3xy + x^2y^3 + T(y)$. Differ-
    entiating with respect to y yields $\dfrac{\partial F}{\partial y} = 3x + 3x^2y^2 + T'(y) = 3x^2y^2 + 3x - 3$ so $T(y) = -3y$.
    Thus $x^2y^3 = 3(c + y - xy)$.

23. The equation is linear in y.

The equation in standard form is $\dfrac{dy}{dx} + (a)y = b$. Thus the integrating factor is

$$\exp\left(a \int dx\right) = \exp ax.$$

Multiplying by the integrating factor we have $\dfrac{d}{dx}(ye^{ax}) = be^{ax}$. Integrating and simplifying yields $ye^{ax} = (b/a)e^{ax} + c$ or $y = b/a + ce^{-ax}$.

The equation is also separable.

Separating the variables gives $\displaystyle\int \dfrac{dy}{b - ay} = \int dx$. Integrating both sides yields $-\frac{1}{a}\ln(b - ay) = x + c$. Thus $y = b/a + ce^{-ax}$.

25. The equation is exact.

Let $\dfrac{\partial F}{\partial x} = \sin y - y\sin x$. Integrating with respect to x we have $F = x\sin y + y\cos x + T(y)$.

Differentiating with respect to y yields $\dfrac{\partial F}{\partial y} = x\cos y + \cos x + T'(y) = \cos x + x\cos y$ so $T(y) = 0$. Thus $x\sin y + y\cos x = c$.

27. The equation is linear in y.

The equation in standard form is $\dfrac{dy}{dx} - (2\cot x)y = \sin^3 x$. Thus the integrating factor is

$$\exp\left(-2 \int \cot x\, dx\right) = \exp\left[-2\ln(\sin x)\right] = \sin^{-2} x.$$

Multiplying by the integrating factor we have $\dfrac{d}{dx}\left[y(\sin^{-2} x)\right] = \sin x$. Integrating and simplifying yields $y(\sin^{-2} x) = -\cos x + c$. Substituting initial conditions gives, $1/\sin^2(\pi/2) = -\cos(\pi/2) + c$ or $c = 1$ so $y = 2\sin^2 x\, \sin^2 \frac{1}{2}x$.

29. The equation is separable.

Separating the variables we have,

$$\int \frac{dx}{\sqrt{1 - x^2}} = -\int \frac{dy}{\sqrt{1 - y^2}}.$$

Integrating both sides yields, $\arcsin x + \arcsin y = c$, or a part of the ellipse $x^2 + 2c_1 xy + y^2 + c_1^2 - 1 = 0$; where $c_1 = \cos c$.

31. Substituting, $\arcsin 0 = -\arcsin \sqrt{3}/2 + c$ or $0 = -\pi/3 + c$. $\arcsin x + \arcsin y = \frac{1}{3}\pi$, or that arc of the ellipse $x^2 + xy + y^2 = \frac{3}{4}$ that is indicated by a light solid line in Figure 2.4.

35. The equation is linear in x.

The equation in standard form is $\dfrac{dx}{dy} - (1/y)x = 2/y^2$. Thus the integrating factor is

$$\exp\left(-\int \frac{dy}{y}\right) = \exp(-\ln y) = y^{-1}.$$

Multiplying by the integrating factor we have $\frac{d}{dy}(xy^{-1}) = 2y^{-3}$. Integrating and simplifying yields $xy^{-1} = -y^{-2} + c$ or $xy = cy^2 - 1$.

37. The equation is linear in y.

The equation in standard form is $\frac{dy}{dx} - (\tan x)y = \cos x$. Thus the integrating factor is

$$\exp\left(-\int \tan x \, dx\right) = \exp\left[\ln\left(\cos x\right)\right] = \cos x.$$

Multiplying by the integrating factor we have $\frac{d}{dx}(y\cos x) = \cos^2 x$. Integrating and simplifying yields $y\cos x = \frac{1}{2}(\cos x \sin x + x) + c$ or $2y = \sin x + (x + c)\sec x$.

39. The equation is linear in y.

The equation in standard form is $\frac{dy}{dx} + [x/(1 - x^2)]y = (1 - 3x^2 + 2x^4)/(1 - x^2)$. Thus the integrating factor is

$$\exp\left(\int \frac{x}{1 - x^2} \, dx\right) = \exp\left[-\frac{1}{2}\ln\left(1 - x^2\right)\right] = (1 - x^2)^{-1/2}.$$

Multiplying by the integrating factor we have $\frac{d}{dx}[y(1 - x^2)^{-1/2}] = \frac{1 - 2x^2}{(1 - x^2)^{1/2}}$. Integrating and simplifying yields $y(1 - x^2)^{-1/2} = \arcsin x + x\sqrt{1 - x^2} - \arcsin x + c$ or $y = x - x^3 + c(1 - x^2)^{1/2}$.

41. The equation is linear in y.

The equation in standard form is $\frac{dy}{dx} + (\tan x)y = \sec x$. Thus the integrating factor is

$$\exp\left(\int \tan x \, dx\right) = \exp\left[\ln\left(\sec x\right)\right] = \sec x.$$

Multiplying by the integrating factor we have $\frac{d}{dx}(y\sec x) = \sec^2 x$. Integrating and simplifying yields $y\cos^{-1}x = \tan x + c$ or $y = \sin x + c\cos x$.

43. The equation is linear in y.

The equation in standard form is $\frac{dy}{dx} + [(1/x) - \tan x]y = 1$. Thus the integrating factor is

$$\exp\left(\int [(1/x) - \tan x] \, dx\right) = \exp\left[\ln x - \ln\left(\sec x\right)\right] = x\cos x.$$

Multiplying by the integrating factor we have $\frac{d}{dx}(yx\cos x) = x\cos x$. Integrating and simplifying yields $xy\cos x = c + \cos x + x\sin x$.

45. The equation is linear in x.

The equation in standard form is $\dfrac{dx}{dy} - (3/y)x = y^2 - y$. Thus the integrating factor is

$$\exp\left(-3\int \frac{dy}{y}\right) = \exp\left(-3\ln y\right) = y^{-3}.$$

Multiplying by the integrating factor we have $\dfrac{d}{dy}(xy^{-3}) = y^{-1} - y^{-2}$. Integrating yields $xy^{-3} = \ln y + y^{-1} + c$. Substituting initial conditions gives $-1 = 0 - 1 + c$ or $c = 0$ so $x = y^2[1 + y\ln(-y)]$.

47. The equation is linear in x.

The equation in standard form is $\dfrac{dx}{dy} + (1/y)x = -1 + 1/y^2$. Thus the integrating factor is

$$\exp\left(\int \frac{dy}{y}\right) = \exp\left(\ln y\right) = y.$$

Multiplying by the integrating factor we have $\dfrac{d}{dy}(xy) = -y + 1/y$.

Integrating yields $xy = -y^2/2 + \ln y + c$. Substituting initial conditions gives $-1 = -1/2 + c$ or $c = -1/2$ so $y^2 + 2xy + 1 = 2\ln y$.

49. The equation is linear in y.

The equation in standard form is $\dfrac{dy}{dx} - y = 3x$. Thus the integrating factor is

$$\exp\left(-\int dx\right) = \exp\left(-x\right).$$

Multiplying by the integrating factor we have $\dfrac{d}{dx}\left[y(e^{-x})\right] = 3xe^{-x}$. Integrating yields $y(e^{-x}) = 3(-xe^{-x} - e^{-x}) + c$. Substituting initial conditions gives $0 = 3[-(-1)e - e] + c$ or $c = 0$ so $y = -3(x + 1)$.

51. The equation is linear in y.

The equation in standard form is $\dfrac{dy}{dx} + [2/(1 - x^2)]y = 1$. Thus the integrating factor is

$$\exp\left(2\int \frac{dx}{1 - x^2}\right) = \exp\left(\ln\left(\frac{x+1}{x-1}\right)\right) = \frac{x+1}{x-1}.$$

Multiplying by the integrating factor we have $\dfrac{d}{dx}\left(y\left(\dfrac{x+1}{x-1}\right)\right) = 1 + \dfrac{2}{x-1}$. Integrating yields $y\left(\dfrac{x+1}{x-1}\right) = x + 2\ln(x - 1) + c$. Substituting initial conditions gives $3/1 = 2 + 0 + c$ or $c = 1$ so $(x + 1)y = (x - 1)[x + 1 + 2\ln(x - 1)]$.

53. The equation is linear in y.

   The equation in standard form is $\dfrac{dy}{dx} + (3/x)y = 1/x^5$. Thus the integrating factor is

   $$\exp\left(3 \int \frac{dx}{x}\right) = \exp\left(3 \ln x\right) = x^3.$$

   Multiplying by the integrating factor we have $\dfrac{d}{dx}\left(yx^3\right) = x^{-2}$. Integrating yields $yx^3 = -x^{-1} + c$. Substituting initial conditions gives $1 = -1 + c$ or $c = 2$ so $x^4 y = 2x - 1$.

55. The equation is linear in y.

   The equation in standard form is $\dfrac{dy}{dx} + \left(\dfrac{3x-4}{x(x-2)}\right)y = \dfrac{1}{x(x-2)}$. Thus the integrating factor is

   $$\exp\left(\int \frac{3x-4}{x(x-2)}\,dx\right) = \exp\left(\int \frac{2}{x} + \frac{1}{x-2}\,dx\right) = \exp[2\ln x + \ln(x-2)] = x^2(x-2).$$

   Multiplying by the integrating factor we have $\dfrac{d}{dx}\left[yx^2(x-2)\right] = x$.

   Integrating yields $yx^2(x-2) = x^2/2 + c$. Substituting initial conditions gives $1(1-2)2 = (1/2) + c$ or $c = -\frac{5}{2}$ so $2x^2(x-2)y = x^2 - 5$.

Chapter 3

Numerical Methods

3.2   Euler's Method

1. The results are found in following table. The correct solution is computed from the solution $y = 2e^x - x - 1$.

Table 3.1:

x	y	$x + y$	dy	Correct
0.0	1.00	1.00	0.10	1.00
0.1	1.10	1.20	0.12	1.11
0.2	1.22	1.42	0.14	1.24
0.3	1.36	1.66	0.17	1.40
0.4	1.53	1.93	0.19	1.58
0.5	1.72	2.22	0.22	1.80
0.6	1.94	2.54	0.25	2.04
0.7	2.19	2.89	0.29	2.33
0.8	2.48	3.28	0.33	2.65
0.9	2.81	3.71	0.37	3.02
1.0	3.18			3.44

3. The approximate value is $y(1) \approx 5.78$.
   The actual value is $y(1) = 6.15$.

5. The approximate value is $y(3) \approx 1.19$.
   The actual value is $y(1) = 1.44$.

7. The approximate value is $y(2) \approx 1.09$.

9. The approximate value is $y(2) \approx 1.03$.

11. The approximate value is $y(2) \approx 2.09$.

13. The approximate value is $y(2) \approx .61$.

3.3 A Modification of Euler's Method

1. The approximate value is $y(1) \approx 3.42$.
   The actual value is $y(1) = 3.44$.

3. The approximate value is $y(1) \approx 6.12$.
   The actual value is $y(1) = 6.15$.

5. The approximate value is $y(3) \approx 1.42$.
   The actual value is $y(1) = 1.44$.

7. The approximate value is $y(2) \approx 1.03$.

9. The approximate value is $y(2) \approx .97$.

11. The approximate value is $y(2) \approx 3.21$.

13. The approximate value is $y(2) \approx .70$.

3.4 A Method of Successive Approximation

1. $y_0 = 1$;
   $$y_1(x) = 1 + \int_0^x (t+1)\,dt = 1 + x + \tfrac{1}{2}x^2;$$
   $$y_2(x) = 1 + \int_0^x (t+1+t+t^2/2)\,dt = 1 + x + x^2 + \tfrac{1}{6}x^3;$$
   $$y_3(x) = 1 + \int_0^x (t+1+t+t^2+t^3/6)\,dt = 1 + x + x^2 + \tfrac{1}{3}x^3 + \tfrac{1}{24}x^4.$$

3. $y_1(x) = 1 + \int_1^x (t+1)\,dt = 1 + \int_1^x 2 + (t-1)\,dt = 1 + 2(x-1) + \tfrac{1}{2}(x-1)^2$;
   $$y_2(x) = 1 + \int_1^x [2 + 3(t-1) + (t-1)^2/2]\,dt = 1 + 2(x-1) + \tfrac{3}{2}(x-1)^2 + \tfrac{1}{6}(x-1)^3;$$
   $$y_3(x) = 1 + \int_1^x [2 + 3(t-1) + \tfrac{3}{2}(t-1)^2 + \tfrac{1}{6}(t-1)^3]\,dt$$
   $$= 1 + 2(x-1) + \tfrac{3}{2}((x-1)^2 + \tfrac{1}{2}(x-1)^3 + \tfrac{1}{24}(x-1)^4.$$

3.5 An Improvement

1. $y_0(x) = 1 + x$;
   $$y_1(x) = 1 + \int_0^x (t+1+t)\,dt = 1 + x + x^2.$$
   $$y_2(x) = 1 + \int_0^x [t+1+t+t^2]\,dt = 1 + x + x^2 + \tfrac{1}{3}x^3.$$
   $$y_3(x) = 1 + \int_0^x [t+1+t+t^2+\tfrac{1}{3}t^3]\,dt = 1 + x + x^2 + \tfrac{1}{3}x^3 + \tfrac{1}{12}x^4.$$

3. $y_0(x) = 2x - 1$;

$$y_1(x) = 1 + \int_1^x (t + 2t - 1)\, dt = 2x - 1 + \tfrac{3}{2}(x-1)^2.$$

$$y_2(x) = 1 + \int_1^x [t + 2t - 1 + \tfrac{3}{2}(t-1)^2]\, dt = 2x - 1 + \tfrac{3}{2}(x-1)^2 + \tfrac{1}{2}(x-1)^3.$$

$$y_3(x) = 1 + \int_1^x [t + 2t - 1 + \tfrac{3}{2}(t-1)^2 + \tfrac{1}{2}(t-1)^3]\, dt$$
$$= 1 + 2(x-1) + \tfrac{3}{2}(x-1)^2 + \tfrac{1}{2}(x-1)^3 + \tfrac{1}{8}(x-1)^4.$$

3.6 The Use of Taylor's Theorem

1. We first find the required derivatives:
   $y' = x + y, \quad y'' = 1 + y', \quad y''' = y'', \quad y'''' = y'''.$
   Then evaluate them at (x_0, y_0).
   $x_0 = 0, \quad y_0 = 1, \quad y_0' = 1, \quad y_0'' = 1 + 1 = 2, \quad y_0''' = y_0'''' = 2.$
   This gives the desired polynomial.
   $y_5(x) = 1 + x + x^2 + \tfrac{1}{3}x^3 + \tfrac{1}{12}x^4 + \tfrac{1}{60}x^5.$

3. Use the same derivatives as in Exercise 1, then evaluate them at (x_0, y_0)
   $x_0 = 1, \quad y_0 = 1, \quad y_0' = 2, \quad y_0'' = 1 + 2 = 3, \quad y_0''' = y_0'''' = 3.$
   This gives the desired polynomial.
   $y_5(x) = 1 + 2(x-1) + \tfrac{3}{2}(x-2)^2 + \tfrac{1}{2}(x-1)^3 + \tfrac{1}{8}(x-1)^4 + \tfrac{1}{40}(x-1)^5.$

5. Use the same derivatives as in Exercise 1, then evaluate them at (x_0, y_0)
   $x_0 = 2, \quad y_0 = -1, \quad y_0' = 1, \quad y_0'' = 1 + 1 = 2, \quad y_0''' = y_0'''' = 2.$
   This gives the desired polynomial.
   $y_5(x) = -1 + (x-2) + (x-2)^2 + \tfrac{1}{3}(x-2)^3 + \tfrac{1}{12}(x-2)^4 + \tfrac{1}{60}(x-2)^5.$

7. Use the same derivatives as in the example in the section, then:
   $y^{(5)} = 2yy^{(4)} + 8y'y'' + 6y''y''',$
   $y^{(6)} = 2yy^{(5)} + 10y'y^{(4)} + 20y''y''',$
   $y^{(7)} = 2yy^{(6)} + 12y'y^{(5)} + 30y''y^{(4)} + 20y'''y'''.$
   Then evaluate them at (x_0, y_0).
   $x_0 = 0, \quad y_0 = 1, \quad y_0' = 1, \quad y_0'' = 1 + 1 = 2, \quad y_0''' = 4,$
   $y_0^{(4)} = 20, \quad y_0^{(5)} = 96, \quad y_0^{(6)} = 552, \quad y_0^{(7)} = 3776.$
   This gives the desired polynomial.
   $y_7(x) = 1 + x + x^2 + \tfrac{2}{3}x^3 + \tfrac{5}{6}x^4 + \tfrac{4}{5}x^5 + \tfrac{23}{30}x^6 + \tfrac{236}{315}x^7.$

3.7   The Runge-Kutta Method

1. The results are found in the following table.

Table 3.2:

x	0.20	0.30	0.40	0.50
y	1.25	1.42	1.64	1.94
K_1	1.52	1.93	2.53	
$x + \frac{1}{2}h$	0.25	0.35	0.45	
$y + \frac{1}{2}hK_1$	1.33	1.52	1.77	
K_2	1.71	2.19	2.93	
$y + \frac{1}{2}hK_2$	1.34	1.53	1.79	
K_3	1.73	2.22	3.00	
$x + h$	0.30	0.40	0.50	
$y + hK_3$	1.42	1.64	1.94	
K_4	1.93	2.53	3.51	
K	1.72	2.21	2.98	

3.8   A Continuing Method

1. $y_5^{(1)} = 1.93$; $y_5^{(2)} = 1.94$; $y_6^{(1)} = 2.36$; $y_6^{(2)} = 2.37$.

Chapter 4

Elementary Applications

4.3 Simple Chemical Conversion

1. From Section 4.1, $v_e = \sqrt{2gR} = \sqrt{2(.165)(.00609)(1080)} = 1.5$ miles/sec.

3. From Section 4.2, $u(t) = 70 + ce^{-kt}$. At $t = 0$, $u(0) = 70 + c = 18$ or $c = -52$. At $t = 1$, $u(1) = 70 - 52e^{-k} = 31$ or $e^{-k} = 3/4$ so $k = .29$. Thus when $t = 5$, $u(5) = 70 - 52\exp[-0.29(5)] = 58°$.

5. From Section 4.2, $u(t) = -10 + ce^{-kt}$. At 1:00 let $t = 0$, $u(0) = -10 + c = 70$ or $c = 80$. At $t = 2$, $u(2) = -10 + 80e^{-2k} = 26$ or $e^{-2k} = .45$ so $k = .4$. Thus when $t = 5$, $u(5) = -10 + 80\exp[-0.4(5)] = 1°$F.
Starting over at 1:05, $u(t) = 70 + ce^{-.4t}$ Let $t = 0$ again $u(0) = 70 + c = 1$ or $c = -69$. Thus at 1:09 when $t = 4$, $u(4) = 70 - 69\exp[-0.4(4)] = 56°$F.

7. From Section 4.2, $u(t) = 20 + ce^{-kt}$. At 2:00 let $t = 0$ so $u(0) = 20 + c = 80$ or $c = 60$. At 2:03, $t = 3$ so $u(3) = 20 + 60e^{-3k} = 42$ or $e^{-3k} = .366$ so $k = 1/3$. If the thermometer is brought in at t_0, it reads $u(t_0) = 20 + 60\exp(-t_0/3)$.
Starting over at $t = t_0$, $u(0) = 80 + c = 20 + 60\exp(-t_0/3)$ or $c = -60[1 - \exp(-t_0/3)]$. At 2:10, $t = 10 - t_0$, thus $u(10 - t_0) = 80 - 60[1 - \exp(-t_0/3)]\exp[-(10 - t_0)/3] = 71$ or $\exp(t_0/3) - 1 = 4.2$ so $t_0 = 5$. The thermometer was brought in at 2:05.

9. From Section 4.3, $x(t) = x_0 e^{-kt}$. At $t = 10$, $x_0/4 = x_0 e^{-10k}$ so $\ln(3/4) = -10k$ or $k = .028$. One-tenth will remain when $x_0/10 = x_0 e^{-.028t}$ or $\ln(1/10) = -(.028)t$ or $t = 80$ sec.

11. The equation is $\dfrac{dx}{dt} = -k\sqrt{x}$. Separating the variables gives $\displaystyle\int x^{-1/2}\,dx = -k\int dt$. Integrating yields $2x^{1/2} = -kt + c$ or $x(t) = (-kt/2 + c)^2$. At $t = 0$, $x(0) = x_0 = c^2$ or $c = \sqrt{x_0}$. The substance will disappear when $0 = (-kt/2 + \sqrt{x_0})^2$ or $t = 2\sqrt{x_0}/k$.

13. If $b \geq a$, $x \to a$; if $b \leq a$, $x \to b$.

15. From Section 4.3, $x(t) = x_0 e^{-kt}$. One-half of the material remains at $t = 38$ so $x_0/2 = x_0 e^{-38k}$ or $\ln(1/2) = -38k$ so $k = .018$. One-tenth of the material reamains when $x_0/10 = x_0 e^{-.018t}$ or $\ln(1/10) = -(.018)t$ so $t = 126$ hr.

19. From Exercise 17, $(w/g)\dfrac{dv}{dt} = w - kv$. This equation is linear in v. Putting it in standard form gives $\dfrac{dv}{dt} + (g/b)v = g$. The integrating factor is $\exp\left(\dfrac{g}{b}\int dt\right) = \exp(gt/b)$. Multiplying

by the integrating factor yields $\frac{d}{dt}\left(e^{gt/b}v\right) = ge^{gt/b}$ or $e^{gt/b}v = be^{gt/b} + c$. At $t = 0$, $v_0 = b + c$ or $c = v_0 - b$ so $v = b + (v_0 - b)\exp\left(-gt/b\right)$.

21. 1.9 sec and 10.5 ft/sec.

23. The rope will always be tangent to the curve, thus $\frac{dy}{dx} = \frac{-y}{\sqrt{a^2 - y^2}}$. Separating the variables

    yields $\int dx = -\int \frac{\sqrt{a^2 - y^2}}{y}\, dy$. Integrating gives $x = a\ln\left(\frac{a + \sqrt{a^2 - y^2}}{y}\right) - \sqrt{a^2 - y^2} + c$.

    At the start, $x = 0$ and $y = a$ so $0 = a\ln(a/a) - 0 + c$ or $c = 0$.

    Thus $x = a\ln\frac{a + \sqrt{a^2 - y^2}}{y} - \sqrt{a^2 - y^2}$.

25. (a) $s = 160$ lb; (b) $t = 64$ min.

27. From Section 4.3, $y(t) = 100e^{.05t}$. At $t = 1$, $y(1) = 100e^{.05} = 100(1.0513) = \$105.13$.
    (a) $\$105.13 - \$100 = \$5.13$;   (b) 5.13%.

4.4   Logistic Growth and the Price of Commodities

1. We want to maximize the growth rate $f(x) = x(b - ax)$. Taking the derivative and setting it equal to zero gives $f'(x) = -2ax + b = 0$ or $x = b/2a$.

3. Taking the derivative and setting it equal to zero gives $t = \dfrac{\ln 9}{\ln 9 - \ln 4} \approx 2.7$ hr.

5. The equation is $\frac{dg}{dt} = c - kg$ which is linear in g. In standard form it becomes $\frac{dg}{dt} + kg = c$.

   The integrating factor is $\exp\left(k\int dt\right) = \exp\left(kt\right)$. Multiplying by the integrating factor yields

   $\frac{d}{dt}\left(e^{kt}g\right) = ce^{kt}$. Integrating gives $e^{kt}g = (c/k)e^{kt} + c$ or $g(t) = (c/k) + ce^{(-kt)}$. As $t \to \infty$, $g(t) \to c/k$.

7.   a. The first dose is given at $t = 0$, so $y(t) = y_0 e^{-kt}$, for $0 \le t \le T$ so $y(T) = y_0 e^{-kT}$.

     b. With $y(T)$ as the new initial condition, $y(t) = y_0(1 + e^{-kT})e^{-kt}$, for $T \le t \le 2T$ so $y(2T) = y_0(e^{-kT} + e^{-k2T})$.

     c. Continuing in this way,
        $$y(nT) = y_0(e^{-kT} + e^{-k2T} + \cdots + e^{-knT}) = y_0(e^{-kT} + (e^{-kT})^2 + \cdots + (e^{-kT})^n).$$
        This is a geometric series with $r = e^{-kt} < 1$ so, $y(nT) = y_0\dfrac{e^{-kt} - (e^{-kt})^{(n+1)}}{1 - e^{-kt}}$.

     d. Taking the limit as $n \to \infty$, $y \to \dfrac{y_0 e^{-kT}}{1 - e^{-kT}}$.

9. From Section 4.4, $\frac{dP}{dt} = k(D - S) = k[(C - dP) - (a + bP + q\sin\beta t)]$ which is linear in P. In standard form it is $\frac{dP}{dt} + k(d + b)P = k(c - a) - kq\sin\beta t$. The integrating factor is

$$\exp\left(k(d + b)\int dt\right) = \exp\left[k(d + b)t\right].$$

Multiplying by the integrating factor yields

$$\frac{d}{dt}\left(Pe^{k(d+b)t}\right) = k(c - a)e^{k(d+b)t} - kqe^{k(d+b)t}\sin\beta t.$$

Integrating gives

$$Pe^{k(d+b)t} = \frac{c - a}{d + b}e^{k(d+b)t} - \frac{kqe^{k(d+b)t}}{Q^2}[k(d + b)\sin\beta t - \beta\cos\beta t] + c_1 \text{ where } Q^2 = k(d + b)^2 + \beta^2.$$

Let $(\beta/Q) = \cos(\beta t + \alpha)$, then $k(d + b)/Q = \sin\alpha$, and

$$P = \frac{c - a}{d + b} + \frac{cq}{Q}[\cos(\beta t + \alpha)] + c_1 e^{-k(d+b)t}.$$

Let $P(0) = P_0$, then $c_1 = P_0 - \frac{c - a}{d + b} - \frac{\beta kq}{Q^2}$ so

$$P(t) = \frac{c - a}{d + b} + \frac{qk}{Q}\cos(\beta t + \alpha) + c_1 e^{-k(d+b)t}.$$

Chapter 5

Additional Topics on Equations of Order One

5.1   Integrating Factors Found by Inspection

1. The terms of the differential equation may be rearranged to give

$$2xy^2 \, dx + y \, dx - x \, dy = 0 \quad \text{or} \quad 2x \, dx + \frac{y \, dx - x \, dy}{y^2} = 0.$$

Integration now yields $x^2 + \dfrac{x}{y} = c \quad \text{or} \quad x(xy + 1) = cy.$

3. The terms of the differential equation may be rearranged to give

$$x^3 y^2 (y \, dx + x \, dy) + dx = 0 \quad \text{or} \quad (xy)^2 (y \, dx + x \, dy) + \frac{dx}{x} = 0.$$

Integration now yields $\dfrac{(xy)^3}{3} + \ln|cx| = 0 \quad \text{or} \quad x^3 y^3 = -3 \ln|cx|.$

5. The terms of the differential equation may be rearranged to give

$$x^4 (y \, dx + x \, dy) + y^2 (x \, dy - y \, dx) = 0 \quad \text{or} \quad (y \, dx + x \, dy) + \frac{y^2}{x^2} \frac{x \, dy - y \, dx}{x^2} = 0.$$

Integration now yields $xy + \dfrac{1}{3} \left(\dfrac{y}{x} \right)^3 = c \quad \text{or} \quad y(3x^4 + y^2) = 3cx^3.$

7. The terms of the differential equation may be rearranged to give

$$y^2 (y \, dx + x \, dy) + y \, dx - x \, dy = 0 \quad \text{or} \quad (y \, dx + x \, dy) + \frac{y \, dx - x \, dy}{y^2} = 0.$$

Integration now yields $xy + \dfrac{x}{y} = c \quad \text{or} \quad x(y^2 + 1) = cy.$

9. The terms of the differential equation may be rearranged to give

$$x^2 y^2 (y \, dx + x \, dy) = y \, dx - x \, dy \quad \text{or} \quad xy(y \, dx + x \, dy) = \frac{y}{x} \frac{y \, dx - x \, dy}{y^2}.$$

Integration now yields $\dfrac{(xy)^2}{2} = \ln \left| \dfrac{x}{y} \right| + \ln|c| \quad \text{or} \quad (xy)^2 = 2 \ln \left| \dfrac{cx}{y} \right|.$

11. The terms of the differential equation may be rearranged to give

$$x^2 y^2 (y\, dx + x\, dy) = my\, dx - nx\, dy \quad \text{or} \quad xy\, d(xy) = m\frac{dx}{x} - n\frac{dy}{y}.$$

Integration now yields $\dfrac{(xy)^2}{2} = m \ln|x| - n \ln|y| + \ln|c| \quad$ or $\quad (xy)^2 = 2 \ln \left| \dfrac{cx^m}{y^n} \right|.$

13. The terms of the differential equation may be rearranged to give

$$y^2 (y\, dx + x\, dy) + 2xy\, dx - x^2\, dy = 0 \quad \text{or} \quad d(xy) + \frac{2xy\, dx - x^2\, dy}{y^2} = 0.$$

Integration now yields $xy + \dfrac{x^2}{y} = c \quad$ or $\quad x(x + y^2) = cy.$

15. The terms of the differential equation may be rearranged to give

$$x(3x^2 y\, dx - x^3\, dy) + x(x\, dy - y\, dx) + y^2\, dx = 0 \text{ or } \frac{3x^2 y\, dx - x^3\, dy}{y^2} + \frac{x\, dy - y\, dx}{y^2} + \frac{dx}{x} = 0.$$

Integration now yields $\dfrac{x^3}{y} - \dfrac{x}{y} + \ln|cx| = 0 \quad$ or $\quad y \ln|cx| = x(1 - x^2).$

17. The terms of the differential equation may be rearranged to give

$$x^2 y(x\, dy - y\, dx) + y(y\, dx + x\, dy) + 2x^2\, dy = 0 \quad \text{or} \quad \frac{y\, dx - x\, dy}{y^2} - \frac{y\, dx + x\, dy}{x^2 y^2} - \frac{2\, dy}{y^3} = 0.$$

Integration now yields $\dfrac{x}{y} + \dfrac{1}{xy} + \dfrac{1}{y^2} = c \quad$ or $\quad x^2 y + y + x = cxy^2.$

19. The terms of the differential equation may be rearranged to give

$$x^2 (y\, dx - x\, dy) + y^2 (x\, dy - y\, dx) + d(xy) = 0 \quad \text{or} \quad d\left(\frac{x}{y}\right) + d\left(\frac{y}{x}\right) + \frac{d(xy)}{x^2 y^2} = 0.$$

Integration now yields $\dfrac{x}{y} + \dfrac{y}{x} - \dfrac{1}{xy} + c = 0 \quad$ or $\quad x^2 + cxy + y^2 = 1.$

21. The terms of the differential equation may be rearranged to give

$$(x^2 + y^2)(y\, dx + x\, dy) + x\, dy - y\, dx = 0 \quad \text{or} \quad y\, dx + x\, dy + \frac{x\, dy - y\, dx}{x^2 + y^2} = 0.$$

Integration now yields $xy + \arctan \dfrac{y}{x} = c.$

23. The terms of the differential equation may be rearranged to give

$$x^3 e^{xy}(y\,dx + x\,dy) + y(x\,dy - y\,dx) = 0 \quad \text{or} \quad e^{xy}\,d(xy) + \frac{y}{x}\,d\left(\frac{y}{x}\right) = 0.$$

Integration now yields $e^{xy} + \dfrac{1}{2}\left(\dfrac{y}{x}\right)^2 = \dfrac{c}{2}$ or $2x^2 e^{xy} + y^2 = cx^2$.

25. The terms of the differential equation may be rearranged to give

$$(x^2 - y^2)(x\,dx - y\,dy) - x^2\,dx = 0 \quad \text{or} \quad \frac{1}{4}(x^2 - y^2)^2 - \frac{1}{3}x^3 = c.$$

When $x = 2$, $y = 0$, so that $c = \frac{4}{3}$. Therefore $3(x^2 - y^2)^2 = 4(x^3 + 4)$.

27. The terms of the differential equation may be rearranged to give

$$x^3 y^3(y\,dx + x\,dy) + x^2(2y\,dx - 2x\,dy) - y^2\,dx = 0 \quad \text{or} \quad xy\,d(xy) + 2\,d\left(\frac{x}{y}\right) - \frac{dx}{x^2} = 0.$$

Integration now yields $\dfrac{1}{2}x^2 y^2 + \dfrac{2x}{y} + \dfrac{1}{x} = c$ or $x^3 y^3 + 4x^2 + 2y = 2cxy$. When $x = 1$, $y = 1$, so that $c = 7/2$. Therefore $x^3 y^3 + 4x^2 - 7xy + 2y = 0$.

29. The terms of the differential equation may be rearranged to give

$$x^n y^n(x\,dx + y\,dy) + a(y\,dx + x\,dy) = 0 \quad \text{or} \quad (x\,dx + y\,dy) + \frac{a(y\,dx + x\,dy)}{x^n y^n} = 0.$$

Integration now yields:

$$\text{If } n \neq 1, \quad \frac{x^2}{2} + \frac{y^2}{2} - \frac{a(xy)^{1-n}}{1-n} - \frac{c}{2} = 0 \quad \text{or} \quad (n-1)(x^2 + y^2 - c)(xy)^{n-1} = 2a;$$

$$\text{If } n = 1, \quad \frac{x^2}{2} + \frac{y^2}{2} - \frac{c}{2} + a\ln|xy| = 0 \quad \text{or} \quad x^2 + y^2 - c = -2a\ln|xy|.$$

5.2   The Determination of Integrating Factors

1. $\dfrac{1}{N}\left(\dfrac{\partial M}{\partial y} - \dfrac{\partial N}{\partial x}\right) = \dfrac{-2}{x}$. Therefore $\exp\left[\displaystyle\int \dfrac{-2}{x}\,dx\right] = \dfrac{1}{x^2}$ is an integrating factor. That is, $\dfrac{x^2 + y^2 + 1}{x^2}\,dx + \dfrac{x - 2y}{x}\,dy = 0$ is exact. The solution is

$$x - \frac{y^2}{x} - \frac{1}{x} + y = c \quad \text{or} \quad x^2 - y^2 + xy - 1 = cx.$$

3. $\dfrac{1}{M}\left(\dfrac{\partial M}{\partial y}-\dfrac{\partial N}{\partial X}\right)=\dfrac{2}{y}$. Therefore $\exp\left[\displaystyle\int\dfrac{-2}{y}\,dy\right]=\dfrac{1}{y^2}$ is an integrating factor. That is,

$\dfrac{4x+y}{y}\,dx-\dfrac{2(x^2-y)}{y^2}\,dy=0$ is exact. The solution is

$$\dfrac{2x^2}{y}+x+2\ln|y|=c \quad\text{or}\quad 2x^2+xy+2y\ln|y|=cy.$$

5. $\dfrac{1}{N}\left(\dfrac{\partial M}{\partial y}-\dfrac{\partial N}{\partial x}\right)=-1$. Therefore $\exp\left[\displaystyle\int -1\,dx\right]=\exp(-x)$ is an integrating factor. That is, $y(y+2x-2)e^{-x}\,dx-2(x+y)e^{-x}\,dy=0$ is exact. The solution is

$$-2xye^{-x}-y^2e^{-x}=-c \quad\text{or}\quad y(2x+y)=ce^{x}.$$

7. $\dfrac{1}{N}\left(\dfrac{\partial M}{\partial y}-\dfrac{\partial N}{\partial x}\right)=\dfrac{2}{x}$. Therefore $\exp\left[\displaystyle\int\dfrac{2}{x}\,dx\right]=x^2$ is an integrating factor. That is, $x^2y(8x-9y)\,dx+2x^3(x-3y)\,dy=0$ is exact. The solution is

$$2x^4y-3x^3y^2=c \quad\text{or}\quad x^3y(2x-y)=c.$$

9. $\dfrac{1}{N}\left(\dfrac{\partial M}{\partial y}-\dfrac{\partial N}{\partial x}\right)=2x$. Therefore $\exp\left[\displaystyle\int 2x\,dx\right]=\exp(x^2)$ is an integrating factor. That is, $\exp(x^2)y(2x^2-xy+1)\,dx+\exp(x^2)(x-y)\,dy=0$ is exact. The solution is

$$\exp(x^2)\left(xy-\dfrac{y^2}{2}\right)=\dfrac{c}{2} \quad\text{or}\quad y(2x-y)=c\exp(-x^2).$$

11. $\dfrac{1}{N}\left(\dfrac{\partial M}{\partial y}-\dfrac{\partial N}{\partial x}\right)=\dfrac{3}{x}$. Therefore $\exp\left[\displaystyle\int\dfrac{3}{x}\,dx\right]=x^3$ is an integrating factor. That is, $2x^3(2y^2+5xy-2y+4)\,dx+x^4(2x+2y-1)\,dy=0$ is exact. The solution is

$$2x^5y+x^4y^2-x^4y+2x^4=c \quad\text{or}\quad x^4(y^2+2xy-y+2)=c.$$

13. $\dfrac{1}{N}\left(\dfrac{\partial M}{\partial y}-\dfrac{\partial N}{\partial x}\right)=\dfrac{1}{x}$. Therefore $\exp\left[\displaystyle\int\dfrac{1}{x}\,dx\right]=x$ is an integrating factor. That is, $(2xy^2+3x^2y-2xy+6x^2)\,dx+(x^3+2x^2y-x^2)\,dy=0$ is exact. The solution is

$$x^3y+x^2y^2-x^2y+2x^3=c \quad\text{or}\quad x^2(y^2+xy-y+2x)=c.$$

15. We consider the equation $\dfrac{M}{Mx+Ny}\,dx+\dfrac{N}{Mx+Ny}=0$ and check to see if it is exact. If we write the equation as $P\,dx+Q\,dy=0$ then

$$\dfrac{\partial P}{\partial y}-\dfrac{\partial Q}{\partial x}=\dfrac{(Mx+Ny)\left(\dfrac{\partial M}{\partial y}-\dfrac{\partial N}{\partial x}\right)+xN\dfrac{\partial M}{\partial x}-xM\dfrac{\partial M}{\partial y}+yN\dfrac{\partial N}{\partial x}-yM\dfrac{\partial N}{\partial y}}{(Mx+Ny)^2}$$

$$=\dfrac{N\left(y\dfrac{\partial M}{\partial y}+x\dfrac{\partial M}{\partial x}\right)-M\left(x\dfrac{\partial N}{\partial x}+y\dfrac{\partial N}{\partial y}\right)}{(Mx+Ny)^2}.$$

Now from Euler's theorem we have $x\dfrac{\partial M}{\partial x} + y\dfrac{\partial M}{\partial y} = kM$ and $x\dfrac{\partial N}{\partial x} + y\dfrac{\partial N}{\partial y} = kN$, so that $\dfrac{\partial P}{\partial y} - \dfrac{\partial Q}{\partial x} = 0$. That is $(Mx + Ny)^{-1}$ is an integrating factor of the original equation.

17. Because the equation is homogeneous, an integrating factor is $\dfrac{1}{x^2y - x^2y - 2y^3} = -\dfrac{1}{2y^3}$. Thus the differential equation $-\dfrac{x}{2y^2}\,dx + \dfrac{x^2 + 2y^2}{2y^3}\,dy = 0$ is exact. The solution is

$$-\frac{x^2}{4y^2} + \ln|y| = \ln|c| \quad \text{or} \quad 4y^2\ln\left|\frac{y}{c}\right| = x^2.$$

19. Because the equation is homogeneous, an integrating factor is $(Mu + Nv)^{-1} = -\dfrac{1}{uv^3}$. Thus the differential equation $-\dfrac{(u^2 + v^2)}{uv^2}\,du + \dfrac{(u^2 + 2v^2)}{v^3}\,dv = 0$ is exact. The solution is

$$-\frac{u^2}{2v^2} + \ln\left|\frac{v^2}{u}\right| + \ln|c| = 0 \quad \text{or} \quad u^2 = 2v^2\ln\left|\frac{cv^2}{u}\right|.$$

21. The general linear equation of order one is $\dfrac{dy}{dx} + P(x)y = Q(x)$. We may rewrite the equation in the form $(Q - Py)\,dx - dy = M\,dx + N\,dy = 0$. Now $\dfrac{1}{N}\left(\dfrac{\partial M}{\partial y} - \dfrac{\partial N}{\partial x}\right) = \dfrac{-P}{-1} = P$ is a function of x alone so that $\exp\left[\displaystyle\int P(x)\,dx\right]$ is an integrating factor. Hence the equation $\exp\left[\displaystyle\int P\,dx\right]\left[\dfrac{dy}{dx} + Py\right] = Q\exp\left[\displaystyle\int P\,dx\right]$ is an exact equation. Integrating we get $y\exp\left[\displaystyle\int P\,dx\right] = \displaystyle\int Q\exp\left[\displaystyle\int P\,dx\right]\,dx + c$.

5.4   Bernoulli's Equation

1. We set $u = 3x - 2y$ and $dy = \dfrac{3\,dx - du}{2}$ and simplify to get $(5u + 11)\,dx - (u + 3)\,du = 0$. The variables now separate and we integrate to get $25x - 5u - 4\ln|5u + 11| = c_1$. Substituting for u and simplifying yields the solution $5(x + y + c) - 2\ln|15x - 10y + 11| = 0$.

3. We set $u = 9x + 4y + 1$ and $\dfrac{dy}{dx} = \dfrac{1}{4}\dfrac{du}{dx} - \dfrac{9}{4}$ and simplify to get $\dfrac{du}{dx} = 4u^2 + 9$. The variables now separate and we integrate to get $\arctan\dfrac{2u}{3} = 6x + c$. Substituting for u and simplifying yields the solution $2(9x + 4y + 1) = 3\tan(6x + c)$.

5. We set $u = x + y$ and $\dfrac{dy}{dx} = \dfrac{du}{dx} - 1$ and simplify to get $\dfrac{du}{dx} = 1 + \sin u$. The variables now separate and we integrate to get $\tan u - \dfrac{1}{\cos u} = x + c$. Substituting for u and simplifying yields the solution $x + c = \tan(x + y) - \sec(x + y)$.

7. We set $u = \sin y$ and $du = \cos y\, dy$ and simplify to get $(3u^2 - 5xu)\, dx + 2x^2\, du = 0$. We solve this homogeneous equation to get $x^3(u - x)^2 = cu^2$. Substituting for u and simplifying yields the solution $x^3(\sin y - x)^2 = c\sin^2 y$.

9. We set $x = u - v$ and $\dfrac{dv}{du} = 1 - \dfrac{dx}{du}$ and simplify to get $dx + (x - 3)(x + 1)\, du = 0$. The variables now separate and we integrate to get $(x - 3)\exp(4u) = c(x + 1)$. Substituting for x and simplifying yields the solution $(u - v - 3)\exp(4u) = c(u - v + 1)$.

11. We set $y = e^{2v}$ and $dy = 2e^{2v}\, dv$ and simplify to get $(ky - u)\, du = (y + ku)\, dy$. We solve this homogeneous equation to get $2k\arctan\dfrac{u}{y} = \ln|c(u^2 + y^2)|$. Substituting for y and simplifying yields the solution $2k\arctan(ue^{-2v}) = \ln|c(u^2 + e^{4v})|$.

13. We set $u = x + 2y$ and $dx = du - 2\, dy$ and simplify to get $(u - 1)\, du - (3u - 7)\, dy = 0$. The variables now separate and we integrate to get $3u + 4\ln|3u - 7| - 9y = -c$. Substituting for u and simplifying yields the solution $3x - 3y + c + 4\ln|3x + 6y - 7| = 0$.

15. A Bernoulli equation. We set $v = y^{1-n}$ and $y^{-n}y' = \dfrac{v'}{1 - n}$ to get $v' - \dfrac{1 - n}{x}v = (1 - n)x^{k-1}$. Solving this linear equation we have $(k + n - 1)v = (1 - n)x^k + cx^{1-n}$. Substituting for v we get the solution $(k + n - 1)y^{1-n} = (1 - n)x^k + cx^{1-n}$.

17. We set $u = x + 2y$ and $dx = du - 2\, dy$ and simplify to get $(u - 1)\, du - dy = 0$. The variables now separate and we integrate to get $(u - 1)^2 = 2y + c$. Substituting for u and simplifying yields the solution $(x + 2y - 1)^2 = 2y + c$. The original equation is also exact.

19. A Bernoulli equation. We set $u = y^{-2}$ and $u' = -2y^{-3}y'$ to get $u' + \dfrac{3}{x}u = -\dfrac{1}{x^3}$. Solving this linear equation we have $ux^3 = c - x$. Substituting for u we get the solution $y^2(c - x) = x^3$.

21. Suppose $n = 1$ and $k = 0$. Then $xy' = 2y$. The variables separate and we get $y = cx^2$.

    Suppose $n = 1$ and $k \neq 0$. Then $xy' = (x^k + 1)y$. The variables separate and we get $x^k = k\ln\left|\dfrac{cy}{x}\right|$.

    Suppose $n \neq 1$ and $k + n = 1$. The solution proceeds as in Exercise 15 down to the linear equation $v' - \dfrac{1 - n}{x}v = (1 - n)x^{k-1}$ which in this case becomes $v' - \dfrac{1 - n}{x}v = (1 - n)x^{-n}$. This linear differential equation has solution $v = (1 - n)x^{1-n}\ln|cx|$. Substituting for v gives us $y^{1-n} = (1 - n)x^{1-n}\ln|cx|$.

23. We set $u = 3x + y$ and $y' = u' - 3$ and simplify to get $u' = 2(u^2 + 1)$. The variables now separate and we integrate to get $\arctan u = 2x + c$. Substituting for u and simplifying yields the solution $\arctan(3x + y) = 2x + c$. When $x = 0$, $y = 1$ implies that $c = \pi/4$. Thus $4\arctan(3x + y) = 8x + \pi$.

25. A Bernoulli equation. We set $u = y^{-3}$ and $\dfrac{du}{dx} = -3y^{-4}\dfrac{dy}{dx}$ to get $\dfrac{du}{dx} + \dfrac{2}{x}u = \dfrac{1}{x^2}$. Solving this linear equation we have $x^2u = x + c$. Substituting for u we get the solution $x^2 = y^3(x + c)$. When $x = 2$, $y = 1$ implies that $c = 2$. Hence $x^2 = y^3(x + 2)$.

27. We set $u = y^2$ and $du = 2y\, dy$ to get $\dfrac{du}{dx} - \dfrac{3}{x}u = \dfrac{x}{2}$. Solving this linear equation we have $u = -\dfrac{x^2}{2} + cx^3$. Substituting for u we get the solution $y^2 = -\dfrac{x^2}{2} + cx^3$. When $x = 1$, $y = 1$ implies that $c = 3/2$. Thus $2y^2 = x^2(3x - 1)$.

5.5   Coefficients Linear in the Two Variables

1. The lines $y = 2$ and $x - y - 1 = 0$ intersect at the point $(3, 2)$ so we set $x = u + 3$ and $y = v + 2$. We get $v\, du - (u - v)\, dv = 0$. We solve this homogeneous equation and get $u = -v \ln |cv|$. Substituting for u and v gives us $x - 3 = (2 - y) \ln |c(y - 2)|$.

3. The lines $y = 2x$ and $4x + y - 6 = 0$ intersect at the point $(1, 2)$ so we set $x = u + 1$ and $y = v + 2$. We get $(2u - v)\, du + (4u + v)\, dv = 0$. We solve this homogeneous equation and get $(u + v)^3 = c(2u + v)^2$. Substituting for u and v gives us $(x + y - 3)^3 = c(2x + y - 4)^2$.

5. We let $u = x + y$ and $du = dx + dy$ and get $(u - 1)\, du + (u + 2)\, dy = 0$. The variables separate and we have $u + y + c = 3 \ln |u + 2|$. Substituting for u we obtain $x + 2y + c = 3 \ln |x + y + 2|$.

7. The lines $x - 3y + 2 = 0$ and $x + 3y - 4 = 0$ intersect at the point $(1, 1)$ so we substitute $x = u + 1$ and $y = v + 1$. We get $(u - 3v)\, du + 3(u + 3v)\, dv = 0$. We solve this homogeneous equation and get $\ln (u^2 + 9v^2) - 2 \arctan \dfrac{u}{3v} = c$. Substituting for u and v gives us the solution $\ln \left[(x - 1)^2 + 9(y - 1)^2 \right] - 2 \arctan \dfrac{x - 1}{3(y - 1)} = c$.

9. The lines $9x - 4y + 4 = 0$ and $2x - y + 1 = 0$ intersect at the point $(0, 1)$ so we need only set $y = u + 1$. We get $(9x - 4u)\, dx - (2x - u)\, du = 0$. We solve this homogeneous equation and get $u = 3(u - 3x) \ln |c(3x - u)|$. Substituting for u gives us $y - 1 = 3(y - 3x - 1) \ln |c(3x - y + 1)|$.

11. The lines $x + 2y - 1 = 0$ and $2x + y - 5 = 0$ intersect at the point $(3, -1)$ so we set $x = u + 3$ and $y = v - 1$. We get $(u + 2v)\, du - (2u + v)\, dv = 0$. We solve this homogeneous equation and get $(uv)^3 = c(v + u)$. Substituting for u and v gives us $(x - y - 4)^3 = c(x + y - 2)$.

13. The lines $3x + 2y + 7 = 0$ and $2x - y = 0$ intersect at the point $(-1, -2)$ so we set $x = u - 1$ and $y = v - 2$. We get $(3u + 2v)\, du + (2u - v)\, dv = 0$. We solve this homogeneous equation and get $3u^2 + 4uv - v^2 = c_1$. Substituting for u and v gives us $3x^2 + 4xy - y^2 + 14x = c$. The original equation is exact, so an easier method of solution is available.

15. We let $u = 2x - y + 3$ and $dy = 2\, dx - du$ and get $(2 + 2u)\, dx - u\, du = 0$. The variables separate and we have $2x - u + \ln |u + 1| = -c - 3$. Substituting for u we obtain $y + c = -\ln |2x - y + 4|$.

17. We let $u = x - y + 2$ and $dy = dx - du$ and get $(u + 3)\, dx - 3\, du = 0$. The variables separate and we have $x + c = 3 \ln |u + 3|$. Substituting for u we obtain $x + c = 3 \ln |x - y + 5|$.

19. The lines $2x - 3y + 4 = 0$ and $x = 1$ intersect at the point $(1, 2)$ so we set $x = u + 1$ and $y = v + 2$. We get $(2u - 3v)\, du + 3u\, dv = 0$. We solve this homogeneous equation and get $2u \ln |u| + 3v = cu$. Substituting for u and v gives us $2(x - 1) \ln |x - 1| + 3(y - 2) = c(x - 1)$. When $x = -1$, $y = 2$ implies $c = 2 \ln 2$, so that $3(y - 2) = -2(x - 1) \ln \left(\dfrac{1 - x}{2} \right)$.

21. The lines $x + y - 4 = 0$ and $3x - y - 4 = 0$ intersect at the point $(2, 2)$ so we set $x = u + 2$ and $y = v + 2$. We get $(u + v)\, du - (3u - v)\, dv = 0$. We solve this homogeneous equation and get $\ln|v - u| + \dfrac{2u}{v - u} = c$. Substituting for u and v gives us $\ln|y - x| + \dfrac{2(x - 2)}{y - x} = c$. When $x = 3$, $y = 7$ implies $c = \frac{1}{2} + \ln 4$. Thus $y - 5x + 8 = 2(y - x)\ln\left(\dfrac{y - x}{4}\right)$.

23. Equation (B) is $\alpha^2 - 5\alpha + 6 = 0$ with roots $\alpha_1 = 2$ and $\alpha_2 = 3$. We substitute $x = 2u + 3v$ and $y = u + v$ and get $(v - 1)\, du + (-2u - 4)\, dv = 0$. The variables separate and we get $(v - 1)^2 + c(u + 2) = 0$. Substituting for u and v we get $(x - 2y - 1)^2 = c(x - 3y - 2)$.

25. Equation (B) is $\alpha^2 + 3\alpha + 2 = 0$ with roots $\alpha_1 = -2$ and $\alpha_2 = -1$. We substitute $x = -2u - v$ and $y = u + v$ and get $3\, du + (-u + 2)\, dv = 0$. The variables separate and we get the solution $v + c = 3\ln|u - 2|$. Substituting for u and v we get $x + 2y + c = 3\ln|x + y + 2|$.

27. Equation (B) is $\alpha^2 - 3\alpha + 2 = 0$ with roots $\alpha_1 = 1$ and $\alpha_2 = 2$. We substitute $x = u + 2v$ and $y = u + v$ and get $(-2v + 4)\, du + (u + 3)\, dv = 0$. The variables separate and we get the solution $(u + 3)^2 = c(2 - v)$. Substituting for u and v we get $(2y - x + 3)^2 = c(y - x + 2)$.

29. We substitute $x = \alpha_1 u + \beta v$ and $y = u + v$. The coefficient of du will be given by the expression $\alpha_1\left[a_1(\alpha_1 u + \beta v) + b_1(u + v) + c_1\right] + \left[a_2(\alpha_1 u + \beta v) + b_2(u + v) + c_2\right]$. In order for the resulting differential equation to be linear in u, the coefficient of u in this polynomial must be zero. That coefficient is $a_1\alpha_1^2 + (a_2 + b_1)\alpha_1 + b_2$. Thus if α_1 is a root of the equation $a_1\alpha^2 + (a_2 + b_1)\alpha + b_2 = 0$ the resulting equation will be linear. The condition that $\beta \neq \alpha_1$ is required so that $x \neq \alpha_1 y$.

31. Equation (B) is now $\alpha^2 + 4\alpha + 4 = 0$. We take $\alpha_1 = -2$ and $\beta = 0$. We substitute $x = -2u$ and $y = u + v$ and get $\dfrac{du}{dv} - \dfrac{u}{v} = \dfrac{1}{v} - 1$. This linear equation has solution $u + 1 - c_1 v = -v\ln|v|$. Substituting for u and v gives $-\dfrac{x}{2} + 1 - \dfrac{c_1(x + 2y)}{2} = -\dfrac{x + 2y}{2}\ln\left|\dfrac{x + 2y}{2}\right|$. Each different choice of β will give a different form for this solution. In this particular form, if we put $c_1 = -1 - \ln 2c$ we get the answer given in the text.

33. Equation (B) is now $\alpha^2 - 3\alpha + 2 = 0$. We take $\alpha_1 = 1$ and $\beta = 0$. We substitute $x = u$ and $y = u + v$ and get $\dfrac{du}{dv} - \dfrac{1}{2v + 4}u = -1 - \dfrac{1}{2(v + 2)}$. This linear equation has solution $(u + 2v + 3)^2 = c(v + 2)$. Substituting for u and v gives $(2y - x + 3)^2 = c(y - x + 2)$.

5.6 Solutions Involving Nonelementary Integrals

1. The variables separate to give $\dfrac{dy}{y} = \left[1 - \exp\left(-x^2\right)\right]\, dx$. Upon integrating we have the solution in the form $\ln|cy| = x - \displaystyle\int_0^x \exp\left(-\beta^2\right)\, d\beta = x - \dfrac{1}{2}\sqrt{\pi}\, \operatorname{erf} x$.

3. The equation is linear with the integrating factor $\exp\left(\int 4x^3\,dx\right) = \exp\left(x^4\right)$. The solution is given by $y\exp\left(x^4\right) = \int_0^x \exp\left(\beta^4\right)\,d\beta + c.$

5. The equation is linear and may be written $\dfrac{dy}{dx} - y = \dfrac{1}{x}$. An integrating factor is e^{-x} so that $ye^{-x} = \int_c^x \dfrac{e^{-\beta}}{\beta}\,d\beta$, where $c > 0$. The condition $y = 0$, $x = 1$ requires that $c = 1$. Therefore $y = e^x \int_1^x \dfrac{e^{-\beta}}{\beta}\,d\beta.$

7. This linear equation may be written $\dfrac{dy}{dx} - 2xy = x^2$ and has an integrating factor $\exp\left(-x^2\right)$. The solution may be written $y\exp\left(-x^2\right) = \int x^2 \exp\left(-x^2\right)\,dx + c$. An integration by parts gives

$$y\exp\left(-x^2\right) = -\frac{x}{2}\exp\left(-x^2\right) + \frac{1}{2}\int_0^x \exp\left(-\beta^2\right)\,d\beta + c.$$

The initial condition yields $c = 1$. Thus $2y = 2\exp\left(x^2\right) - x + \dfrac{\sqrt{\pi}}{2}\exp\left(x^2\right)\operatorname{erf} x.$

Miscellaneous Exercises

1. We set $u = y^2 - 3y$, and $du = (2y - 3)\,dy$ and simplify to get $\dfrac{du}{dx} + u = x$. We solve this linear equation and obtain $u = x - 1 + ce^{-x}$. Substituting for u and simplifying yields the solution $y^2 - 3y - x + 1 = ce^{-x}$.

3. The lines $x + 3y - 5 = 0$ and $x - y - 1 = 0$ intersect at the point $(2, 1)$ so we set $x = u + 2$ and $y = v + 1$. We get $(u + 3v)\,du - (u - v)\,dv = 0$. We solve this homogeneous equation and get $2v = (u+v)\ln|c(u+v)|$. Substituting for u and v gives us $2(y-1) = (x+y-3)\ln|c(x + y - 3)|$.

5. This differential equation is exact as it stands. The standard procedure for solving exact equations yields the solution $2x^2 + 2xy - 3y^2 - 8x + 24y = c$.

7. We set $u = x^4$, and $du = 4x^3\,dx$ and simplify to get $\dfrac{du}{dy} + \dfrac{12}{y}u = 4y^2$. We solve this linear equation and obtain $15uy^{12} = 4y^{15} + c$. Substituting for u and simplifying yields the solution $15x^4 y^{12} = 4y^{15} + c$.

9. We set $u = x^2$, and $du = 2x\,dx$ and simplify to get $\dfrac{du}{dy} - \dfrac{6}{y}u = -2y^3$. We solve this linear equation and obtain $u = y^4 + cy^6$. Substituting for u and simplifying yields the solution $x^2 = y^4(1 + cy^2)$.

11. We set $u = e^y$, and $du = e^y\,dy$ and simplify to get $\dfrac{du}{dx} + \dfrac{3}{2x}u = -\dfrac{5}{2}$. We solve this linear equation and obtain $x^3(x + u)^2 = c$. Substituting for u and simplifying yields the solution $x^3(x + e^y)^2 = c$.

13. The lines $x - 3y + 4 = 0$ and $x - y - 2 = 0$ intersect at the point $(5, 3)$ so we set $x = u + 5$ and $y = v + 3$. We get $(u - 3v)\,du + 2(u - v)\,dv = 0$. We solve this homogeneous equation and get $(u + v)^4 = c(u - 2v)$. Substituting for u and v gives us $(x + y - 8)^4 = c(x - 2y + 1)$.

15. A Bernoulli equation. We set $u = x^{-1}$ and $\dfrac{du}{dy} = -x^{-2}\dfrac{dx}{dy}$ to get $\dfrac{du}{dy} + \dfrac{1}{y}u = -y^3$. Solving this linear equation we have $y(5u + y^4) = c$. Substituting for u we get the solution $y(5 + xy^4) = cx$.

17. A Bernoulli equation. We set $u = x^{-2}$ and $\dfrac{du}{dy} = -2x^{-3}\dfrac{dx}{dy}$ to get $\dfrac{du}{dy} + \dfrac{2}{y}u = 2$. Solving this linear equation we have $y^2(2y - 3u) = c$. Substituting for u we get $y^2(2x^2y - 3) = cx^2$.

19. We set $u = 4x + 3y$, and $dy = \dfrac{du - 4\,dx}{3}$ and simplify to get $-(u + 25)\,dx + (u + 1)\,du = 0$. The variables now separate and we integrate to get $u - x + 3c = 24\ln|u + 25|$. Substituting for u and simplifying yields the solution $x + y + c = 8\ln|4x + 3y + 25|$.

21. We set $u = x - y$, and $dx = du + dy$ and simplify to get $(3u - 2)\,du + (2u - 3)\,dy = 0$. The variables now separate and we integrate to get $-6u - 4y + 2c = 5\ln|2u - 3|$. Substituting for u and simplifying yields the solution $2(y - 3x + c) = 5\ln|2x - 2y - 3|$.

23. The lines $x - y - 1 = 0$ and $y = 2$ intersect at the point $(3, 2)$ so we set $x = u + 3$ and $y = v + 2$. We get $(u - v)\,du - 2v\,dv = 0$. We solve this homogeneous equation and get $(u + v)^2(u - 2v) = c$. Substituting for u and v gives us $(x + y - 5)^2(x - 2y + 1) = c$.

25. We set $u = x + 2y$, and $dx = du - 2\,dy$ and simplify to get $(2u - 1)\,du - 5(u - 1)\,dy = 0$. The variables now separate and we integrate to get $\ln|u - 1| = 5y - 2u + c$. Substituting for u and simplifying yields the solution $\ln|x + 2y - 1| = y - 2x + c$.

27. $\dfrac{1}{N}\left(\dfrac{\partial M}{\partial y} - \dfrac{\partial N}{\partial x}\right) = 3$. Therefore e^{3x} is an integrating factor. That is, the differential equation $e^{3x}(6xy - 3y^2 + 2y)\,dx + 2e^{3x}(x - y)\,dy = 0$ is exact. The solution is $y(2x - y) = ce^{-3x}$.

29. We set $u = ax + by + c$ and $du = a\,dx + b\,dy$ and simplify to get $\dfrac{du}{dx} - bu = a$. We solve this linear equation and obtain $bu = -a + c_1e^{bx}$. Substituting for u and simplifying yields the solution $b^2y = c_1e^{bx} - abx - a - cb$.

31. A Bernoulli equation. We set $u = v^{-2}$ and $\dfrac{du}{dx} = -2v^{-3}\dfrac{dv}{dx}$ to get $\dfrac{du}{dx} - \dfrac{2}{x}u = 1$. Solving this linear equation we have $u = cx^2 - x$. Substituting for u we get the solution $v^2(cx^2 - x) = 1$. When $x = 1$, $v = \frac{1}{2}$ implies that $c = 5$. Therefore $xv^2(5x - 1) = 1$.

33. We set $u = x + y$, and $dx = du - dy$ and simplify to get $(1 + u^2)\,du - yu\,dy = 0$. The variables now separate and we integrate to get $y^2 = u^2 + 2\ln|u| + c$. Substituting for u and simplifying yields the solution $y^2 = (x + y)^2 + 2\ln|x + y| + c$.

35. This differential equation is exact as it stands. The standard procedure for solving exact equations yields the solution $x^2 + x - 3xy - y^2 + 4y = c$.

37. This equation is homogeneous. We set $y = vx$ and get $\dfrac{dx}{x} - \dfrac{4v}{v^4 - 1}\, dv = 0$. Integrating we get $x(v^2 + 1) = c(v^2 - 1)$. Substituting for v gives $x(y^2 + x^2) = c(y^2 - x^2)$. When $x = 1$, $y = 2$ implies that $c = 5/3$. We finally have $3x(y^2 + x^2) = 5(y^2 - x^2)$ which may be written $y^2(5 - 3x) = x^2(5 + 3x)$.

39. We set $u = x - y + 2$, and $u' = 1 - y'$ and simplify to get $du + (u - 1)\, dx = 0$. The variables now separate and we integrate to get $\ln|u - 1| = c - x$. Substituting for u and simplifying yields the solution $\ln|x - y + 1| = c - x$.

Chapter 6

Linear Differential Equations

6.2 An Existence and Uniqueness Theorem

1. All the coefficients and the function $R(x)$ are continuous for all x. Moreover $b_0(x) = x - 1$ is zero for only $x = 1$. Hence the equation is normal for $x > 1$ and for $x < 1$.

3. All the coefficients are continuous for all x. $R(x) = \ln x$ is continuous for $x > 0$. $b_0(x) = x^2$ is not zero for $x > 0$. Hence the equation is normal for $x > 0$.

5. Since e^x and e^{-x} are solutions of the differential equation, $y = c_1 e^x + c_2 e^{-x}$ is a solution also. For this solution $y(0) = 4$ and $y'(0) = 2$ implies that $c_1 + c_2 = 4$ and $c_1 - c_2 = 2$. Hence $c_1 = 3$ and $c_2 = 1$. It follows that $y = 3e^x + e^{-x}$ is the unique solution of the initial value problem.

7. Since e^x and xe^x are solutions of the differential equation, $y = c_1 e^x + c_2 x e^x$ is a solution also. For this solution $y(0) = 7$ and $y'(0) = 4$ implies that $c_1 = 7$ and $c_1 + c_2 = 4$. Hence $c_1 = 7$ and $c_2 = -3$. It follows that $y = 7e^x - 3xe^x$ is the unique solution of the initial value problem.

9. Consider the general homogeneous linear equation $b_0(x)\dfrac{d^n y}{dx^n} + b_1(x)\dfrac{d^{n-1}y}{dx^{n-1}} + \cdots b_n(x)y = 0$ that is normal on the interval I. The function $y(x) = 0$ that has value zero for all x on the interval I is clearly a solution of this differential equation. Moreover for this zero function $y(x_0) = y'(x_0) = \cdots = y^{(n-1)}(x_0) = 0$ for any x on the interval I. By Theorem 6.2 this zero function is the unique solution that has these properties.

6.4 The Wronskian

1. $W(x) = \begin{vmatrix} 1 & x & x^2 & \cdots & x^{n-1} \\ 0 & 1 & 2x & \cdots & (n-1)x^{n-2} \\ \vdots & \vdots & \vdots & & \vdots \\ 0 & 0 & 0 & \cdots & (n-1)! \end{vmatrix} = 0!\,1!\,2! \cdots (n-1)!$.

3. $W(x) = \begin{vmatrix} e^x & \cos x & \sin x \\ e^x & -\sin x & \cos x \\ e^x & -\cos x & -\sin x \end{vmatrix} = \begin{vmatrix} e^x & \cos x & \sin x \\ e^x & -\sin x & \cos x \\ 2e^x & 0 & 0 \end{vmatrix} = 2e^x \begin{vmatrix} \cos x & \sin x \\ -\sin x & \cos x \end{vmatrix} = 2e^x$. Since $W(x)$ is never zero these three functions are linearly independent on any interval.

5. The trigonometric identity $\cos{(\omega t - \beta)} = \cos{\omega t}\cos{\beta} + \sin{\omega t}\sin{\beta}$ may be written in the form $1 \cdot \cos{(\omega t - \beta)} + (-\cos{\beta}) \cdot \cos{\omega t} + (-\sin{\beta}) \cdot \sin{\omega t} = 0$ for all t. Hence the three functions are linearly dependent on any interval.

7. The trigonometric identity $\sin^2{x} + \cos^2{x} = 1$ may be written $1 \cdot 1 + (-1) \cdot \sin^2{x} + (-1) \cdot \cos^2{x} = 0$ for all x. Hence the three functions are linearly dependent on any interval.

9. $W(x) = \begin{vmatrix} f & xf & x^2f \\ f' & xf' + f & x^2f' + 2xf \\ f'' & xf'' + 2f' & x^2f'' + 4xf' + 2f \end{vmatrix} = \begin{vmatrix} f & 0 & 0 \\ f' & f & 2xf \\ f'' & 2f' & 4xf' + 2f \end{vmatrix} = f\begin{vmatrix} f & 0 \\ 2f' & 2f \end{vmatrix} = 2f^3(x)$

Since $f(x)$ is never zero on the interval the Wronskian is never zero on the interval and the functions are linearly independent by Theorem 6.3.

11. Since the equation is homogeneous the zero function, $f(x) = 0$ for all x in the interval, is a solution of the differential equation. Moreover the zero function satisfies the initial conditions $f(x_0) = f'(x_0) = 0$. By the uniqueness theorem the zero function is the only function that has these properties. Therefore $y = f$.

13. Since y_1 and y_2 are solutions of the homogeneous equation (A), by Theorem 6.1 the linear combination $y(x) = \bar{c}_1 y_1(x) + \bar{c}_2 y_2(x)$ is also a solution. Moreover $y(x_0)$ and $y'(x_0)$ are both zero. Hence by the result of Exercise 11, $y(x) = 0$ for all x in the interval.

6.8   The Fundamental Laws of Operation

1. From the definitions we have

$$(4D + 1)(D - 2)y = (4D + 1)\big[(D - 2)y\big] = (4D + 1)(y' - 2y) = (4D)(y' - 2y) + (y' - 2y)$$
$$= 4(y'' - 2y') + (y' - 2y) = 4y'' - 7y' - 2y = (4D^2 - 7D - 2)y.$$

Hence $(4D + 1)(D - 2) = 4D^2 - 7D - 2$.

3. From the definitions we have

$$(D + 2)(D^2 - 2D + 5)y = (D + 2)(y'' - 2y' + 5y) = D(y'' - 2y' + 5y) + 2(y'' - 2y' + 5y)$$
$$= (y''' - 2y'' + 5y') + 2(y'' - 2y' + 5y)$$
$$= y''' + y' + 10y = (D^3 + D + 10)y.$$

Hence $(D + 2)(D^2 - 2D + 5) = D^3 + D + 10$.

5. $2D^2 + 3D - 2 = (D + 2)(2D - 1)$.

7. $D^3 - 2D^2 - 5D + 6 = (D - 1)(D^2 - D - 6) = (D - 1)(D + 2)(D - 3)$.

9. $D^4 - 4D^2 = D^2(D^2 - 4) = D^2(D - 2)(D + 2)$.

11. $D^3 - 21D + 20 = (D - 1)(D^2 + D - 20) = (D - 1)(D - 4)(D + 5)$.

13. $2D^4 + 11D^3 + 18D^2 + 4D - 8 = (D + 2)(2D^3 + 7D^2 + 4D - 4) = (D + 2)^2(2D^2 + 3D - 2) = (D + 2)^3(2D - 1)$.

15. $D^4 + D^3 - 2D^2 + 4D - 24 = (D-2)(D^3 + 3D^2 + 4D + 12) = (D-2)(D+3)(D^2+4)$.

17. From the definitions we have

$$(D-x)(D+x)y = (D-x)(y' + xy) = D(y' + xy) - x(y' + xy)$$
$$= y'' + xy' + y - xy' - x^2 y = y'' + (1 - x^2)y = [D^2 + (1 - x^2)]y.$$

Hence $(D-x)(D+x) = D^2 + 1 - x^2$.

19. $D(xD - 1)y = D(xy' - y) = xy'' = (xD^2)y$. Hence $D(xD - 1) = xD^2$.

21. From the definitions we have

$$(xD + 2)(xD - 1)y = (xD + 2)(xy' - y) = (xD)(xy' - y) + 2(xy' - y)$$
$$= x(xy'') + 2xy' - 2y = x^2 y'' + 2xy' - 2y = (x^2 D^2 + 2xD - 2)y.$$

Hence $(xD + 2)(xD - 1) = x^2 D^2 + 2xD - 2$.

6.9   Some Properties of Differential Operators

1. Given the equation $(D - 2)^3 y = 0$ we multiply by the factor e^{-2x} and use the exponential shift $e^{-2x}(D - 2)^3 y = D^3(e^{-2x}y) = 0$. We now obtain $e^{-2x}y = c_1 + c_2 x + c_3 x^2$ so that $y = (c_1 + c_2 x + c_3 x^2)e^{2x}$.

3. Given the equation $(2D - 1)^2 y = 0$ we multiply by the factor $e^{-x/2}$ and use the exponential shift $e^{-x/2}(2D - 1)^2 y = (2D)^2(e^{-x/2}y) = 0$. We now obtain $e^{-x/2}y = c_1 + c_2 x$ so that $y = (c_1 + c_2 x)e^{x/2}$.

5. Suppose the four functions are linearly dependent on an interval I. Then there exist four constants, not all zero, such that $c_1 e^{-3x} + c_2 x e^{-3x} + c_3 x^2 e^{-3x} + c_4 x^3 e^{-3x} = 0$ for all x in I. But e^{-3x} is never zero so we can divide both sides by e^{-3x} and obtain $c_1 + c_2 x + c_3 x^2 + c_4 x^3 = 0$. It follows that $1, x, x^2, x^3$ are linearly dependent on I. This contradicts the results obtained in Exercise 1 of Section 6.4. Therefore the supposition that the four functions are linearly dependent is false.

Chapter 7

Linear Equations with Constant Coefficients

7.2   The Auxiliary Equation: Distinct Roots

1. The auxiliary equation is $m^2 + 2m - 3 = 0$ and its roots are $m = -3,\ 1$. The solution is
   $y = c_1 e^{-3x} + c_2 e^x$.

3. The auxiliary equation is $m^2 + m - 6 = 0$ and its roots are $m = -3,\ 2$. The solution is
   $y = c_1 e^{-3x} + c_2 e^{2x}$.

5. The auxiliary equation is $m^3 + 3m^2 - 4m = 0$ and its roots are $m = 0,\ 1,\ -4$. The solution
   is $y = c_1 + c_2 e^x + c_3 e^{-4x}$.

7. The auxiliary equation is $m^3 + 6m^2 + 11m + 6 = 0$ and its roots are $m = -1,\ -2,\ -3$. The
   solution is $y = c_1 e^{-x} + c_2 e^{-2x} + c_3 e^{-3x}$.

9. The auxiliary equation is $4m^3 - 7m + 3 = 0$ and its roots are $m = 1,\ \frac{1}{2},\ -\frac{3}{2}$. The solution is
   $y = c_1 e^x + c_2 \exp\left(\frac{1}{2}x\right) + c_3 \exp\left(-\frac{3}{2}x\right)$.

11. The auxiliary equation is $m^3 + m^2 - 2m = 0$ and its roots are $m = 0,\ 1,\ -2$. The solution is
    $y = c_1 + c_2 e^t + c_3 e^{-2t}$.

13. The auxiliary equation is $9m^3 - 7m + 2 = 0$ and its roots are $m = -1,\ \frac{1}{3},\ \frac{2}{3}$. The solution is
    $y = c_1 e^{-x} + c_2 \exp\left(\frac{1}{3}x\right) + c_3 \exp\left(\frac{2}{3}x\right)$.

15. The auxiliary equation is $m^3 - 14m + 8 = 0$ and its roots are $m = -4,\ 2 \pm \sqrt{2}$. The solution
    is $y = c_1 e^{-4x} + c_2 \exp\left[(2 + \sqrt{2})x\right] + c_3 \exp\left[(2 - \sqrt{2})x\right]$.

17. The auxiliary equation is $4m^4 - 8m^3 - 7m^2 + 11m + 6 = 0$ and its roots are $m = -1,\ 2,\ -\frac{1}{2},\ \frac{3}{2}$.
    The solution is $y = c_1 e^{-x} + c_2 e^{2x} + c_3 \exp\left(-\frac{1}{2}x\right) + c_4 \exp\left(\frac{3}{2}x\right)$.

19. The auxiliary equation is $4m^4 + 4m^3 - 13m^2 - 7m + 6 = 0$ and its roots are $m = -1,\ -2,\ \frac{1}{2},\ \frac{3}{2}$.
    The solution is $y = c_1 e^{-x} + c_2 e^{-2x} + c_3 \exp\left(\frac{1}{2}x\right) + c_4 \exp\left(\frac{3}{2}x\right)$.

21. The auxiliary equation is $m^2 - 4am + 3a^2 = 0$ and its roots are $m = 3a,\ a$. The solution is
    $y = c_1 e^{3ax} + c_2 e^{ax}$.

23. The auxiliary equation is $m^2 - 2m - 3 = 0$ and its roots are $m = -1,\ 3$. The general solution is
    $y = c_1 e^{-x} + c_2 e^{3x}$ and $y' = -c_1 e^{-x} + 3c_2 e^{3x}$. But $y(0) = c_1 + c_2 = 0$ and $y'(0) = -c_1 + 3c_2 = -4$,
    so that $c_1 = 1,\ c_2 = -1$. The particular solution is $y = e^{-x} - e^{3x}$.

25. The auxiliary equation is $m^2 - 2m - 3 = 0$ and its roots are $m = 3, -1$. The general solution is $y = c_1 e^{3x} + c_2 e^{-x}$ and $y' = 3c_1 e^{3x} - c_2 e^{-x}$. But $y(0) = c_1 + c_2 = 4$ and $y'(0) = 3c_1 - c_2 = 0$, so that $c_1 = 1$, $c_2 = 3$. The particular solution is $y = e^{3x} + 3e^{-x}$. Thus $y(1) = e^3 + e^{-1}$.

27. The auxiliary equation is $m^2 - m - 6 = 0$ and its roots are $m = 3, -2$. The general solution is $y = c_1 e^{3x} + c_2 e^{-2x}$ and $y' = 3c_1 e^{3x} - 2c_2 e^{-2x}$. But $y(0) = c_1 + c_2 = 3$ and $y'(0) = 3c_1 - 2c_2 = -1$, so that $c_1 = 1$, $c_2 = 2$. The particular solution is $y = e^{3x} + 2e^{-2x}$. Thus $y(1) = e^3 + 2e^{-2}$.

29. The auxiliary equation is $m^3 - 2m^2 - 5m + 6 = 0$ and its roots are $m = 1, 3, -2$. The general solution is $y = c_1 e^x + c_2 e^{3x} + c_3 e^{-2x}$, so that $y' = c_1 e^x + 3c_2 e^{3x} - 2c_3 e^{-2x}$, and $y'' = c_1 e^x + 9c_2 e^{3x} + 4c_3 e^{-2x}$. But $y(0) = c_1 + c_2 + c_3 = 1$, $y'(0) = c_1 + 3c_2 - 2c_3 = -7$, and $y''(0) = c_1 + 9c_2 + 4c_3 = -1$. Thus $c_1 = 0$, $c_2 = -1$, $c_3 = 2$. The particular solution is $y = -e^{3x} + 2e^{-2x}$ and $y(1) = -e^3 + 2e^{-2}$.

7.3 The Auxiliary Equation: Repeated Roots

1. The auxiliary equation is $m^2 - 6m + 9 = 0$ and its roots are $m = 3, 3$. The general solution is $y = (c_1 + c_2 x)e^{3x}$.

3. The auxiliary equation is $4m^3 + 4m^2 + m = 0$ and its roots are $m = 0, -\frac{1}{2}, -\frac{1}{2}$. The general solution is $y = c_1 + (c_2 + c_3 x)\exp\left(-\frac{1}{2}x\right)$.

5. The auxiliary equation is $m^4 + 6m^3 + 9m^2 = 0$ and its roots are $m = 0, 0, -3, -3$. The general solution is $y = c_1 + c_2 x + (c_3 + c_4 x)e^{-3x}$.

7. The auxiliary equation is $4m^3 - 3m + 1 = 0$ and its roots are $m = -1, \frac{1}{2}, \frac{1}{2}$. The general solution is $y = (c_1 + c_2 x)\exp\left(\frac{1}{2}x\right) + c_3 e^{-x}$.

9. The auxiliary equation is $m^3 + 3m^2 + 3m + 1 = 0$ and its roots are $m = -1, -1, -1$. The general solution is $y = (c_1 + c_2 x + c_3 x^2)e^{-x}$.

11. The auxiliary equation is $m^5 - m^3 = 0$ and its roots are $m = 0, 0, 0, 1, -1$. The general solution is $y = c_1 + c_2 x + c_3 x^2 + c_4 e^x + c_5 e^{-x}$.

13. The auxiliary equation is $4m^4 + 4m^3 - 3m^2 - 2m + 1 = 0$ and its roots are $m = -1, -1, \frac{1}{2}, \frac{1}{2}$. The general solution is $y = (c_1 + c_2 x)e^{-x} + (c_3 + c_4 x)\exp\left(\frac{1}{2}x\right)$.

15. The auxiliary equation is $m^4 + 3m^3 - 6m^2 - 28m - 24 = 0$ and its roots are $m = -2, -2, -2, 3$. The general solution is $y = (c_1 + c_2 x + c_3 x^2)e^{-2x} + c_4 e^{3x}$.

17. The auxiliary equation is $4m^5 - 23m^3 - 33m^2 - 17m - 3 = 0$ with roots $m = -1, -1, 3, -\frac{1}{2}, -\frac{1}{2}$. The general solution is $y = (c_1 + c_2 x)e^{-x} + c_3 e^{3x} + (c_4 + c_5 x)\exp\left(-\frac{1}{2}x\right)$.

19. The auxiliary equation is $m^4 - 5m^2 - 6m - 2 = 0$ and its roots are $m = 1 \pm \sqrt{3}, -1, -1$. The general solution is $y = (c_1 + c_2 x)e^{-x} + c_3 \exp\left[(1 + \sqrt{3})x\right] + c_4 \exp\left[(1 - \sqrt{3})x\right]$.

21. The auxiliary equation is $m^2 + 4m + 4 = 0$ and its roots are $m = -2, -2$. The general solution is $y = (c_1 + c_2 x)e^{-2x}$ and $y' = (-2c_1 + c_2 - 2c_2 x)e^{-2x}$. But $y(0) = c_1 = 1$ and $y'(0) = -2c_1 + c_2 = -1$, so that $c_1 = c_2 = 1$. The particular solution is $y = (1 + x)e^{-2x}$.

23. The auxiliary equation is $m^3 - 3m - 2 = 0$ and its roots are $m = -1, -1, 2$. The general solution is $y = (c_1 + c_2 x)e^{-x} + c_3 e^{2x}$, so that we have $y' = (-c_1 + c_2 - c_2 x)e^{-x} + 2c_3 e^{2x}$, and $y'' = (c_1 - 2c_2 + c_2 x)e^{-x} + 4c_3 e^{2x}$. But $y(0) = c_1 + c_3 = 0$, $y'(0) = -c_1 + c_2 + 2c_3 = 9$, and $y''(0) = c_1 - 2c_2 + 4c_3 = 0$, so that $c_1 = -2$, $c_2 = 3$, and $c_3 = 2$. The particular solution is $y = (-2 + 3x)e^{-x} + 2e^{2x}$.

25. From Exercise 24 we have $y = c_1 + c_2 x + c_3 e^{-x} + c_4 e^{-2x}$, so that $y' = c_2 - c_3 e^{-x} - 2c_4 e^{-2x}$, $y'' = c_3 e^{-x} + 4c_4 e^{-2x}$, and $y''' = -c_3 e^{-x} - 8c_4 e^{-2x}$. We have $y(0) = c_1 + c_3 + c_4 = 0$, $y'(0) = c_2 - c_3 - 2c_4 = 3$, $y''(0) = c_3 + 4c_4 = -5$, and $y'''(0) = -c_3 - 8c_4 = 9$. Solving these equations gives us $c_1 = 2$, $c_2 = 0$, and $c_3 = c_4 = -1$. The particular solution is $y = 2 - e^{-x} - e^{-2x}$.

27. The auxiliary equation is $4m^2 - 4m + 1 = 0$ and its roots are $m = \frac{1}{2}, \frac{1}{2}$. The general solution is $y = (c_1 + c_2 x)\exp\left(\frac{1}{2}x\right)$ and $y' = \left(\frac{1}{2}c_1 + c_2 + c_2 x\right)\exp\left(\frac{1}{2}x\right)$. But $y(0) = c_1 = -2$ and $y'(0) = \frac{1}{2}c_1 + c_2 = 2$, so that $c_1 = -2$, $c_2 = 3$. The particular solution is $y = (-2 + 3x)\exp\left(\frac{1}{2}x\right)$ and $y(2) = 4e$.

29. The auxiliary equation is $m^3 + 5m^2 + 3m - 9 = 0$ and its roots are $m = 1, -3, -3$. The general solution is $y = c_1 e^x + (c_2 + c_3 x)e^{-3x}$ and $y' = c_1 e^x + (-3c_2 + c_3 + c_3 x)e^{-3x}$. But $y(0) = c_1 + c_2 = 1$ and $y(1) = c_1 e + (c_2 + c_3)e^{-3} = 0$. Moreover, $\lim\limits_{x \to \infty} y = 0$ implies that $c_1 = 0$, so that $c_1 = 0$, $c_2 = -1$, and $c_3 = 1$. The particular solution is $y = (-1 + x)e^{-3x}$, and $y(2) = e^{-6}$.

7.6   A Note on Hyperbolic Functions

1. In what follows we will use equation (5) on page 114 of the text.

$$\begin{aligned}
\left[(D-a)^2 + b^2\right] y &= (D-a)^2 y + b^2 y \\
&= c_3 (D-a)^2 \left[e^{ax}\cos bx\right] + c_4 (D-a)^2 \left[e^{ax}\sin bx\right] \\
&\quad + b^2 \left[c_3 e^{ax}\cos bx + c_4 e^{ax}\sin bx\right] \\
&= c_3 e^{ax} D^2 \left[\cos bx\right] + c_4 e^{ax} D^2 \left[\sin bx\right] + c_3 b^2 e^{ax}\cos bx + c_4 b^2 e^{ax}\sin bx \\
&= 0.
\end{aligned}$$

3. The auxiliary equation is $m^2 - 2m + 2 = 0$ and its roots are $m = 1 \pm i$. The general solution is $y = e^x(c_1 \cos x + c_2 \sin x)$.

5. The auxiliary equation is $m^2 - 9 = 0$ and its roots are $m = \pm 3$. The general solution is $y = c_1 \cosh 3x + c_2 \sinh 3x$.

7. The auxiliary equation is $m^2 - 4m + 7 = 0$ and its roots are $m = 2 \pm \sqrt{3}i$. The general solution is $y = e^{2x}(c_1 \cos\sqrt{3}x + c_2 \sin\sqrt{3}x)$.

9. The auxiliary equation is $m^4 + 2m^3 + 10m^2 = 0$ and its roots are $m = 0, 0, -1 \pm 3i$. The general solution is $y = c_1 + c_2 x + e^{-x}(c_1 \cos 3x + c_2 \sin 3x)$.

11. The auxiliary equation is $m^4 + 18m^2 + 81 = 0$ with roots $m = 3i, \; 3i, \; -3i, \; -3i$. The general solution is $y = (c_1 + c_2 x)\cos 3x + (c_3 + c_4 x)\sin 3x$.

13. The auxiliary equation is $m^6 + 9m^4 + 24m^2 + 16 = 0$ with roots $m = i, \; -i, \; 2i, \; 2i, \; -2i, \; -2i$. The general solution is $y = c_1 \cos x + c_2 \sin x + (c_3 + c_4 x)\cos 2x + (c_5 + c_6 x)\sin 2x$.

15. The auxiliary equation is $m^2 - 1 = 0$ and its roots are $m = \pm 1$. The general solution is $y = c_1 \cosh x + c_2 \sinh x$ and $y' = c_1 \sinh x + c_2 \cosh x$. But $y(0) = c_1 = y_0$ and $y'(0) = c_2 = 0$, so that $c_1 = y_0$ and $c_2 = 0$. The particular solution is $y = y_0 \cosh x$.

17. Here $m^3 + 7m^2 + 19m + 13 = 0$ with roots $m = -1, \; -3 \pm 2i$. The general solution is $y = c_1 e^{-x} + c_2 e^{-3x}\cos 2x + c_3 e^{-3x}\sin 2x$. Thus $y' = -c_1 e^{-x} + (-3c_2 + 2c_3)e^{-3x}\cos 2x + (-2c_2 - 3c_3)e^{-3x}\sin 2x$, and $y'' = c_1 e^{-x} + (5c_2 - 12c_3)e^{-3x}\cos 2x + (12c_2 + 5c_3)e^{-3x}\sin 2x$. But $y(0) = c_1 + c_2 = 0$, $y'(0) = -c_1 - 3c_2 + 2c_3 = 2$, and $y''(0) = c_1 + 5c_2 - 12c_3 = -12$, so that $c_1 = c_2 = 0$, and $c_3 = 1$. The particular solution is $y = e^{-3x}\sin 2x$.

19. The auxiliary equation is $m^2 + k^2 = 0$ with roots $m = \pm ki$. Thus, the general solution is $x = c_1 \cos kt + c_2 \sin kt$, $\dfrac{dx}{dt} = -kc_1 \sin kt + kc_2 \cos kt$, and $\dfrac{d^2 x}{dt^2} = -k^2 c_1 \cos kt - k^2 c_2 \sin kt$. But $x(0) = c_1 = 0$ and $\dfrac{dx}{dt}(0) = kc_2 = v_0$, so that $c_1 = 0$, $c_2 = \dfrac{v_0}{k}$. The particular solution is $y = \left(\dfrac{v_0}{k}\right)\sin kt$.

21. The auxiliary equation is $m^2 + 2bm + k^2 = 0$ with roots $m = -b \pm i\sqrt{k^2 - b^2}$, where $k > b > 0$. The general solution is $x = c_1 e^{-bt}\cos\sqrt{k^2 - b^2}\,t + c_2 e^{-bt}\cos\sqrt{k^2 - b^2}\,t$. But $x(0) = c_1 = 0$ and $x'(0) = -bc_1 + \sqrt{k^2 - b^2}\,c_2$, so that $c_1 = 0$, $c_2 = \dfrac{v_0}{\sqrt{k^2 - b^2}}$. The particular solution is $x = \dfrac{v_0}{\sqrt{k^2 - b^2}}e^{-bt}\sin\left(\sqrt{k^2 - b^2}\,t\right)$.

Miscellaneous Exercises

1. The auxiliary equation is $m^2 + 3m = 0$ and its roots are $m = 0, \; -3$. The general solution is $y = c_1 + c_2 e^{-3x}$.

3. The auxiliary equation is $m^2 + m - 6 = 0$ and its roots are $m = -3, \; 2$. The general solution is $y = c_1 e^{-3x} + c_2 e^{2x}$.

5. The auxiliary equation is $m^3 - 3m^2 + 4 = 0$ and its roots are $m = -1, \; 2, \; 2$. The general solution is $y = c_1 e^{-x} + (c_2 + c_3 x)e^{2x}$.

7. The auxiliary equation is $4m^3 - 3m + 1 = 0$ and its roots are $m = -1, \; \frac{1}{2}, \; \frac{1}{2}$. The general solution is $y = c_1 e^{-x} + (c_2 + c_3 x)\exp\left(\frac{1}{2}x\right)$.

9. The auxiliary equation is $m^3 + 3m^2 + 3m + 1 = 0$ and its roots are $m = -1, \; -1, \; -1$. The general solution is $y = (c_1 + c_2 x + c_3 x^2)e^{-x}$.

11. The auxiliary equation is $4m^3 - 7m + 3 = 0$ and its roots are $m = 1$, $\frac{1}{2}$, $-\frac{3}{2}$. The general solution is $y = c_1 e^x + c_2 \exp\left(\frac{1}{2}x\right) + c_3 \exp\left(-\frac{3}{2}x\right)$.

13. The auxiliary equation is $8m^3 - 4m^2 - 2m + 1 = 0$ and its roots are $m = \frac{1}{2}$, $\frac{1}{2}$, $-\frac{1}{2}$. The general solution is $y = c_1 \exp\left(-\frac{1}{2}x\right) + (c_2 + c_3 x) \exp\left(\frac{1}{2}x\right)$.

15. The auxiliary equation is $m^4 - 2m^3 + 5m^2 - 8m + 4 = 0$ and its roots are $m = 1$, 1, $\pm 2i$. The general solution is $y = (c_1 + c_2 x)e^x + c_3 \cos 2x + c_4 \sin 2x$

17. The auxiliary equation is $m^4 + 5m^2 + 4 = 0$ and its roots are $m = \pm i$, $\pm 2i$. The general solution is $y = c_1 \cos x + c_2 \sin x + c_3 \cos 2x + c_4 \sin 2x$

19. The auxiliary equation is $m^4 - 11m^3 + 36m^2 - 16m - 64 = 0$ and $m = -4$, -4, -4, -1. The general solution is $y = c_1 e^{-x} + (c_2 + c_3 x + c_4 x^2)e^{-4x}$.

21. The auxiliary equation is $m^4 + 4m^3 + 2m^2 - 8m - 8 = 0$ and its roots are $m = -2$, -2, $\pm\sqrt{2}$. The general solution is $y = (c_1 + c_2 x)e^{-2x} + c_3 e^{\sqrt{2}x} + c_4 e^{-\sqrt{2}x}$.

23. The auxiliary equation is $4m^4 + 20m^3 + 35m^2 + 25m + 6 = 0$ and $m = -1$, -2, $-\frac{1}{2}$, $-\frac{3}{2}$. The general solution is $y = c_1 e^{-x} + c_2 e^{-2x} + c_3 \exp\left(-\frac{1}{2}x\right) + c_4 \exp\left(-\frac{3}{2}x\right)$.

25. The auxiliary equation is $m^3 + 5m^2 + 7m + 3 = 0$ and its roots are $m = -1$, -1, -3. The general solution is $y = (c_1 + c_2 x)e^{-x} + c_3 e^{-3x}$.

27. The auxiliary equation is $m^3 - m^2 + m - 1 = 0$ and its roots are $m = 1$, $\pm i$. The general solution is $y = c_1 e^x + c_2 \cos x + c_3 \sin x$.

29. The auxiliary equation is $m^4 - 13m^2 + 36 = 0$ and its roots are $m = \pm 3$, ± 2. The general solution is $y = c_1 e^{3x} + c_2 e^{-3x} + c_3 e^{2x} + c_4 e^{-2x}$.

31. The auxiliary equation is $4m^3 + 8m^2 - 11m + 3 = 0$ and its roots are $m = -3$, $\frac{1}{2}$, $\frac{1}{2}$. The general solution is $y = c_1 e^{-3x} + (c_2 + c_3 x) \exp\left(\frac{1}{2}x\right)$.

33. The auxiliary equation is $m^4 - m^3 - 3m^2 + m + 2 = 0$ and its roots are $m = 1$, 2, -1, -1. The general solution is $y = c_1 e^x + +c_2 e^{2x} + (c_3 + c_4 x)e^{-x}$.

35. The auxiliary equation is $m^5 + m^4 - 6m^3 = 0$ and its roots are $m = 0$, 0, 0, -3, 2. The general solution is $y = c_1 + c_2 x + c_3 x^2 + c_4 e^{-3x} + c_5 e^{2x}$.

37. The auxiliary equation is $4m^3 + 12m^2 + 13m + 10 = 0$ and its roots are $m = -2$, $-\frac{1}{2} \pm i$. The general solution is $y = c_1 e^{-2x} + \exp\left(-\frac{1}{2}x\right)(c_2 \cos x + c_3 \sin x)$.

39. The auxiliary equation is $m^5 - 2m^3 - 2m^2 - 3m - 2 = 0$ and its roots are $m = -1$, -1, 2, $\pm i$. The general solution is $y = (c_1 + c_2 x)e^{-x} + c_3 e^{2x} + c_4 \cos x + c_5 \sin x$.

41. The auxiliary equation is $m^5 - 15m^3 + 10m^2 + 60m - 72 = 0$ and $m = 2$, 2, 2, -3, -3. The general solution is $y = (c_1 + c_2 x + c_3 x^2)e^{2x} + (c_4 + c_5 x)e^{-3x}$.

43. The auxiliary equation is $m^4 + 3m^3 - 6m^2 - 28m - 24 = 0$ and its roots are $m = -2$, -2, -2, 3. The general solution is $y = (c_1 + c_2 x + c_3 x^2)e^{-2x} + c_4 e^{3x}$.

45. The auxiliary equation is $4m^5 - 23m^3 - 33m^2 - 17m - 3 = 0$ and $m = 3, -1, -1, -\frac{1}{2}, -\frac{1}{2}$. The general solution is $y = c_1 e^{3x} + (c_2 + c_3 x)e^{-x} + (c_4 + c_5 x)\exp\left(-\frac{1}{2}x\right)$.

47. The auxiliary equation is $m^4 + 6m^3 + 9m^2 = 0$ and its roots are $m = 0, 0, -3, -3$. The general solution is $y = c_1 + c_2 x + (c_3 + c_4 x)e^{-3x}$. Hence, $y' = c_2 + (-3c_3 + c_4 - 3c_4 x)e^{-3x}$, and $y'' = (9c_3 - 6c_4 + 9c_4 x)e^{-3x}$. But $y(0) = c_1 + c_3 = 0$, $y'(0) = c_2 - 3c_3 + c_4 = 0$, $y''(0) = 9c_3 - 6c_4 = 6$, and $\lim\limits_{x \to \infty} y' = c_2 = 1$. It follows that $c_1 = 0$, $c_2 = 1$, $c_3 = 0$, and $c_4 = -1$. The particular solution is $y = x - xe^{-3x}$, so that $y(1) = 1 - e^{-3}$.

49. The auxiliary equation is $m^5 + m^4 - 9m^3 - 13m^2 + 8m + 12 = 0$, the roots of which are $m = 1, 3, -1, -2, -2$. The general solution is $y = c_1 e^x + c_2 e^{3x} + c_3 e^{-x} + (c_4 + c_5 x)e^{-2x}$.

51. The auxiliary equation is $m^5 + m^4 - 7m^3 - 11m^2 - 8m - 12 = 0$, the roots of which are $m = -2, -2, 3, \pm i$. The general solution is $y = c_1 e^{3x} + (c_2 + c_3 x)e^{-2x} + c_4 \cos x + c_5 \sin x$.

Chapter 8

Nonhomogeneous Equations: Undetermined Coefficients

8.1 Construction of a Homogeneous Equation from a Specific Solution

1. From $y = 4e^{2x} + 3e^{-x}$ an auxiliary equation is $(m-2)(m+1) = 0$. Hence a linear differential equation is $(D-2)(D+1)y = 0$.

3. From $y = -2x + \frac{1}{2}e^{4x}$ an auxiliary equation is $m^2(m-4) = 0$. Hence a linear differential equation is $D^2(D-4)y = 0$.

5. From $y = 2e^x \cos 3x$ an auxiliary equation is $(m-1)^2 + 9 = 0$. Hence a linear differential equation is $\left[(D-1)^2 + 9\right]y = 0$.

7. From $y = -2e^{3x} \cos x$ an auxiliary equation is $(m-3)^2 + 1 = 0$. Hence a linear differential equation is $\left[(D-3)^2 + 1\right]y = 0$.

9. From $y = xe^{-x} \sin 2x + 3e^{-x} \cos 2x$ an auxiliary equation is $\left[(m+1)^2 + 4\right]^2 = 0$. Hence a linear differential equation is $\left[(D+1)^2 + 4\right]^2 y = 0$.

11. From $y = \cos kx$ an auxiliary equation is $m^2 + k^2 = 0$. Hence a linear differential equation is $(D^2 + k^2)y = 0$.

13. From $y = 4\sinh x$ an auxiliary equation is $m^2 - 1 = 0$. Hence a linear differential equation is $(D^2 - 1)y = 0$.

15. From $y = 3xe^{2x}$ an auxiliary equation is $(m-2)^2 = 0$. Hence $m = 2,\ 2$.

17. From $y = e^{-x} \cos 4x$ an auxiliary equation is $(m+1)^2 + 4^2 = 0$. Hence $m = -1 \pm 4i$.

19. From $y = x\left(e^{2x} + 4\right)$ an auxiliary equation is $m^2(m-2)^2 = 0$. Hence $m = 0,\ 0,\ 2,\ 2$.

21. From $y = xe^x$ an auxiliary equation is $(m-1)^2 = 0$. Hence $m = 1,\ 1$.

23. From $y = 4\cos 2x$ an auxiliary equation is $m^2 + 4 = 0$. Hence $m = \pm 2i$.

25. From $y = x\cos 2x$ an auxiliary equation is $(m^2 + 4)^2 = 0$. Hence $m = \pm 2i,\ \pm 2i$.

27. From $y = x\cos 2x - 3\sin 2x$ an auxiliary equation is $(m^2 + 4)^2 = 0$. Hence $m = \pm 2i,\ \pm 2i$.

29. From $y = \sin^3 x = \frac{1}{4}(3\sin x - \sin 3x)$, $(m^2 + 1)(m^2 + 9) = 0$. Hence $m = \pm i,\ \pm 3i$.

31. From $y = x^2 - x + e^{-x}(x + \cos x)$ an auxiliary equation is $m^3\big[(m+1)^2 + 1\big](m+1)^2 = 0$. Hence $m = 0,\ 0,\ 0,\ -1 \pm i,\ -1,\ -1$.

33. From $y = x^2 \sin x + x \cos x$ an auxiliary equation is $(m^2 + 1)^3 = 0$. Hence $m = \pm i,\ \pm i,\ \pm i$.

8.3 The Method of Undetermined Coefficieents

1. From $(D^2 + D)y = -\cos x$ we have $m = 0,\ -1$ and $m' = \pm i$. Hence $y_c = c_1 + c_2 e^{-x}$ and $y_p = A\cos x + B\sin x$. Substituting y_p into the differential equation and simplifying we have $(-A + B)\cos x + (-A - B)\sin x = -\cos x$, so that

$$-A + B = -1, \qquad\qquad A = \tfrac{1}{2},$$
$$-A - B = 0, \qquad\qquad B = -\tfrac{1}{2}.$$

The general solution is $y = c_1 + c_2 e^{-x} + \frac{1}{2}\cos x - \frac{1}{2}\sin x$.

3. From $(D^2 + 3D + 2)y = 12x^2$ we have $m = -2,\ -1$ and $m' = 0,\ 0,\ 0$. Hence $y_c = c_1 e^{-2x} + c_2 e^{-x}$ and $y_p = A + Bx + Cx^2$. Substituting y_p into the differential equation and simplifying we have $(2A + 3B + 2C) + (2B + 6C)x + 2Cx^2 = 12x^2$, so that

$$2A + 3B + 2C = 0, \qquad\qquad A = 21,$$
$$2B + 6C = 0, \qquad\qquad B = -18,$$
$$2C = 12, \qquad\qquad C = 6.$$

The general solution is $y = c_1 e^{-2x} + c_2 e^{-x} + 21 - 18x + 6x^2$.

5. From $(D^2 + 9)y = 5e^x - 162x$ we have $m = \pm 3i$ and $m' = 1,\ 0,\ 0$. Hence $y_c = c_1 \cos 3x + c_2 \sin 3x$ and $y_p = Ae^x + B + Cx$. Substituting y_p into the differential equation and simplifying we have $10Ae^x + 9B + 9Cx = 5e^x - 162x$, so that $A = \frac{1}{2}$, $B = 0$, and $C = -18$. The general solution is $y = c_1 \cos 3x + c_2 \sin 3x + \frac{1}{2}e^x - 18x$.

7. From $y'' - 3y' - 4y = 30e^x$ we have $m = 4,\ -1$ and $m' = 1$. Hence $y_c = c_1 e^{4x} + c_2 e^{-x}$ and $y_p = Ae^x$. Substituting y_p into the differential equation and simplifying we have $-6Ae^x = 30e^x$, so that $A = -5$. The general solution is $y = c_1 e^{4x} + c_2 e^{-x} - 5e^x$.

9. From $(D^2 - 4)y = e^x + 2$ we have $m = \pm 2$ and $m' = 0,\ 2$. Hence $y_c = c_1 e^{2x} + c_2 e^{-2x}$ and $y_p = A + Bxe^{2x}$. Substituting y_p into the differential equation and simplifying we obtain the equation $4Be^{2x} - 4A = e^{2x} + 2$, so that $A = -\frac{1}{2}$, $B = \frac{1}{4}$. The general solution is $y = c_1 e^{2x} + c_2 e^{-2x} - \frac{1}{2} + \frac{1}{4}xe^{2x}$.

11. From $y'' - 4y' + 3y = 20\cos x$ we have $m = 3,\ 1$ and $m' = \pm i$. Hence $y_c = c_1 e^{3x} + c_2 e^x$ and $y_p = A\cos x + B\sin x$. Substituting y_p into the differential equation and simplifying we have $(2A - 4B)\cos x + (4A + 2B)\sin x = 20\cos x$, so that

$$2A - 4B = 20, \qquad\qquad A = 2,$$
$$4A + 2B = 0, \qquad\qquad B = -4.$$

The general solution is $y = c_1 e^{3x} + c_2 e^x + 2\cos x - 4\sin x$.

13. From $y'' + 2y' + y = 7 + 75\sin 2x$ we have $m = -1, -1$ and $m' = 0, \pm 2i$. It follows that $y_c = (c_1 + c_2 x)e^{-x}$ and $y_p = A + B\cos 2x + C\sin 2x$. Substituting y_p into the differential equation and simplifying we have $A + (-3B + 4C)\cos 2x + (-4B - 3C)\sin 2x = 7 + 75\sin 2x$, so that

$$A = 7, \qquad\qquad A = 7,$$
$$-3B + 4C = 0, \qquad\qquad B = -12,$$
$$-4B - 3C = 75, \qquad\qquad C = -9.$$

The general solution is $y = (c_1 + c_2 x)e^{-x} + 7 - 12\cos 2x - 9\sin 2x$.

15. From $(D^2 + 1)y = \cos x$ we have $m = \pm i$ and $m' = \pm i$. Hence $y_c = c_1\cos x + c_2\sin x$ and $y_p = Ax\cos x + Bx\sin x$. Substituting y_p into the differential equation and simplifying we obtain the equation $-2A\sin x + 2B\cos x = \cos x$, so that $A = 0$, $B = \frac{1}{2}$. The general solution is $y = c_1\cos x + c_2\sin x + \frac{1}{2}x\sin x$.

17. $(D^2 - 1)y = e^{-x}(2\sin x + 4\cos x)$ yields $m = 1, -1$ and $m' = -1 \pm i$. Hence $y_c = c_1 e^x + c_2 e^{-x}$ and $y_p = Ae^{-x}\sin x + Be^{-x}\cos x$. Substituting y_p into the differential equation and simplifying we have $(-A + 2B)e^{-x}\sin x + (-2A - B)e^{-x}\cos x = 2e^{-x}\sin x + 4e^{-x}\cos x$, so that

$$-A + 2B = 2, \qquad\qquad A = -2,$$
$$-2A - B = 4, \qquad\qquad B = 0.$$

The general solution is $y = c_1 e^x + c_2 e^{-x} - 2e^{-x}\sin x$.

19. From $(D^3 - D)y = x$ we have $m = 0, \pm 1$ and $m' = 0, 0$. It follows that $y_c = c_1 + c_2 e^x + c_3 e^{-x}$ and $y_p = Ax + Bx^2$. Substituting y_p into the differential equation and simplifying we obtain $-A - 2Bx = x$, so that $A = 0$, $B = -\frac{1}{2}$. The general solution is $y = c_1 + c_2 e^x + c_3 e^{-x} - \frac{1}{2}x^2$.

21. From $(D^3 + D^2 - 4D - 4)y = 3e^{-x} - 4x - 6$ we have $m = 2, -2, -1$ and $m' = 0, 0, -1$. It follows that $y_c = c_1 e^{2x} + c_2 e^{-2x} + c_3 e^{-x}$ and $y_p = A + Bx + Cxe^{-x}$. Substituting y_p into the differential equation and simplifying we have $-3Ce^{-x} - 4Bx + (-4A - 4B) = 3e^{-x} - 4x - 6$, so that

$$-3C = 3, \qquad\qquad C = -1,$$
$$-4B = -4, \qquad\qquad B = 1,$$
$$-4A - 4B = -6, \qquad\qquad A = \tfrac{1}{2}.$$

The general solution is $y = c_1 e^{2x} + c_2 e^{-2x} + c_3 e^{-x} + \frac{1}{2} + x - xe^{-x}$.

23. From the equation $(D^4 - 1)y = e^{-x}$ we have $m = 1, -1, i, -i$ and $m' = -1$. We therefore obtain $y_c = c_1 e^x + c_2 e^{-x} + c_3\cos x + c_4\sin x$ and $y_p = Axe^{-x}$. Substituting y_p into the differential equation and simplifying we have $-4Ae^{-x} = e^{-x}$, so that $A = -\frac{1}{4}$. The general solution is $y = c_1 e^x + c_2 e^{-x} + c_3\cos x + c_4\sin x - \frac{1}{4}xe^{-x}$.

25. From $(D^2 + 1)y = 12\cos^2 x = 6 + 6\cos 2x$ we have $m = \pm i$ and $m' = 0, \pm 2i$. It follows that $y_c = c_1\cos x + c_2\sin x$ and $y_p = A + B\cos 2x + C\sin 2x$. Substituting y_p into the differential equation and simplifying we have $A - 3B\cos 2x - 3C\sin 3x = 6 + 6\cos 2x$, so that $A = 6$, $B = -2$, and $C = 0$. The general solution is $y = c_1\cos x + c_2\sin x + 6 - 2\cos 2x$.

27. From $y'' - 3y' - 4y = 16x - 50\cos 2x$ we have $m = 4, -1$ and $m' = 0, 0, \pm 2i$. Hence $y_c = c_1 e^{4x} + c_2 e^{-x}$ and $y_p = A + Bx + C\cos 2x + D\sin 2x$. Substituting y_p into the differential equation we get $(-4A - 3B) - 4Bx + (-8C - 6D)\cos 2x + (6C - 8D)\sin 2x = 16x - 50\cos 2x$, so that

$$
\begin{array}{ll}
-4A - 3B = 0, & A = 3, \\
-4B = 16, & B = -4, \\
-8C - 6D = -50, & C = 4, \\
6C - 8D = 0, & D = 3.
\end{array}
$$

The general solution is $y = c_1 e^{4x} + c_2 e^{-x} + 3 - 4x + 4\cos 2x + 3\sin 2x$.

29. From the equation $y'' + 4y' + 3y = 15e^{2x} + e^{-x}$ we have $m = -3, -1$ and $m' = 2, -1$. Hence $y_c = c_1 e^{-x} + c_2 e^{-3x}$ and $y_p = Ae^{2x} + Bxe^{-x}$. Substituting y_p into the differential equation and simplifying we have $15Ae^{2x} + 2Bxe^{-x} = 15e^{2x} + e^{-x}$, so that $A = 1$, $B = \frac{1}{2}$. The general solution is $y = c_1 e^{-x} + c_2 e^{-3x} + e^{2x} + \frac{1}{2}xe^{-x}$.

31. From $y'' - y' - 2y = 6x + 6e^{-x}$ we have $m = 2, -1$ and $m' = 0, 0, -1$. Hence $y_c = c_1 e^{2x} + c_2 e^{-x}$ and $y_p = A + Bx + Cxe^{-x}$. Substituting y_p into the differential equation and simplifying we have $(-2A - B) - 2Bx - 3Ce^{-x} = 6x + 6e^{-x}$, so that

$$
\begin{array}{ll}
-2A - B = 0, & A = \frac{3}{2}, \\
-2B = 6, & B = -3, \\
-3C = 6, & C = -2.
\end{array}
$$

The general solution is $y = c_1 e^{2x} + c_2 e^{-x} + \frac{3}{2} - 3x - 2xe^{-x}$.

33. From $(D^3 - 3D^2 + 4)y = 6 + 80\cos 2x$ we have $m = -1, 2, 2$ and $m' = 0, \pm 2i$. Hence $y_c = c_1 e^{-x} + (c_2 + c_3 x)e^{2x}$ and $y_p = A + B\cos 2x + C\sin 2x$. Substituting y_p into the differential equation and simplifying we have $4A + (16B - 8C)\cos 2x + (8B + 16C)\sin 2x = 6 + 80\cos 2x$, so that

$$
\begin{array}{ll}
4A = 6, & A = \frac{3}{2}, \\
16B - 8C = 80, & B = 4, \\
8B + 16C = 0, & C = -2.
\end{array}
$$

The general solution is $y = c_1 e^{-x} + (c_2 + c_3 x)e^{2x} + \frac{3}{2} + 4\cos 2x - 2\sin 2x$.

35. From $(D^3 + D^2 - 4D - 4)y = 8x + 8 + 6e^{-x}$ we have $m = \pm 2, -1$ and $m' = 0, 0, -1$. Hence $y_c = c_1 e^{2x} + c_2 e^{-2x} + c_3 e^{-x}$ and $y_p = A + Bx + Cxe^{-x}$. Substituting y_p into the differential equation and simplifying we have $(-4A - 4B) - 4Bx - 3Ce^{-x} = 8x + 8 + 6e^{-x}$, so that

$$
\begin{array}{ll}
-4A - 4B = 8, & A = 0, \\
-4B = 8, & B = -2, \\
-3C = 6, & C = -2.
\end{array}
$$

The general solution is $y = c_1 e^{2x} + c_2 e^{-2x} + c_3 e^{-x} - 2x - 2xe^{-x}$.

37. From $(D^2 - 4)y = 2 - 8x$ we have $m = 2, -2$ and $m' = 0, 0$. Hence $y_c = c_1 e^{2x} + c_2 e^{-2x}$ and $y_p = A + Bx$. Substituting this y_p into the differential equation and simplifying we have $-4A - 4Bx = 2 - 8x$, so that $A = -\frac{1}{2}$, $B = 2$. The general solution is $y = c_1 e^{2x} + c_2 e^{-2x} - \frac{1}{2} + 2x$. But

$$y(0) = c_1 + c_2 - \tfrac{1}{2} = 0, \qquad\qquad c_1 = 1,$$
$$y'(0) = 2c_1 - 2c_2 + 2 = 5, \qquad\qquad c_2 = -\tfrac{1}{2}.$$

The particular solution is $y = e^{2x} - \frac{1}{2}e^{-2x} - \frac{1}{2} + 2x.$.

39. From the differential equation $(D^2 + 4D + 5)y = 10e^{-3x}$ we have $m = -2 \pm i$ and $m' = -3$. Hence $y_c = e^{-2x}(c_1 \cos x + c_2 \sin x)$ and $y_p = Ae^{-3x}$. Substituting this y_p into the differential equation and simplifying we have $2Ae^{-3x} = 10e^{-3x}$, so that $A = 5$. The general solution is $y = e^{-2x}(c_1 \cos x + c_2 \sin x) + 5e^{-3x}$. But

$$y(0) = c_1 + 5 = 4, \qquad\qquad c_1 = -1,$$
$$y'(0) = -2c_1 + c_2 - 15 = 0, \qquad\qquad c_2 = 13.$$

The desired solution is $y = e^{-2x}(-\cos x + 13 \sin x) + 5e^{-3x}$.

41. From $\ddot{x} + 4\dot{x} + 5x = 8 \sin t$ we have $m = -2 \pm i$ and $m' = \pm i$. Hence $x_c = e^{-2t}(c_1 \cos t + c_2 \sin t)$ and $x_p = A \cos t + B \sin t$. Substituting x_p into the differential equation and simplifying we have $(4A + 4B) \cos t + (-4A + 4B) \sin t = 8 \sin t$, so that

$$4A + 4B = 0, \qquad\qquad A = -1,$$
$$-4A + 4B = 8, \qquad\qquad B = 1.$$

The general solution is $x = e^{-2t}(c_1 \cos t + c_2 \sin t) - \cos t + \sin t$. But

$$x(0) = c_1 - 1 = 0, \qquad\qquad c_1 = 1,$$
$$\dot{x}(0) = -2c_1 + c_2 + 1 = 0, \qquad\qquad c_2 = 1.$$

The desired solution is $x = e^{-2t}(\cos t + \sin t) - \cos t + \sin t$.

43. From $(D^3 + 4D^2 + 9D + 10)y = -24e^x$ we have $m = -2, -1 \pm 2i$ and $m' = 1$. Hence $y_c = c_1 e^{-2x} + e^{-x}(c_2 \cos 2x + c_3 \sin 2x)$ and $y_p = Ae^x$. Substituting y_p into the differential equation and simplifying we have $24Ae^x = -24e^x$, so that $A = -1$. The general solution is $y = c_1 e^{-2x} + e^{-x}(c_2 \cos 2x + c_3 \sin 2x) - e^x$. But

$$y(0) = c_1 + c_2 - 1 = 0, \qquad\qquad c_1 = 2,$$
$$y'(0) = -2c_1 - c_2 + 2c_3 - 1 = -4, \qquad\qquad c_2 = -1,$$
$$y''(0) = 4c_1 - 3c_2 - 4c_3 - 1 = 10, \qquad\qquad c_3 = 0.$$

The particular solution is $y = 2e^{-2x} - e^{-x} \cos 2x - e^x$.

45. From $y'' + 2y' + y = x$ we have $m = -1$, -1 and $m' = 0$, 0. Hence $y_c = (c_1 + c_2 x)e^{-x}$ and $y_p = A + Bx$. Substituting y_p into the differential equation and simplifying we have $A + 2B + Bx = x$, so that $A = -2$, $B = 1$. The general solution is $y = (c_1 + c_2 x)e^{-x} + x - 2$. But

$$y(0) = c_1 - 2 = -3, \qquad\qquad c_1 = -1,$$
$$y(1) = (c_1 + c_2)e^{-1} - 1 = -1, \qquad\qquad c_2 = 1.$$

The particular solution is $y = (-1 + x)e^{-x} + x - 2$. It follows that $y(2) = e^{-2}$ and $y'(2) = 1$.

47. From $4y'' + y = 2$ we have $m = \pm\frac{1}{2}i$ and $m' = 0$. Hence $y_c = c_1 \cos\frac{1}{2}x + c_2 \sin\frac{1}{2}x$ and $y_p = A$. Substituting y_p into the differential equation and simplifying we get $A = 2$. The general solution is $y = c_1 \cos\frac{1}{2}x + c_2 \sin\frac{1}{2}x + 2$. But

$$y(\pi) = c_2 + 2 = 0, \qquad\qquad c_1 = -2,$$
$$y'(\pi) = -\tfrac{1}{2}c_1 = 1, \qquad\qquad c_2 = -2.$$

The particular solution is $y = -2\cos\frac{1}{2}x - 2\sin\frac{1}{2}x + 2$. It follows that $y(2) = -2\cos 1 - 2\sin 1 + 2$ and $y'(2) = \sin 1 - \cos 1$.

49. From $(D^2 + D)y = x + 1$ we have $m = 0$, -1 and $m' = 0$, 0. Hence $y_c = c_1 + c_2 e^{-x}$ and $y_p = Ax + Bx^2$. Substituting y_p into the differential equation and simplifying we get $(A + 2B) + 2Bx = x + 1$, so that $A = 0$, $B = \frac{1}{2}$. The general solution is $y = c_1 + c_2 e^{-x} + \frac{1}{2}x^2$. But

$$y(0) = c_1 + c_2 = 1, \qquad\qquad c_1 = \frac{-e^{-1}}{1 - e^{-1}},$$
$$y(1) = c_1 + c_2 e^{-1} + \tfrac{1}{2} = \tfrac{1}{2}, \qquad\qquad c_2 = \frac{1}{1 - e^{-1}}.$$

The particular solution is $y = \dfrac{-e^{-1}}{1 - e^{-1}} + \dfrac{1}{1 - e^{-1}}e^{-x} + \dfrac{1}{2}x^2$, and $y(4) = 8 - e^{-1} - e^{-2} - e^{-3}$.

51. From $(D^2 + 1)y = 2\cos x$ we have $m = \pm i$ and $m' = \pm i$. Hence $y_c = c_1 \cos x + c_2 \sin x$ and $y_p = Ax\cos x + Bx\sin x$. Substituting y_p into the differential equation and simplifying we get $-2A\sin x + 2B\cos x = 2\cos x$, so that $A = 0$, $B = 1$. The general solution of the differential equation is $y = c_1 \cos x + c_2 \sin x + x\sin x$. But $y(0) = c_1 = 0$ and $y(\pi) = -c_1 = 0$, so that $c_1 = 0$. There are an unlimited number of solutions, $y = c_2 \sin x + x\sin x$.

53. From $(D^2 - D)y = 2 - 2x$ we have $m = 0$, 1 and $m' = 0$, 0. Hence $y_c = c_1 + c_2 e^x$ and $y_p = Ax + Bx^2$. Substituting y_p into the differential equation and simplifying we get $-A + 2B - 2Bx = 2 - 2x$, so that $A = 0$, $B = 1$. The general solution of the differential equation is $y = c_1 + c_2 e^x + x^2$. We want $y'(a) = c_2 e^a + 2a = 0$ and $y''(a) = c_2 e^a + 2 = 0$, so that $a = 1$. But

$$y(1) = c_1 + c_2 e + 1 = 0, \qquad\qquad c_1 = 1,$$
$$y'(1) = c_2 e + 2 = 0, \qquad\qquad c_2 = -2e^{-1}.$$

The desired solution is $y = 1 - 2e^{x-1} + x^2$.

8.4   Solution by Inspection

1. If $b \neq a$ then

$$(D^2 + a^2)\left[(a^2 - b^2)^{-1} \sin bx\right] = -b^2(a^2 - b^2)^{-1} \sin bx + a^2(a^2 - b^2)^{-1} \sin bx$$
$$= \sin bx.$$

Hence $(D^2 + a^2)y = \sin bx$ has the particular solution $y = (a^2 - b^2)^{-1} \sin bx$ if $a \neq b$.

3. If $y = 3$ then $(D^2 + 4)y = 4 \cdot 3 = 12$. Hence $y = 3$ is a particular solution.

5. If $y = 2$ then $(D^2 + 4D + 4)y = 4 \cdot 2 = 8$. Hence $y = 2$ is a particular solution.

7. If $y = -\frac{7}{2}$ then $(D^3 - 3D + 2)y = 2 \cdot (-\frac{7}{2}) = -7$. Hence $y = -\frac{7}{2}$ is a particular solution.

9. If $y = 3x$ then $(D^2 + 4D)y = 4 \cdot 3 = 12$. Hence $y = 3x$ is a particular solution.

11. If $y = 3x$ then $(D^3 + 5D)y = 5 \cdot 3 = 15$. Hence $y = 3x$ is a particular solution.

13. If $y = -3x^2$ then $(D^4 - 4D^2)y = (-6) \cdot (-4) = 24$. Hence $y = -3x^2$ is a particular solution.

15. If $y = -4x^3$ then $(D^5 - D^3)y = 4! = 24$. Hence $y = -4x^3$ is a particular solution.

17. If $y = 2\sin x$ then $(D^2 + 4)y = -2\sin x + 8\sin x = 6\sin x$. Hence $y = 2\sin x$ is a particular solution.

19. If $y = 2x + \frac{1}{4} - 3e^x$ then $(D^2 + 4)y = -3e^x + 8x + 1 - 12e^x = -15e^x + 8x + 1$. Hence $y = 2x + \frac{1}{4} - 3e^x$ is a particular solution.

21. If $y = 3e^{2x}$ then $(D^2 + 3D - 4)y = 12e^{2x} + 18e^{2x} - 12e^{2x} = 18e^{2x}$. Hence $y = 3e^{2x}$ is a particular solution.

23. If $y = \frac{1}{4}e^{3x}$ then $(D^2 - 1)y = \frac{9}{4}e^{3x} - \frac{1}{4}e^{3x} = 2e^{3x}$. Hence $y = \frac{1}{4}e^{3x}$ is a particular solution.

25. If $y = -\frac{1}{5}\cos 2x$ then $(D^2 - 1)y = \frac{4}{5}\cos 2x + \frac{1}{5}\cos 2x = \cos 2x$. Hence $y = -\frac{1}{5}\cos 2x$ is a particular solution.

27. If $y = \frac{1}{2}e^x + 3x$ then $(D^2 + 1)y = \frac{1}{2}e^x + \frac{1}{2}e^x + 3x = e^x + 3x$. Hence $y = \frac{1}{2}e^x + 3x$ is a particular solution.

29. If $y = -\frac{1}{3}\cos 2x - 2x$ then $(D^2 + 1)y = \frac{4}{3}\cos 2x - \frac{1}{3}\cos 2x - 2x = \cos 2x - 2x$. It follows that $y = -\frac{1}{3}\cos 2x - 2x$ is a particular solution.

31. If $y = -\frac{2}{3}\sin 4x$ then $(D^2 + 1)y = \frac{32}{3}\sin 4x - \frac{2}{3}\sin 4x = 10\sin 4x$. Hence $y = -\frac{2}{3}\sin 4x$ is a particular solution.

33. If $y = 3e^x$ then $(D^2 + 2D + 1)y = 3e^x + 6e^x + 3e^x = 12e^x$. Hence $y = 3e^x$ is a particular solution.

35. If $y = 3e^{-x}$ then $(D^2 - 2D + 1)y = 3e^{-x} + 6e^{-x} + 3e^{-x} = 12e^{-x}$. Hence $y = 3e^{-x}$ is a particular solution.

37. If $y = -\frac{1}{4}e^x$ then $(D^2 - 2D - 3)y = -\frac{1}{4}e^x + \frac{1}{2}e^x + \frac{3}{4}e^x = e^x$. Hence $y = -\frac{1}{4}e^x$ is a particular solution.

39. If $y = -4\sin x$ then $(4D^2 + 1)y = 16\sin x - 4\sin x = 12\sin x$. Hence $y = -4\sin x$ is a particular solution.

41. If $y = 2e^x - 5$ then $(4D^2 + +4D + 1)y = 8e^x + 8e^x + 2e^x - 5 = 18e^x - 5$. Hence $y = 2e^x - 5$ is a particular solution.

43. If $y = -\frac{1}{2}e^{-x}$ then $(D^3 - 1)y = \frac{1}{2}e^{-x} + \frac{1}{2}e^{-x} = e^{-x}$. Hence $y = -\frac{1}{2}e^{-x}$ is a particular solution.

45. If $y = \frac{1}{6}e^{2x}$ then $(D^3 - D)y = \frac{8}{6}e^{2x} - \frac{2}{6}e^{2x} = e^{2x}$. Hence $y = \frac{1}{6}e^{2x}$ is a particular solution.

47. If $y = \frac{3}{10}\sin 2x$ then $(D^4 + 4)y = \frac{48}{10}\sin 2x + \frac{12}{10}\sin 2x = 6\sin 2x$. Hence $y = \frac{3}{10}\sin 2x$ is a particular solution.

49. If $y = \frac{1}{2}\cos 2x$ then $(D^3 - D)y = 4\sin 2x + \sin 2x = 5\sin 2x$. Hence $y = \frac{1}{2}\cos 2x$ is a particular solution.

Chapter 9

Variation of Parameters

9.2   Reduction of Order

1. One solution of the homogeneous equation is e^x. We make the substitution $y = ve^x$ and simplify to obtain

$$v'' + 2v' = (x - 1)e^{-x}.$$

Multiplying by the integrating factor e^{2x} and integrating we get

$$e^{2x}v' = xe^x - 2e^x + c_1.$$

Dividing by e^{2x} and integrating again yields

$$v = -xe^{-x} + e^{-x} + c_2e^{-2x} + c_3,$$
$$y = ve^x = c_2e^{-x} + c_3e^x - x + 1.$$

3. One solution of the homogeneous equation is e^{2x}. We make the substitution $y = ve^{2x}$ and simplify to obtain $v'' = e^{-x}$. Integrating twice gives $v = e^{-x} + c_1 + c_2x$. It follows that $y = ve^{2x} = (c_1 + c_2x)e^{2x} + e^x$.

5. One solution of the homogeneous equation is $\sin x$. We make the substitution $y = v\sin x$ and simplify to obtain

$$v'' + 2\cot x v' = \sec x \csc x.$$

Multiplying by the integrating factor $\sin^2 x$ and integrating we get

$$\sin^2 x \, v' = -\ln|\cos x| + c_1.$$

Dividing by $\sin^2 x$ and integrating again yields

$$v = \cot x \ln|\cos x| + x + c_2 \cot x + c_3,$$
$$y = v\sin x = c_3 \sin x + c_2 \cos x + x \sin x + \cos x \ln|\cos x|.$$

7. One solution of the homogeneous equation is e^{-x}. We make the substitution $y = ve^{-x}$ and simplify to obtain $v'' = e^x(e^x - 1)^{-2}$. Integrating twice gives $v = c_1x + c_2 - \ln|1 - e^{-x}|$. It follows that $y = ve^{-x} = e^{-x}[c_1x + c_2 - \ln|1 - e^{-x}|]$.

9. One solution of the homogeneous equation is $\cos x$. We make the substitution $y = v \cos x$ and simplify to obtain

$$v'' - 2 \tan x v' = \sec x \csc x.$$

Multiplying by the integrating factor $\cos^2 x$ and integrating we get

$$\cos^2 x\, v' = \ln|\sin x| + c_2.$$

Dividing by $\cos^2 x$ and integrating again yields

$$v = -x + \tan x \ln|\sin x| + c_2 \tan x + c_1,$$
$$y = ve^x = c_1 \cos x + c_2 \sin x - x \cos x + \sin x \ln|\sin x|.$$

11. One solution of the homogeneous equation is $\cos x$. We make the substitution $y = v \cos x$ and simplify to obtain

$$v'' - 2 \tan x v' = \sec x \csc^3 x.$$

Multiplying by the integrating factor $\cos^2 x$ and integrating we get

$$\cos^2 x\, v' = -\frac{1}{2} \cot^2 x + c_1.$$

Dividing by $\cos^2 x$ and integrating again yields

$$v = \frac{1}{2} \cot x + c_1 \tan x + c_2,$$
$$y = v \cos x = \frac{1}{2} \cos x \cot x + c_1 \sin x + c_2 \cos x,$$
$$= c_3 \sin x + c_2 \cos x + \frac{1}{2} \csc x.$$

13. One solution of the homogeneous equation is x. We make the substitution $y = vx$ and simplify to obtain

$$v'' + \frac{5}{2x} v' = 0.$$

Multiplying by the integrating factor $x^{5/2}$ and integrating we get

$$x^{5/2} v' = c_3.$$

Dividing by $x^{5/2}$ and integrating again yields

$$v = c_2 x^{-3/2} + c_1,$$
$$y = vx = c_1 x + c_2 x^{-1/2}.$$

15. One solution of the homogeneous equation is x. We make the substitution $y = vx$ and simplify to obtain

$$x(1 - x^2)v'' + (2 - 4x^2)v' = 0.$$

This may be written

$$\frac{v''}{v'} + \frac{2}{x} - \frac{1}{1-x} + \frac{1}{1+x} = 0.$$

Upon integrating we obtain

$$v' = \frac{2c_1}{x^2(1-x^2)} = c_1\left[\frac{2}{x^2} + \frac{1}{1-x} + \frac{1}{1+x}\right].$$

Integrating again yields

$$v = c_1\left[\frac{-2}{x} + \ln\left|\frac{1+x}{1-x}\right|\right] + c_2,$$

$$y = vx = c_1\left[-2 + x\ln\left|\frac{1+x}{1-x}\right|\right] + c_2 x.$$

9.4   Solution of $y'' + y = f(x)$

1. We consider

$$y = Ae^x + Be^{-x},$$
$$y' = Ae^x - Be^{-x} + A'e^x + B'e^{-x}.$$

Setting $A'e^x + B'e^{-x} = 0$ gives

$$y'' = Ae^x + Be^{-x} + A'e^x - B'e^{-x},$$
$$y'' - y = A'e^x - B'e^{-x} = e^x + 1.$$

We must solve the system

$$A'e^x + B'e^{-x} = 0,$$
$$A'e^x - B'e^{-x} = e^x + 1,$$

for A' and B' and then integrate to find A and B. We get

$$A' = \tfrac{1}{2} + \tfrac{1}{2}e^{-x}, \qquad\qquad A = \tfrac{1}{2}x - \tfrac{1}{2}e^{-x},$$
$$B' = -\tfrac{1}{2}e^{2x} - \tfrac{1}{2}e^{x}, \qquad\qquad B = -\tfrac{1}{4}e^{2x} - \tfrac{1}{2}e^{x}.$$

Thus

$$y_p = \left(\tfrac{1}{2}x - \tfrac{1}{2}e^{-x}\right)e^x + \left(-\tfrac{1}{4}e^{2x} - \tfrac{1}{2}e^x\right)e^{-x}$$
$$= \tfrac{1}{2}xe^x - \tfrac{1}{4}e^x - 1,$$
$$y = y_c + \tfrac{1}{2}xe^x - \tfrac{1}{4}e^x - 1.$$

3. We consider

$$y = A\cos x + B\sin x,$$
$$y' = -A\sin x + B\cos x + A'\cos x + B'\sin x.$$

Setting $A'\cos x + B'\sin x = 0$ gives

$$y'' = -A\cos x - B\sin x - A'\sin x + B'\cos x,$$
$$y'' + y = -A'\sin x + B'\cos x = \csc x.$$

We must solve the system

$$A'\cos x + B'\sin x = 0,$$
$$-A'\sin x + B'\cos x = \csc x,$$

for A' and B' and then integrate to find A and B. We get

$$A' = -1, \qquad\qquad\qquad A = -x,$$
$$B' = \cot x, \qquad\qquad\qquad B = \ln|\sin x|.$$

Thus

$$y_p = -x\cos x + \sin x\ln|\sin x|,$$
$$y = y_c - x\cos x + \sin x\ln|\sin x|.$$

5. We consider

$$y = A\cos x + B\sin x,$$
$$y' = -A\sin x + B\cos x + A'\cos x + B'\sin x.$$

Setting $A'\cos x + B'\sin x = 0$ gives

$$y'' = -A\cos x - B\sin x - A'\sin x + B'\cos x,$$
$$y'' + y = -A'\sin x + B'\cos x = \sec^3 x.$$

We must solve the system

$$A'\cos x + B'\sin x = 0,$$
$$-A'\sin x + B'\cos x = \sec^3 x,$$

for A' and B' and then integrate to find A and B. We get

$$A' = -\sin x\sec^3 x, \qquad\qquad A = -\tfrac{1}{2}\tan^2 x,$$
$$B' = \sec^2 x, \qquad\qquad\qquad B = \tan x.$$

Thus

$$y_p = \left(-\tfrac{1}{2}\tan^2 x\right)\cos x + \tan x\sin x$$
$$= \tfrac{1}{2}\sec x - \tfrac{1}{2}\cos x,$$
$$y = y_c + \tfrac{1}{2}\sec x.$$

7. We consider

$$y = A \cos x + B \sin x,$$
$$y' = -A \sin x + B \cos x + A' \cos x + B' \sin x.$$

Setting $A' \cos x + B' \sin x = 0$ gives

$$y'' = -A \cos x - B \sin x - A' \sin x + B' \cos x,$$
$$y'' + y = -A' \sin x + B' \cos x = \tan x.$$

We must solve the system

$$A' \cos x + B' \sin x = 0,$$
$$-A' \sin x + B' \cos x = \tan x,$$

for A' and B' and then integrate to find A and B. We get

$$A' = -\sin x \tan x, \qquad\qquad A = \sin x - \ln|\sec x + \tan x|,$$
$$B' = \sin x, \qquad\qquad B = -\cos x.$$

Thus

$$y_p = (\sin x - \ln|\sec x + \tan x|) \cos x - \cos x \sin x$$
$$= -\cos x \ln|\sec x + \tan x|,$$
$$y = y_c - \cos x \ln|\sec x + \tan x|.$$

9. We consider

$$y = A \cos x + B \sin x,$$
$$y' = -A \sin x + B \cos x + A' \cos x + B' \sin x.$$

Setting $A' \cos x + B' \sin x = 0$ gives

$$y'' = -A \cos x - B \sin x - A' \sin x + B' \cos x,$$
$$y'' + y = -A' \sin x + B' \cos x = \sec x \csc x.$$

We must solve the system

$$A' \cos x + B' \sin x = 0,$$
$$-A' \sin x + B' \cos x = \sec x \csc x,$$

for A' and B' and then integrate to find A and B. We get

$$A' = -\sec x, \qquad\qquad A = -\ln|\sec x + \tan x|,$$
$$B' = \csc x, \qquad\qquad B = -\ln|\csc x + \cot x|.$$

Thus

$$y_p = -\cos x \ln|\sec x + \tan x| - \sin x \ln|\csc x + \cot x|,$$
$$y = y_c - \cos x \ln|\sec x + \tan x| - \sin x \ln|\csc x + \cot x|.$$

11. We consider

$$y = (A + Bx)e^x,$$
$$y' = Ae^x + B(e^x + xe^x) + A'e^x + B'xe^x.$$

Setting $A'e^x + B'xe^x = 0$ gives

$$y'' = Ae^x + B(2e^x + xe^x) + A'e^x + B'(e^x + xe^x),$$
$$y'' - 2y' + 1 = B'e^x = e^{2x}(e^x + 1)^{-2}.$$

We must solve the system

$$A'e^x + B'xe^x = 0,$$
$$B'e^x = e^{2x}(e^x + 1)^{-2},$$

for A' and B' and then integrate to find A and B. We get

$$A' = -xe^x(e^x + 1)^{-2}, \qquad A = x(e^x + 1)^{-1} + \ln(1 + e^{-x}),$$
$$B' = e^x(e^x + 1)^{-2}, \qquad B = -(e^x + 1)^{-1}.$$

Thus

$$\begin{aligned} y_p &= \left[x(e^x + 1)^{-1} + \ln(1 + e^{-x})\right]e^x + \left[-(e^x + 1)^{-1}\right]xe^x \\ &= e^x \ln(1 + e^{-x}) = e^x\left[\ln(1 + e^x) - x\right], \\ y &= y_c + e^x \ln(1 + e^x). \end{aligned}$$

13. We consider

$$y = Ae^x + Be^{2x},$$
$$y' = Ae^x + 2Be^{2x} + A'e^x + B'e^{2x}.$$

Setting $A'e^x + B'e^{2x} = 0$ gives

$$y'' = Ae^x + 4Be^{2x} + A'e^x + 2B'e^{2x},$$
$$y'' - 3y' + 2y = A'e^x + 2B'e^{2x} = \cos(e^{-x}).$$

We must solve the system

$$A'e^x + B'e^{2x} = 0,$$
$$A'e^x + 2B'e^{2x} = \cos(e^{-x}),$$

for A' and B' and then integrate to find A and B. We get

$$A' = -e^{-x}\cos(e^{-x}), \qquad A = \sin(e^{-x}),$$
$$B' = e^{-2x}\cos(e^{-x}), \qquad B = -e^{-x}\sin(e^{-x}) - \cos(e^{-x}).$$

Thus

$$y_p = e^x \sin\left(e^{-x}\right) + \left[-e^{-x} \sin\left(e^{-x}\right) - \cos\left(e^{-x}\right)\right] e^{2x}$$
$$= -e^{2x} \cos\left(e^{-x}\right),$$
$$y = y_c - e^{2x} \cos\left(e^{-x}\right).$$

15. We consider

$$y = Ae^x + Be^{-x},$$
$$y' = Ae^x - Be^{-x} + A'e^x + B'e^{-x}.$$

Setting $A'e^x + B'e^{-x} = 0$ gives

$$y'' = Ae^x + Be^{-x} + A'e^x - B'e^{-x},$$
$$y'' - y = A'e^x - B'e^{-x} = e^{-2x} \sin\left(e^{-x}\right).$$

We must solve the system

$$A'e^x + B'e^{-x} = 0,$$
$$A'e^x - B'e^{-x} = e^{-2x} \sin\left(e^{-x}\right),$$

for A' and B' and then integrate to find A and B. We get

$$A' = \tfrac{1}{2} e^{-3x} \sin\left(e^{-x}\right), \qquad A = \tfrac{1}{2} e^{-2x} \cos\left(e^{-x}\right) - e^{-x} \sin\left(e^{-x}\right) - \cos\left(e^{-x}\right),$$
$$B' = -\tfrac{1}{2} e^{-x} \sin\left(e^{-x}\right), \qquad B = -\tfrac{1}{2} \cos\left(e^{-x}\right).$$

Thus

$$y_p = -\sin\left(e^{-x}\right) - e^x \cos\left(e^{-x}\right)$$
$$y = y_c - \sin\left(e^{-x}\right) - e^x \cos\left(e^{-x}\right).$$

17. We consider

$$y = A + Be^x + Ce^{-x},$$
$$y' = Be^x - Ce^{-x} + A' + B'e^x + C'e^{-x},$$
$$y'' = Be^x + Ce^{-x} + B'e^x - C'e^{-x},$$
$$y''' = Be^x - Ce^{-x} + B'e^x + C'e^{-x}$$

In the first two differentiations we have let the terms involving A', B', and C' equal zero. If we now substitute back into the differential equation we have

$$y''' - y' = B'e^x + C'e^{-x} = x,$$
$$A' + B'e^x + C'e^{-x} = 0,$$
$$B'e^x - C'e^{-x} = 0.$$

Solving this system and integrating we obtain

$$A' = -x, \qquad\qquad A = -\tfrac{1}{2}x^2,$$
$$B' = \tfrac{1}{2}xe^{-x}, \qquad\qquad B = -\tfrac{1}{2}xe^{-x} - \tfrac{1}{2}e^{-x},$$
$$C' = \tfrac{1}{2}xe^{x}, \qquad\qquad C = -\tfrac{1}{2}xe^{x} - \tfrac{1}{2}e^{x}.$$

Thus

$$y_p = -\tfrac{1}{2}x^2 + \left(-\tfrac{1}{2}xe^{-x} - \tfrac{1}{2}e^{-x}\right)e^{x} + \left(\tfrac{1}{2}xe^{x} - \tfrac{1}{2}e^{x}\right)e^{-x}$$
$$= -\tfrac{1}{2}x^2 - 1,$$
$$y = y_c - \tfrac{1}{2}x^2 - 1.$$

19. We consider

$$y = Ax + Be^{x},$$
$$y' = A + Be^{x} + A'x + B'e^{x}.$$

Setting $A'x + B'e^{x} = 0$ gives

$$y'' = Be^{x} + A' + B'e^{x},$$
$$(1 - x)y'' + xy - y = A'(1 - x) + B'(1 - x)e^{x} = 2(x - 1)^2 e^{-x}.$$

We must solve the system

$$A'x + B'e^{x} = 0,$$
$$A' + B'e^{x} = 2(1 - x)e^{-x},$$

for A' and B' and then integrate to find A and B. We get

$$A' = 2e^{-x}, \qquad\qquad A = -2e^{-x},$$
$$B' = -2xe^{-2x}, \qquad\qquad B = xe^{-2x} + \tfrac{1}{2}e^{-2x}.$$

Thus

$$y_p = \left(-2e^{-x}\right)x + \left(xe^{-2x} + \tfrac{1}{2}e^{-2x}\right)e^{x}$$
$$= -e^{-x}(x - \tfrac{1}{2}),$$
$$y = y_c - e^{-x}(x - \tfrac{1}{2}).$$

21. We consider

$$y = A\cos x + B\sin x,$$
$$y' = -A\sin x + B\cos x + A'\cos x + B'\sin x.$$

Setting $A' \cos x + B' \sin x = k$ gives

$$y'' = -A \cos x - B \sin x - A' \sin x + B' \cos x,$$
$$y'' + y = -A' \sin x + B' \cos x = \sec^3 x.$$

We must solve the system

$$A' \cos x + B' \sin x = k,$$
$$-A' \sin x + B' \cos x = \sec^3 x,$$

for A' and B' and then integrate to find A and B. We get

$$A' = k \cos x - \sin x \sec^3 x, \qquad\qquad A = k \sin x - \tfrac{1}{2} \tan^2 x,$$
$$B' = \sec^2 x + k \sin x, \qquad\qquad B = \tan x - k \cos x.$$

Note that when we form $y_p = A \cos x + B \sin x$ the terms involving k will cancel leaving us with the same solution as we had in Exercise 5.

23. If we let $W = y_2 y_1' - y_1 y_2'$ then the differential equation can be written $W' + pW = 0$. Multiplying by the integrating factor $\exp \left[\int P \, dx \right]$ gives us

$$\exp \left[\int P \, dx \right] W(x) = c,$$
$$W(x) = c \exp \left[-\int P \, dx \right].$$

25. We choose $a = x_0$ in the solution given in equation (10).Thus

$$y = c_1 \cos x + c_2 \sin x + \int_{x_0}^{x} f(\beta) \sin (x - \beta) \, d\beta,$$
$$y' = -c_1 \sin x + c_2 \cos x,$$
$$y(x_0) = c_1 \cos x_0 + c_2 \sin x_0 = y_0,$$
$$y'(x_0) = -c_1 \sin x_0 + c_2 \cos x_0 = y_0'.$$

We must solve the last two equations for c_1 and c_2. We obtain

$$c_1 = y_0 \cos x_0 - y_0' \sin x_0,$$
$$c_2 = y_0' \cos x_0 + y_0 \sin x_0.$$

Finally we have

$$y = [y_0 \cos x_0 - y_0' \sin x_0] \cos x + [y_0' \cos x_0 + y_0 \sin x_0] \sin x + \int_{x_0}^{x} f(\beta) \sin (x - \beta) \, d\beta$$
$$= y_0 \cos (x - x_0) + y_0' \sin (x - x_0) + \int_{x_0}^{x} f(\beta) \sin (x - \beta) \, d\beta.$$

Miscellaneous Exercises

1. We consider

$$y = Ae^x + Be^{-x},$$
$$y' = Ae^x - Be^{-x} + A'e^x + B'e^{-x}.$$

Setting $A'e^x + B'e^{-x} = 0$ gives

$$y'' = Ae^x + Be^{-x} + A'e^x - B'e^{-x},$$
$$y'' - y = A'e^x - B'e^{-x} = 2e^{-x}(1 + e^{-2x})^{-2}.$$

We must solve the system

$$A'e^x + B'e^{-x} = 0,$$
$$A'e^x - B'e^{-x} = 2e^{-x}(1 + e^{-2x})^{-2},$$

for A' and B' and then integrate to find A and B. We get

$$A' = e^{-2x}(1 + e^{-2x})^{-2}, \qquad\qquad A = \frac{1/2}{1 + e^{-2x}},$$

$$B' = -(1 + e^{-2x})^{-2} = -1 + \frac{e^{-2x}}{(1 + e^{-2x})^2} + \frac{e^{-2x}}{1 + e^{-2x}}, \quad B = -x - \tfrac{1}{2}\ln\left(1 + e^{-2x}\right) + \frac{1/2}{1 + e^{-2x}}.$$

Thus

$$y_p = \left[\frac{1/2}{1 + e^{-2x}}\right]e^x + \left[-x - \tfrac{1}{2}\ln\left(1 + e^{-2x}\right) + \frac{1/2}{1 + e^{-2x}}\right]e^{-x}$$
$$= -xe^{-x} - \tfrac{1}{2}e^{-x}\ln\left(1 + e^{-2x}\right) + \tfrac{1}{2}e^x,$$
$$y = y_c - xe^{-x} - \tfrac{1}{2}e^{-x}\ln\left(1 + e^{-2x}\right).$$

3. We consider

$$y = A\cos x + B\sin x,$$
$$y' = -A\sin x + B\cos x + A'\cos x + B'\sin x.$$

Setting $A'\cos x + B'\sin x = 0$ gives

$$y'' = -A\cos x - B\sin x - A'\sin x + B'\cos x,$$
$$y'' + y = -A'\sin x + B'\cos x = \sec^2 x \tan x.$$

We must solve the system

$$A'\cos x + B'\sin x = 0,$$
$$-A'\sin x + B'\cos x = \sec^2 x \tan x,$$

for A' and B' and then integrate to find A and B. We get

$$A' = -\sec x \tan^2 x, \qquad A = -\tfrac{1}{2} \sec x \tan x + \tfrac{1}{2} \ln|\sec x + \tan x|,$$
$$B' = \sec x \tan x, \qquad B = \sec x.$$

Thus

$$y_p = \left[-\tfrac{1}{2}\sec x \tan x + \tfrac{1}{2}\ln|\sec x + \tan x|\right]\cos x + \sec x \sin x$$
$$= \tfrac{1}{2}\tan x + \tfrac{1}{2}\cos x \ln|\sec x + \tan x|,$$
$$y = y_c + \tfrac{1}{2}\tan x + \tfrac{1}{2}\cos x \ln|\sec x + \tan x|.$$

5. We consider

$$y = A\cos x + B\sin x,$$
$$y' = -A\sin x + B\cos x + A'\cos x + B'\sin x.$$

Setting $A'\cos x + B'\sin x = 0$ gives

$$y'' = -A\cos x - B\sin x - A'\sin x + B'\cos x,$$
$$y'' + y = -A'\sin x + B'\cos x = \cot x.$$

We must solve the system

$$A'\cos x + B'\sin x = 0,$$
$$-A'\sin x + B'\cos x = \cot x,$$

for A' and B' and then integrate to find A and B. We get

$$A' = -\cos x, \qquad A = -\sin x,$$
$$B' = \cos x \cot x, \qquad B = \cos x - \ln|\csc x + \cot x|.$$

Thus

$$y_p = -\sin x \cos x + [\cos x - \ln|\csc x + \cot x|]\sin x$$
$$= -\sin x \ln|\csc x + \cot x|,$$
$$y = y_c - \sin x \ln|\csc x + \cot x|.$$

7. We use equation (10) of Section 9.4 with $a = 0$ and $f(x) = \sec x$.

$$y = y_c + \int_0^x \sec\beta \sin(x-\beta)\,d\beta$$
$$= y_c + \int_0^x (\sin x - \cos x \tan\beta)\,d\beta$$
$$= y_c + \left[\beta\sin x + \cos x \ln|\cos\beta|\right]_0^x$$
$$= y_c + x\sin x + \cos x \ln|\cos x|.$$

9. We consider

$$y = Ae^x + Be^{-x},$$
$$y' = Ae^x - Be^{-x} + A'e^x + B'e^{-x}.$$

Setting $A'e^x + B'e^{-x} = 0$ gives

$$y'' = Ae^x + Be^{-x} + A'e^x - B'e^{-x},$$
$$y'' - y = A'e^x - B'e^{-x} = \frac{2}{e^x - e^{-x}}.$$

We must solve the system

$$A'e^x + B'e^{-x} = 0,$$
$$A'e^x - B'e^{-x} = \frac{2}{e^x - e^{-x}},$$

for A' and B' and then integrate to find A and B. We get

$$A' = \frac{e^{-x}}{e^x - e^{-x}} = \frac{e^{-2x}}{1 - e^{-2x}}, \qquad\qquad A = \tfrac{1}{2}\ln|1 - e^{-2x}|,$$
$$B' = \frac{-e^x}{e^x - e^{-x}} = \frac{-e^{2x}}{e^{2x} - 1}, \qquad\qquad B = -\tfrac{1}{2}\ln|e^{2x} - 1|.$$

Thus

$$y_p = \left[\tfrac{1}{2}\ln|1 - e^{-2x}|\right] e^x + \left[-\tfrac{1}{2}\ln|e^{2x} - 1|\right] e^{-x}$$
$$= -xe^{-x} + \tfrac{1}{2}(e^x - e^{-x})\ln|1 - e^{-2x}|,$$
$$y = y_c - xe^{-x} + \tfrac{1}{2}(e^x - e^{-x})\ln|1 - e^{-2x}|.$$

11. We consider

$$y = Ae^x + Be^{-x},$$
$$y' = Ae^x - Be^{-x} + A'e^x + B'e^{-x}.$$

Setting $A'e^x + B'e^{-x} = 0$ gives

$$y'' = Ae^x + Be^{-x} + A'e^x - B'e^{-x},$$
$$y'' - y = A'e^x - B'e^{-x} = \frac{1}{e^{2x} + 1}.$$

We must solve the system

$$A'e^x + B'e^{-x} = 0,$$
$$A'e^x - B'e^{-x} = \frac{1}{e^{2x} + 1},$$

for A' and B' and then integrate to find A and B. We get

$$A' = \frac{\frac{1}{2}e^{-x}}{e^{2x}+1} = \frac{1}{2}e^{-x} - \frac{\frac{1}{2}e^{-x}}{1+e^{-2x}}, \qquad A = -\frac{1}{2}e^{-x} - \frac{1}{2}\arctan\left(e^{-x}\right),$$

$$B' = \frac{-\frac{1}{2}e^{x}}{e^{2x}+1} = -\frac{\frac{1}{2}e^{-x}}{1+e^{-2x}}, \qquad B = -\frac{1}{2}\arctan\left(e^{-x}\right).$$

Thus

$$y_p = \left[-\tfrac{1}{2}e^{-x} - \tfrac{1}{2}\arctan\left(e^{-x}\right)\right]e^{x} + \left[-\tfrac{1}{2}\arctan\left(e^{-x}\right)\right]e^{-x}$$

$$= -\frac{1}{2} - \frac{e^{x}+e^{-x}}{2}\arctan e^{-x},$$

$$y = y_c - \tfrac{1}{2} - \cosh x \arctan e^{-x}.$$

13. We consider

$$y = Ae^{-x} + Be^{-3x},$$

$$y' = -Ae^{-x} - 3Be^{-3x} + A'e^{-x} + B'e^{-3x}.$$

Setting $A'e^{-x} + B'e^{-3x} = 0$ gives

$$y'' = Ae^{-x} + 9Be^{-3x} - A'e^{-x} - 3B'e^{-3x},$$

$$y'' + 4y' + 3y = -A'e^{-x} - 3B'e^{-3x} = \sin\left(e^{x}\right).$$

We must solve the system

$$A'e^{-x} + B'e^{-3x} = 0,$$

$$-A'e^{-x} - 3B'e^{-3x} = \sin\left(e^{x}\right),$$

for A' and B' and then integrate to find A and B. We get

$$A' = \tfrac{1}{2}e^{x}\sin\left(e^{x}\right), \qquad A = -\tfrac{1}{2}\cos\left(e^{x}\right),$$

$$B' = -\tfrac{1}{2}e^{3x}\sin\left(e^{x}\right), \qquad B = \tfrac{1}{2}e^{2x}\cos\left(e^{x}\right) - e^{x}\sin\left(e^{x}\right) - \cos\left(e^{x}\right).$$

Thus

$$y_p = -e^{-2x}\sin\left(e^{x}\right) - e^{-3x}\cos\left(e^{x}\right),$$

$$y = y_c - e^{-2x}\sin\left(e^{x}\right) - e^{-3x}\cos\left(e^{x}\right).$$

15. We consider

$$y = Ae^{x} + Be^{-x},$$

$$y' = Ae^{x} - Be^{-x} + A'e^{x} + B'e^{-x}.$$

Setting $A'e^x + B'e^{-x} = 0$ gives

$$y'' = Ae^x + Be^{-x} + A'e^x - B'e^{-x},$$
$$y'' - y = A'e^x - B'e^{-x} = e^{2x}\left[3\tan(e^x) + e^x\sec^2(e^x)\right].$$

We must solve the system

$$A'e^x + B'e^{-x} = 0,$$
$$A'e^x - B'e^{-x} = e^{2x}\left[3\tan(e^x) + e^x\sec^2(e^x)\right],$$

for A' and B' and then integrate to find A and B. We get

$$A' = \tfrac{1}{2}e^x\left[3\tan(e^x) + e^x\sec^2(e^x)\right], \qquad A = -\ln|\cos(e^x)| + \tfrac{1}{2}e^x\tan(e^x),$$
$$B' = -\tfrac{1}{2}e^{3x}\left[3\tan(e^x) + e^x\sec^2(e^x)\right], \qquad B = -\tfrac{1}{2}e^{3x}\tan(e^x).$$

Thus

$$y_p = -e^x\ln|\cos(e^x)| = e^x\ln|\sec(e^x)|,$$
$$y = y_c + e^x\ln|\sec(e^x)|.$$

17. We consider

$$y = A\cos x + B\sin x,$$
$$y' = -A\sin x + B\cos x + A'\cos x + B'\sin x.$$

Setting $A'\cos x + B'\sin x = 0$ gives

$$y'' = -A\cos x - B\sin x - A'\sin x + B'\cos x,$$
$$y'' + y = -A'\sin x + B'\cos x = \sec x\tan^2 x.$$

We must solve the system

$$A'\cos x + B'\sin x = 0,$$
$$-A'\sin x + B'\cos x = \sec x\tan^2 x,$$

for A' and B' and then integrate to find A and B. We get

$$A' = -\tan^3 x, \qquad A = -\ln|\cos x| - \tfrac{1}{2}\tan^2 x,$$
$$B' = \tan^2 x, \qquad B = \tan x - x.$$

Thus

$$y_p = \left[-\ln|\cos x| - \tfrac{1}{2}\tan^2 x\right]\cos x + [\tan x - x]\sin x$$
$$= \tfrac{1}{2}\sin x\tan x - x\sin x - \cos x\ln|\cos x|,$$
$$y = y_c + \tfrac{1}{2}\sin x\tan x - x\sin x - \cos x\ln|\cos x|.$$

Chapter 10

Applications

10.3   Resonance

1. The spring is stretched $\frac{1}{2}$ ft by the 5-lb weight so $5 = \frac{1}{2}k$ or $k = 10$ lb/ft. The initial value problem is therefore

$$\tfrac{5}{32}x''(t) + 10x(t) = 0; \; x(0) = \tfrac{1}{4}, \; x'(0) = -6.$$

We therefore have

$$
\begin{aligned}
x(t) &= c_1 \sin 8t + c_2 \cos 8t, & x(0) &= c_2 = \tfrac{1}{4}, & c_2 &= \tfrac{1}{4}, \\
x'(t) &= 8c_1 \cos 8t - 8c_2 \sin 8t, & x'(0) &= 8c_1 = -6, & c_1 &= -\tfrac{3}{4}.
\end{aligned}
$$

The position of the weight is given by

$$x(t) = \tfrac{1}{4} \cos 8t - \tfrac{3}{4} \sin 8t.$$

3. The initial value problem is

$$\tfrac{2}{32}x''(t) + 16x(t) = 0; \; x(0) = \tfrac{1}{3}, \; x'(0) = 8.$$

We therefore have

$$
\begin{aligned}
x(t) &= c_1 \sin 16t + c_2 \cos 16t, & x(0) &= c_2 = \tfrac{1}{3}, & c_2 &= \tfrac{1}{3}, \\
x'(t) &= 16c_1 \cos 16t - 16c_2 \sin 16t, & x'(0) &= 16c_1 = 8, & c_1 &= \tfrac{1}{2}.
\end{aligned}
$$

The position of the weight is given by

$$x(t) = \tfrac{1}{2} \sin 16t + \tfrac{1}{3} \cos 16t.$$

5. The initial value problem is

$$\tfrac{4}{32}x''(t) + 8x(t) = \tfrac{1}{2}\cos 8t; \; x(0) = 0, \; x'(0) = -4.$$

The differential equation

$$x''(t) + 64x(t) = 4\cos 8t$$

has the general solution

$$x(t) = c_1 \sin 8t + c_2 \cos 8t + \tfrac{1}{4}t \sin 8t, \qquad\qquad x(0) = c_2 = 0, \qquad c_2 = 0,$$
$$x'(t) = 8c_1 \cos 8t - 8c_2 \sin 8t + 2t \cos 8t + \tfrac{1}{4} \sin 8t, \qquad x'(0) = 8c_1 = -4, \qquad c_1 = -\tfrac{1}{2}.$$

The position of the weight is given by

$$x(t) = \tfrac{1}{4}t \sin 8t - \tfrac{1}{2} \sin 8t.$$

7. From Exercise 6 we have

$$\begin{aligned}
x(t) &= \tfrac{1}{4} \cos 8t - \tfrac{1}{4} \sin 8t + \tfrac{1}{2} \sin 4t \\
&= \tfrac{1}{4}\sqrt{2}(\tfrac{1}{2}\sqrt{2} \cos 8t - \tfrac{1}{2}\sqrt{2} \sin 8t) + \tfrac{1}{2} \sin 4t \\
&= \tfrac{1}{4}\sqrt{2}(\cos 8t \cos \tfrac{1}{4}\pi - \sin 8t \sin \tfrac{1}{4}\pi) + \tfrac{1}{2} \sin 4t \\
&= \tfrac{1}{4}\sqrt{2} \cos(8t + \tfrac{1}{4}\pi) + \tfrac{1}{2} \sin 4t.
\end{aligned}$$

9. From Exercise 8 we have $x'(t) = -2 \sin 8t + 2t \sin 8t = -2(1 - t) \sin 8t$. A stop will occur whenever $x'(t) = 0$. That will happen when $t = \tfrac{1}{8}\pi,\ \tfrac{1}{4}\pi,\ 1,\ \tfrac{3}{8}\pi$ (sec). The position of the weight at each of these times is given by $x = -0.15,\ +0.05,\ +0.03,\ +0.04$ (ft), respectively.

11. We must solve the initial value problem

$$\tfrac{16}{32}x''(t) + 128x(t) = 360 \cos 4t; \quad x(0) = \tfrac{1}{3},\ x'(0) = 4.$$

The differential equation becomes $x''(t) + 256x(t) = 720 \cos 4t$ and its general solution can be found to be

$$x(t) = c_1 \cos 16t + c_2 \sin 16t + 3 \cos 4t, \qquad\qquad x(0) = c_1 + 3 = \tfrac{1}{3}, \qquad c_1 = -\tfrac{8}{3},$$
$$x'(t) = -16c_1 \sin 16t + 16c_2 \cos 16t - 12 \sin 4t, \qquad x'(0) = 16c_2 = 4, \qquad c_2 = \tfrac{1}{4}.$$

The motion is described by

$$x(t) = -\tfrac{8}{3} \cos 16t + \tfrac{1}{4} \sin 16t + 3 \cos 4t.$$

Finally, $x(\tfrac{1}{8}\pi) = -\tfrac{8}{3}$, and $x'(\tfrac{1}{8}\pi) = -8$.

13. We need to solve the initial value problem

$$x''(t) + 128x(t) = 0; \quad x(0) = \tfrac{1}{3},\ x'(0) = -8.$$

The general solution and its derivative are given by

$$x(t) = c_1 \cos 8\sqrt{2}t + c_2 \sin 8\sqrt{2}t, \qquad\qquad x(0) = c_1 = \tfrac{1}{3}, \qquad c_1 = \tfrac{1}{3},$$
$$x'(t) = -8\sqrt{2}c_1 \sin 8\sqrt{2}t + 8\sqrt{2}c_2 \cos 8\sqrt{2}t, \qquad x'(0) = 8\sqrt{2}c_2 = -8, \qquad c_2 = -\tfrac{1}{2}\sqrt{2}.$$

Hence the motion can be described as

$$x(t) = \tfrac{1}{3} \cos 8\sqrt{2}t - \tfrac{1}{2}\sqrt{2} \sin 8\sqrt{2}t.$$

15. The initial value problem is

$$\frac{20}{32}x''(t) + 24x(t) = 0; \ x(0) = -\frac{14}{12}, \ x'(0) = 8.$$

We therefore have

$$x(t) = c_1 \sin\sqrt{\frac{192}{5}}t + c_2 \cos\sqrt{\frac{192}{5}}t, \qquad x(0) = c_2 = -\frac{7}{6}, \qquad c_2 = -\frac{7}{6},$$

$$x'(t) = \sqrt{\frac{192}{5}}c_1 \cos\sqrt{\frac{192}{5}}t - \sqrt{\frac{192}{5}}c_2 \sin\sqrt{\frac{192}{5}}t, \quad x'(0) = \sqrt{\frac{192}{5}}c_1 = 8, \quad c_1 = 8\sqrt{\frac{5}{192}}.$$

The position of the weight is given by

$$x(t) = -\frac{7}{6}\cos\sqrt{\frac{192}{5}}t + 8\sqrt{\frac{5}{192}}\sin\sqrt{\frac{192}{5}}t.$$

The amplitude of this motion is $\sqrt{\frac{49}{36} + 64\cdot\frac{5}{192}} = \frac{1}{6}\sqrt{109}$ ft or $2\sqrt{109}$ in. Because the weight began to fall 14 in. above the equilibrium and ends its fall $2\sqrt{109}$ in. below equilibrium, the answer to the question is $14 + 2\sqrt{109}$ in. or approximately 35 in.

17. We need to solve the initial value problem

$$\frac{4}{32}x''(t) + 16x(t) = 0; \ x(0) = -\frac{1}{6}, \ x'(0) = -15.$$

The general solution and its derivative are given by

$$x(t) = c_1 \cos 8\sqrt{2}t + c_2 \sin 8\sqrt{2}t, \qquad x(0) = c_1 = -\frac{1}{6}, \qquad c_1 = -\frac{1}{6},$$

$$x'(t) = -8\sqrt{2}c_1 \sin 8\sqrt{2}t + 8\sqrt{2}c_2 \cos 8\sqrt{2}t, \qquad x'(0) = 8\sqrt{2}c_2 = -15, \qquad c_2 = -\frac{15}{16}\sqrt{2}.$$

It follows that

$$x'(t) = \frac{4}{3}\sqrt{2}\sin 8\sqrt{2}t - 15\cos 8\sqrt{2}t.$$

The first stop will occur when the weight reaches its highest point. The second stop will occur half a period later, when the weight reaches its lowest point. The first stop occurs the first time $x'(t) = 0$. That will happen when $\tan(8\sqrt{2}t) = \frac{45}{8}\sqrt{2}$; that is, at $t = \dfrac{\arctan\left(\frac{45}{8}\sqrt{2}\right)}{8\sqrt{2}}$. Since half of a period of the motion is $\frac{1}{16}\pi\sqrt{2}$, the weight will reach its lowest point when $t = \dfrac{\arctan\left(\frac{45}{8}\sqrt{2}\right)}{8\sqrt{2}} + \frac{1}{16}\pi\sqrt{2} = 0.4$ (sec).

19. We need to solve the initial value problem

$$x''(t) + 96x(t) = 0; \ x(0) = \frac{1}{2}, \ x'(0) = -8.$$

The general solution and its derivative are given by

$$x(t) = c_1 \cos 4\sqrt{6}t + c_2 \sin 4\sqrt{6}t, \qquad x(0) = c_1 = \frac{1}{2}, \qquad c_1 = \frac{1}{2},$$

$$x'(t) = -4\sqrt{6}c_1 \sin 4\sqrt{6}t + 4\sqrt{6}c_2 \cos 4\sqrt{6}t, \qquad x'(0) = 4\sqrt{6}c_2 = -8, \qquad c_2 = -\frac{1}{3}\sqrt{6}.$$

Hence the motion can be described as

$$x(t) = \frac{1}{2}\cos 4\sqrt{6}t - \frac{1}{3}\sqrt{6}\sin 4\sqrt{6}t.$$

21. We need to solve the initial value problem

$$x''(t) + 64x(t) = 8\sin 8t; \quad x(0) = \tfrac{1}{4}, \ x'(0) = 0.$$

The method of undetermined coefficients may be used to obtain $x_p = -\tfrac{1}{2}t\cos 8t$ as a particular solution of this differential equation. The general solution and its derivative are given by

$$x(t) = c_1\sin 8t + c_2\cos 8t - \tfrac{1}{2}t\cos 8t, \qquad x(0) = c_2 = \tfrac{1}{4}, \qquad c_2 = \tfrac{1}{4},$$
$$x'(t) = 8c_1\cos 8t - 8c_2\sin 8t + 4t\sin 8t - \tfrac{1}{2}\cos 8t, \qquad x'(0) = 8c_1 - \tfrac{1}{2} = 0, \qquad c_1 = \tfrac{1}{16}.$$

It follows that

$$x'(t) = -2\sin 8t + 4t\sin 8t = 4(t - \tfrac{1}{2})\sin 8t.$$

Stops will occur when $t = \tfrac{1}{2}, \ n\pi/8, \ n = 1, 2, \cdots$. The first four stops in order are: $\tfrac{1}{8}\pi, \ \tfrac{1}{2}, \ \tfrac{1}{4}\pi, \ \tfrac{3}{8}\pi$ (sec).

23. From Exercise 22 we have

$$x(t) = \tfrac{1}{4}(1 - 8t)\cos 16t + \tfrac{1}{8}\sin 16t,$$
$$x'(t) = -4(1 - 8t)\sin 16t.$$

Stops will occur when $t = \tfrac{1}{8}, \ \tfrac{1}{16}n\pi$. The first four stops will occur at $t = \tfrac{1}{8}, \ \tfrac{1}{16}\pi, \ \tfrac{1}{8}\pi, \ \tfrac{3}{16}\pi$ (sec).

10.4 Damped Vibrations

1. The auxiliary equation is $m^2 + 2\gamma m + 169 = 0$ and its roots are $-\gamma \pm \sqrt{\gamma^2 - 169}$. Critical damping will occur when $\gamma = 13$. For this value of γ we have $m = -13, -13$ and the general solution of the differential equation is

$$x(t) = (c_1 + c_2 t)e^{-13t}, \qquad x(0) = c_1 = 0, \qquad c_1 = 0,$$
$$x'(t) = \big[-13(c_1 + c_2 t) + c_2\big]e^{-13t}, \qquad x'(0) = -13c_1 + c_2 = 8, \qquad c_2 = 8.$$

Thus the solution with critical damping is $x(t) = 8te^{-13t}$.

If $\gamma = 12$, the auxiliary equation is $m^2 + 24m + 169 = 0$ with roots $m = -12 \pm 5i$. The general solution of the differential equation is

$$x(t) = e^{-12t}(c_1\sin 5t + c_2\cos 5t), \qquad x(0) = c_2 = 0, \qquad c_2 = 0,$$
$$x'(t) = e^{-12t}\big[(5c_1 - 12c_2)\cos 5t + (-12c_1 - 5c_2)\sin 5t\big], \quad x'(0) = 5c_1 - 12c_2 = 8, \quad c_1 = \tfrac{8}{5}.$$

Thus we have the damped oscillatory solution $x(t) = \tfrac{8}{5}e^{-12t}\sin 5t$.

If $\gamma = 14$, the auxiliary equation is $m^2 + 28m + 169 = 0$ and $m = -7 \pm 3\sqrt{3} = -8.8, \ -19.2$. The general solution is

$$x(t) = c_1 e^{-8.8t} + c_2 e^{-19.2t}, \qquad x(0) = c_1 + c_2 = 0, \qquad c_1 = 0.77,$$
$$x'(t) = -8.8c_1 e^{-8.8t} - 19.2c_2 e^{-19.2t}, \qquad x'(0) = -8.8c_1 - 19.2c_2 = 8, \qquad c_2 = -0.77.$$

The overdamped motion is described by $x(t) = 0.77\left(e^{-8.8t} - e^{-19.2t}\right)$.

3. The initial value problem is

$$\frac{4}{32}x''(t) + \frac{1}{4}x'(t) + \frac{4}{.64}x(t) = 0; \ x(0) = -\tfrac{1}{3}, \ x'(0) = 5.$$

The differential equation may be written $x''(t) + 2x'(t) + 50x(t) = 0$ with auxiliary equation $m^2 + 2m + 50 = 0$. Here $m = -1 \pm 7i$. The general solution is

$$x(t) = e^{-t}(c_1 \sin 7t + c_2 \cos 7t), \qquad\qquad x(0) = c_2 = -\tfrac{1}{3}, \qquad c_2 = -\tfrac{1}{3},$$
$$x'(t) = e^{-t}\big[(7c_1 - c_2)\cos 7t + (-c_1 - 7c_2)\sin 7t\big], \quad x'(0) = 7c_1 - c_2 = 5, \quad c_1 = \tfrac{2}{3}.$$

The motion is described by $x(t) = \tfrac{1}{3}e^{-t}(2\sin 7t - \cos 7t)$.

5. The initial value problem is

$$\frac{4}{32}x''(t) + x'(t) + 10x(t) = 0; \ x(0) = 0, \ x'(0) = -2.$$

The differential equation can be written $x''(t) + 8x'(t) + 80x(t) = 0$ with auxiliary equation $m^2 + 8m + 80 = 0$. Here $m = -4 \pm 8i$. The general solution is

$$x(t) = e^{-4t}(c_1 \cos 8t + c_2 \sin 8t), \qquad\qquad x(0) = c_1 = 0, \qquad c_1 = 0,$$
$$x'(t) = e^{-4t}\big[(-8c_1 - 4c_2)\sin 8t + (-4c_1 + 8c_2)\cos 8t\big], \quad x'(0) = -4c_1 + 8c_2 = -2, \quad c_2 = -\tfrac{1}{4}.$$

The motion is described by $x(t) = -\tfrac{1}{4}e^{-4t}\sin 8t$.

7. The initial value problem is

$$\frac{2}{32}x''(t) + x'(t) + 4x(t) = 2\sin 8t; \ x(0) = \tfrac{1}{4}, \ x'(0) = 0.$$

The differential equation can be written $x''(t) + 16x'(t) + 64x(t) = 32\sin 8t$ with auxiliary equation $m^2 + 16m + 64 = 0$. Here $m = -8, \ -8$. The method of undetermined coefficients can be used to find a particular solution $y_p = -\tfrac{1}{4}\cos 8t$. The general solution is

$$x(t) = (c_1 + c_2 t)e^{-8t} - \tfrac{1}{4}\cos 8t, \qquad\qquad x(0) = c_1 - \tfrac{1}{4} = \tfrac{1}{4}, \qquad c_1 = \tfrac{1}{2},$$
$$x'(t) = (-8c_1 + c_2 - 8c_2 t)e^{-8t} + 2\sin 8t, \qquad x'(0) = -8c_1 + c_2 = 0, \qquad c_2 = 4.$$

The motion is described by $x(t) = \big(\tfrac{1}{2} + 4t\big)e^{-8t} - \tfrac{1}{4}\cos 8t$.

9. Here

$$x(t) = e^{-t}(0.50\cos 6.1t - 1.23\sin 6.1t),$$
$$x'(t) = e^{-t}(-8\cos 6.1t - 1.82\sin 6.1t).$$

The motion will stop when $x'(t) = 0$. This will occur when $\tan 6.1t = \dfrac{-8}{1.82}$. One solution of this equation, a negative solution, is $t = \dfrac{\arctan\left(-\frac{8}{1.82}\right)}{6.1} = -0.22$. The period of the motion is $\dfrac{2\pi}{6.1}$. The stopping places will occur every half-period. The first three stopping places, with positive values of t, and the corresponding values of x are

$$t_1 = -0.22 + \frac{\pi}{6.1} = 0.3 \text{ (sec)}, \quad t_2 = -0.22 + \frac{2\pi}{6.1} = 0.8 \text{ (sec)}, \quad t_3 = -0.22 + \frac{3\pi}{6.1} = 1.3 \text{ (sec)},$$

$$x_1 = -12 \text{ (in.)}, \qquad\qquad x_2 = 7 \text{ (in.)}, \qquad\qquad x_3 = -4 \text{ (in.)}.$$

11. From Exercise 10 we have $x(t) = 1.22e^{-0.16t}\sin 9.8t$. The initial value of the damping factor$1.22e^{-0.16t}$ is 1.22. We wish to determine the value of t such that $1.22e^{-0.16t} = \frac{1}{10}(1.22)$. Solving this equation for t gives us $t = \dfrac{\ln 10}{0.16} = 14.4 \text{ (sec)}$.

13. The initial value problem is

$$\frac{2}{32}x''(t) + .6x'(t) + 4x(t) = \tfrac{1}{4}\sin 8t; \ x(0) = \tfrac{1}{4}, \ x'(0) = 0.$$

The differential equation can be written $x''(t) + 9.6x'(t) + 64x(t) = 4\sin 8t$ with auxiliary equation $m^2 + 9.6m + 64 = 0$. Here $m = -4.8 \pm 6.4i$. The method of undetermined coefficients can be used to find a particular solution $y_p = -0.05\cos 8t$. The general solution is

$$x(t) = e^{-4.8t}(c_1\cos 6.4t + c_2\sin 6.4t) - 0.05\cos 8t,$$
$$x'(t) = e^{-4.8t}\big[(-6.4c_1 - 4.8c_2)\sin 6.4t + (-4.8c_1 + 6.4c_2)\cos 6.4t\big] + 0.4\sin 8t.$$

Therefore $x(0) = c_1 - 0.05 = 0.25$ and $x'(0) = -4.8c_1 + 6.4c_2 = 0$, so that $c_1 = 0.30$ and $c_2 = 0.22$. The motion is described by $x(t) = e^{-4.8t}(0.30\cos 6.4t + 0.22\sin 6.4t) - 0.05\cos 8t$.

15. The initial value problem is

$$\frac{2}{32}x''(t) + x'(t) + 4x(t) = \tfrac{1}{4}\sin 8t; \ x(0) = \tfrac{1}{4}, \ x'(0) = 0.$$

The differential equation can be written $x''(t) + 16x'(t) + 64x(t) = 4\sin 8t$ with auxiliary equation $m^2 + 16m + 64 = 0$. Here $m = -8, -8$. The method of undetermined coefficients can be used to find a particular solution $y_p = -\frac{1}{32}\cos 8t$. The general solution is

$$x(t) = (c_1 + c_2 t)e^{-8t} - \tfrac{1}{32}\cos 8t, \qquad\qquad x(0) = c_1 - \tfrac{1}{32} = \tfrac{1}{4}, \qquad c_1 = \tfrac{9}{32},$$
$$x'(t) = (-8c_1 + c_2 - 8c_2 t)e^{-8t} + \tfrac{1}{4}\sin 8t, \qquad x'(0) = -8c_1 + c_2 = 0, \qquad c_2 = \tfrac{9}{4}.$$

The motion is described by $x(t) = \frac{9}{32}(8t + 1)e^{-8t} - \frac{1}{32}\cos 8t$. Note that when $t = 1$, the first term in the solution has value $\frac{81}{32}e^{-8} = 0.001$, so that $x(t) = -\frac{1}{32}\cos 8t$ for $t > 1$, to the nearest 0.01 ft.

17. The initial value problem is

$$\frac{12}{32}x''(t) + \tfrac{1}{2}x'(t) + 24x(t) = 9\sin 4t; \ x(0) = \tfrac{1}{4}, \ x'(0) = 0.$$

The differential equation can be written $3x''(t) + 4x'(t) + 192x(t) = 72 \sin 4t$ with auxiliary equation $3m^2 + 4m + 192 = 0$. Here $m = -\frac{2}{3} \pm 8.0i$. The method of undetermined coefficients can be used to find a particular solution $y_p = \frac{81}{164} \sin 4t - \frac{9}{164} \cos 4t$. The general solution is

$$x(t) = e^{-\frac{2}{3}t}(c_1 \cos 8.0t + c_2 \sin 8.0t) + \frac{81}{164} \sin 4t - \frac{9}{164} \cos 4t,$$

$$x'(t) = e^{-\frac{2}{3}t}\left[\left(-\frac{2}{3}c_1 + 8c_2\right) \cos 8.0t + \left(-8c_1 - \frac{2}{3}c_2\right) \sin 8.0t\right] + \frac{81}{41} \cos 4t + \frac{9}{41} \sin 4t.$$

Therefore $x(0) = c_1 - \frac{9}{164} = \frac{1}{4}$ and $x'(0) = -\frac{2}{3}c_1 + 8c_2 + \frac{81}{41} = 0$, so that $c_1 = 0.30$ and $c_2 = -0.22$. Finally, $x(t) = e^{-\frac{2}{3}t}(0.30 \cos 8.0t - 0.22 \sin 8.0t) - 0.05 \cos 4t + 0.49 \sin 4t$.

19. We assume that the acceleration due to gravity is a constant g. Then Newton's law, $F = ma$, becomes $\dfrac{w}{g}\dfrac{d^2x}{dt^2} = w$ for $x < h$. The initial value problem may be written

$$\frac{d^2x}{dt^2} = g; \ x(0) = 0, \ x'(0) = 0.$$

We get

$$x(t) = \tfrac{1}{2}gt^2 + c_1 + c_2t, \qquad\qquad x(0) = c_1 = 0,$$
$$x'(t) = gt + c_2, \qquad\qquad\qquad x'(0) = c_2 = 0.$$

Finally, $x(t) = \frac{1}{2}gt^2$.

21. From $v = gt + v_0$ we get $t = \dfrac{v - v_0}{g}$. Thus

$$x = \frac{1}{2}\frac{(v - v_0)^2}{g} + \frac{v_0(v - v_0)}{g},$$
$$2gx = (v - v_0)^2 + 2v_0(v - v_0) = (v - v_0)(v + v_0).$$

Finally, $v^2 = v_0^2 + 2gx$.

23.

$$x = \frac{gt}{a} + \frac{(av_0 - g)(1 - e^{-at})}{a^2}$$
$$= \frac{gt}{a} + \left(\frac{v_0}{a} - \frac{g}{a^2}\right)\left(at - \tfrac{1}{2}a^2t^2 + \tfrac{1}{6}a^3t^3 - \tfrac{1}{24}a^4t^4 + \cdots\right)$$
$$= \frac{gt}{a} + \left(v_0t - \tfrac{1}{2}v_0at^2 + \tfrac{1}{6}v_0a^2t^3 + \cdots\right) + \left(-\frac{gt}{a} + \tfrac{1}{2}gt^2 - \tfrac{1}{6}gat^3 + \tfrac{1}{24}ga^2t^4 + \cdots\right)$$
$$= \tfrac{1}{2}gt^2 + v_0t - \tfrac{1}{6}at^2(3v_0 + gt) + \tfrac{1}{24}a^2t^3(4v_0 + gt) + \cdots.$$

25. The initial value problem is

$$x''(t) + 6x'(t) + 25x(t) = 0; \ x(0) = 0, \ x'(0) = -12.$$

The auxiliary equation is $m^2 + 6m + 25 = 0$ with roots $m = -3 \pm 4i$. The general solution is

$$x(t) = e^{-3t}(c_1 \sin 4t + c_2 \cos 4t), \qquad\qquad x(0) = c_2 = 0, \qquad\qquad c_2 = 0,$$
$$x'(t) = e^{-3t}\left[(-3c_1 - 4c_2)\sin 4t + (4c_1 - 3c_2)\cos 4t\right], \quad x'(0) = 4c_1 - 3c_2 = -12, \quad c_1 = -3.$$

We have $x(t) = -3e^{-3t}\sin 4t$. Stops will occur when $x'(t) = -3e^{-3t}(4\cos 4t - 3\sin 4t) = 0$. The first stop will be at $t = \arctan\frac{4}{3}$. Since the period is $\pi/2$, stops occur at intervals of length $\pi/4$. Thus $t_n = \arctan\frac{4}{3} + \frac{1}{4}n\pi$, for $n = 0, 1, \cdots$. For these values of t

$$x(t_n) = -3\exp\left[-3t_n\right]\sin 4(\arctan\tfrac{4}{3} + \tfrac{1}{4}n\pi)$$
$$= -3\exp\left[-3t_n\right]\sin 4(\arctan\tfrac{4}{3}).$$

Thus

$$\frac{x(t_{n+1})}{x(t_n)} = \frac{\exp\left[-3t_{n+1}\right]}{\exp\left[-3t_n\right]} = \exp\left[-3(t_{n+1} - t_n)\right] = \exp\left[-3\pi/4\right].$$

This ratio is independent of n.

10.5   The Simple Pendulum

1. Using the linear approximation to the pendulum problem the motion is governed by the differential equation $\dfrac{d^2\theta}{dt^2} + 64\theta = 0$. Thus the period of the motion is $\pi/4$ (sec). If the clock ticks once every period, in 30 seconds it will tick $30 \cdot \dfrac{4}{\pi} = \dfrac{120}{\pi} = 38$ times.

3. Here $\theta = 0.1\cos 8t$ (radians). The maximum value of θ' will occur when $\theta''(t) = -6.4\cos 8t = 0$. That will first happen when $t = \pi/16 = 0.2$ (sec). The maximum angular speed is therefore $|\theta'(\pi/16)| = 0.8$ (radians/sec).

5. The motion is described by $\theta = \frac{1}{10}\cos 8t - \frac{1}{8}\sin 8t$. This is a simple harmonic motion with amplitude $\sqrt{(\frac{1}{10})^2 = (\frac{1}{8})^2} = 0.16$ (radians). Converting to degrees gives $9°$.

7. The motion is simple harmonic with amplitude $\sqrt{\theta_0^2 + \beta^{-2}\omega_0^2}$.

Chapter 11

Linear Systems of Equations

11.2   First-Order Systems with Constant Coefficients

1. If we let $u = y'$ then we get the system $y' = u$, $u' = -8y + 6u + x + 2$.

3. If we let $u = y'$ then we get the system $y' = u$, $u' = -qy - pu + f(x)$.

5. If we let $u = y'$, $v = u'$, and $w = v'$ then we get the system $y' = u$, $u' = v$, $v' = w$, $w' = y$.

7. We can write the given system in the form

$$3v' + w' = -2v + 6w + 5e^x,$$
$$4v' + w' = -2v + 8w + 5e^x + 2x - 3.$$

   Solving for v' and w' yields the system

$$v' = 2w + 2x - 3,$$
$$w' = -2v + 5e^x - 6x + 9.$$

9. If we let $u = y'$ and $w = v'$ then we get the system

$$y' = u,$$
$$v' = w,$$
$$u' = -v + 2w + 1,$$
$$w' = -y + 4v - 2u.$$

11.4   Some Matrix Algebra

1. $A + 2B = \begin{pmatrix} 1 & 2 \\ 3 & 1 \end{pmatrix} + 2\begin{pmatrix} 2 & 0 \\ 1 & -1 \end{pmatrix} = \begin{pmatrix} 1 & 2 \\ 3 & 1 \end{pmatrix} + \begin{pmatrix} 4 & 0 \\ 2 & -2 \end{pmatrix} = \begin{pmatrix} 5 & 2 \\ 5 & -1 \end{pmatrix}.$

3. $AB + 2I = \begin{pmatrix} 1 & 2 \\ 3 & 1 \end{pmatrix}\begin{pmatrix} 2 & 0 \\ 1 & -1 \end{pmatrix} + 2\begin{pmatrix} 1 & 0 \\ 0 & 1 \end{pmatrix} = \begin{pmatrix} 4 & -2 \\ 7 & -1 \end{pmatrix} + \begin{pmatrix} 2 & 0 \\ 0 & 2 \end{pmatrix} = \begin{pmatrix} 6 & -2 \\ 7 & 1 \end{pmatrix}.$

5. $C - 2I = \begin{pmatrix} 1 & -1 \\ 1 & 2 \end{pmatrix} - 2\begin{pmatrix} 1 & 0 \\ 0 & 1 \end{pmatrix} = \begin{pmatrix} 1 & -1 \\ 1 & 2 \end{pmatrix} - \begin{pmatrix} 2 & 0 \\ 0 & 2 \end{pmatrix} = \begin{pmatrix} -1 & -1 \\ 1 & 0 \end{pmatrix}.$

7. $AC - B = \begin{pmatrix} 1 & 2 \\ 3 & 1 \end{pmatrix} \begin{pmatrix} 1 & -1 \\ 1 & 2 \end{pmatrix} - \begin{pmatrix} 2 & 0 \\ 1 & -1 \end{pmatrix} = \begin{pmatrix} 3 & 3 \\ 4 & -1 \end{pmatrix} - \begin{pmatrix} 2 & 0 \\ 1 & -1 \end{pmatrix} = \begin{pmatrix} 1 & 3 \\ 3 & 0 \end{pmatrix}.$

9. Let d_{ij} and e_{ij} be the elements of D and E that are in the ith row and the jth column. Then the corresponding elements in $D + E$ and $E + D$ are $d_{ij} + e_{ij}$ and $e_{ij} + d_{ij}$. Since these two numbers are always equal, we have $D + E = E + D$.

11. Let a_{ij} be the element of A that is in the ith row and the jth column, The corresponding element in O is 0. But $a_{ij} + 0 = a_{ij}$ for all values of i and j. Hence $A + O = A$.

13. Since $k(a_{ij} + b_{ij}) = ka_{ij} + kb_{ij}$, it follows that $k(A + B) = kA + kB$.

15. Since $\det A = -3$, the only solution of the system is the trivial solution, O.

17. Here the two equations $2x + y = 0$ and $-4x - 2y = 0$ have the same solutions. Hence $y = -2x$ and $\begin{pmatrix} x \\ y \end{pmatrix} = \begin{pmatrix} x \\ -2x \end{pmatrix} = x \begin{pmatrix} 1 \\ -2 \end{pmatrix}$ where x is arbitrary.

19. Since $\det A = 0$, the only solution is $\begin{pmatrix} x \\ y \\ z \end{pmatrix} = O.$

21. Here $3x + 2y = 0$ so that if $y = 3c$ then $x = -2c$. That is $\begin{pmatrix} x \\ y \end{pmatrix} = \begin{pmatrix} -2c \\ 3c \end{pmatrix} = c \begin{pmatrix} -2 \\ 3 \end{pmatrix}.$

23. Here $x + y + z = 0$ so that if $x = b$ and $z = -c$ then $y = -b + c$. We therefore have the solutions $\begin{pmatrix} x \\ y \\ z \end{pmatrix} = \begin{pmatrix} b \\ -b + c \\ -c \end{pmatrix} = b \begin{pmatrix} 1 \\ -1 \\ 0 \end{pmatrix} + c \begin{pmatrix} 0 \\ 1 \\ -1 \end{pmatrix}.$

25. $X' = AX + B$ where $\begin{pmatrix} 1 & -1 & 1 \\ 1 & 2 & -1 \\ 2 & -1 & 1 \end{pmatrix}$ and $B = \begin{pmatrix} t \\ 1 \\ e^t \end{pmatrix}.$

27. $X' = AX + B$ where $\begin{pmatrix} t & 1 & 1 \\ t^2 & t & 0 \\ 2 & 1 & t \end{pmatrix}$ and $B = \begin{pmatrix} \sin t \\ 1 \\ 0 \end{pmatrix}.$

11.5 First-Order Systems Revisited

1. The characteristic equation is $\begin{vmatrix} 8 - m & -3 \\ 16 & -8 - m \end{vmatrix} = m^2 - 16 = (m - 4)(m + 4) = 0.$

If $m = 4$, $\begin{pmatrix} 4 & -3 \\ 16 & -12 \end{pmatrix} \begin{pmatrix} c_1 \\ c_2 \end{pmatrix} = O$, so that $4c_1 = 3c_2$. One such eigenvector is $\begin{pmatrix} 3 \\ 4 \end{pmatrix}.$

If $m = -4$, $\begin{pmatrix} 12 & -3 \\ 16 & -4 \end{pmatrix} \begin{pmatrix} c_1 \\ c_2 \end{pmatrix} = O$, so that $12c_1 = 4c_2$. One such eigenvector is $\begin{pmatrix} 1 \\ 4 \end{pmatrix}$. Two

solutions of the system are therefore $\begin{pmatrix} 3 \\ 4 \end{pmatrix} e^{4t}$ and $\begin{pmatrix} 1 \\ 4 \end{pmatrix} e^{-4t}$. The linear independence is checked

by computing $W(0) = \begin{vmatrix} 3 & 1 \\ 4 & 4 \end{vmatrix} = 8 \neq 0$.

3. The characteristic equation is $\begin{vmatrix} 4 - m & 3 \\ -4 & -4 - m \end{vmatrix} = m^2 - 4 = (m - 2)(m + 2) = 0$.

If $m = 2$, $\begin{pmatrix} 2 & 3 \\ -4 & -6 \end{pmatrix} \begin{pmatrix} c_1 \\ c_2 \end{pmatrix} = O$, so that $2c_1 + 3c_2 = 0$. One such eigenvector is $\begin{pmatrix} 3 \\ -2 \end{pmatrix}$.

If $m = -2$, $\begin{pmatrix} 6 & 3 \\ -4 & -2 \end{pmatrix} \begin{pmatrix} c_1 \\ c_2 \end{pmatrix} = O$, so that $6c_1 + 3c_2 = 0$. One such eigenvector is $\begin{pmatrix} 1 \\ -2 \end{pmatrix}$. Two

solutions of the system are therefore $\begin{pmatrix} 3 \\ -2 \end{pmatrix} e^{2t}$ and $\begin{pmatrix} 1 \\ -2 \end{pmatrix} e^{-2t}$. The linear independence is

checked by computing $W(0) = \begin{vmatrix} 3 & 1 \\ -2 & -2 \end{vmatrix} = 8 \neq 0$.

5. The characteristic equation is $\begin{vmatrix} 3 - m & 3 \\ -1 & -1 - m \end{vmatrix} = m^2 - 2m = m(m - 2) = 0$.

If $m = 0$, $\begin{pmatrix} 3 & 3 \\ -1 & -1 \end{pmatrix} \begin{pmatrix} c_1 \\ c_2 \end{pmatrix} = O$, so that $3c_1 + 3c_2 = 0$. One such eigenvector is $\begin{pmatrix} 1 \\ -1 \end{pmatrix}$.

If $m = 2$, $\begin{pmatrix} 1 & 3 \\ -1 & -3 \end{pmatrix} \begin{pmatrix} c_1 \\ c_2 \end{pmatrix} = O$, so that $c_1 + 3c_2 = 0$. One such eigenvector is $\begin{pmatrix} 3 \\ -1 \end{pmatrix}$. Two

solutions of the system are therefore $\begin{pmatrix} 1 \\ -1 \end{pmatrix}$ and $\begin{pmatrix} 3 \\ -1 \end{pmatrix} e^{2t}$. The linear independence is checked

by computing $W(0) = \begin{vmatrix} 1 & 3 \\ -1 & -1 \end{vmatrix} = 2 \neq 0$.

7. The characteristic equation is $\begin{vmatrix} 12 - m & -15 \\ 4 & -4 - m \end{vmatrix} = m^2 - 8m + 12 = (m - 2)(m - 6) = 0$.

If $m = 2$, $\begin{pmatrix} 10 & -15 \\ 4 & -6 \end{pmatrix} \begin{pmatrix} c_1 \\ c_2 \end{pmatrix} = O$, so that $10c_1 - 15c_2 = 0$. One such eigenvector is $\begin{pmatrix} 3 \\ 2 \end{pmatrix}$.

If $m = 6$, $\begin{pmatrix} 6 & -15 \\ 4 & -10 \end{pmatrix} \begin{pmatrix} c_1 \\ c_2 \end{pmatrix} = O$, so that $6c_1 - 15c_2 = 0$. One such eigenvector is $\begin{pmatrix} 5 \\ 2 \end{pmatrix}$. Two

solutions of the system are therefore $\begin{pmatrix} 3 \\ 2 \end{pmatrix} e^{2t}$ and $\begin{pmatrix} 5 \\ 2 \end{pmatrix} e^{6t}$. The linear independence is checked

by computing $W(0) = \begin{vmatrix} 3 & 5 \\ 2 & 2 \end{vmatrix} = -4 \neq 0$.

11.6 Complex Eigenvalues

1. The characteristic equation is $\begin{vmatrix} 4-m & 5 \\ -4 & -4-m \end{vmatrix} = m^2 + 4 = (m-2i)(m+2i) = 0.$

If $m = 2i$, $\begin{pmatrix} 4-2i & 5 \\ -4 & -4-2i \end{pmatrix} \begin{pmatrix} b_1 \\ b_2 \end{pmatrix} = O$, so that $(4-2i)b_1 + 5b_2 = 0$. One such eigenvector is $\begin{pmatrix} 5 \\ -4+2i \end{pmatrix} = \begin{pmatrix} 5 \\ -4 \end{pmatrix} + i\begin{pmatrix} 0 \\ 2 \end{pmatrix}$. The general solution is therefore

$$X = c_1\left[\begin{pmatrix} 5 \\ -4 \end{pmatrix}\cos 2t - \begin{pmatrix} 0 \\ 2 \end{pmatrix}\sin 2t\right] + c_2\left[\begin{pmatrix} 0 \\ 2 \end{pmatrix}\cos 2t + \begin{pmatrix} 5 \\ -4 \end{pmatrix}\sin 2t\right].$$

3. The characteristic equation is $\begin{vmatrix} 4-m & -13 \\ 2 & -6-m \end{vmatrix} = m^2 + 2m + 2 = (m+1-i)(m+1+i) = 0.$

If $m = -1+i$, $\begin{pmatrix} 5-i & -13 \\ 2 & -5-i \end{pmatrix} \begin{pmatrix} b_1 \\ b_2 \end{pmatrix} = O$, so that $(5-i)b_1 - 13b_2 = 0$. One such eigenvector is $\begin{pmatrix} 13 \\ 5-i \end{pmatrix} = \begin{pmatrix} 13 \\ 5 \end{pmatrix} + i\begin{pmatrix} 0 \\ -1 \end{pmatrix}$. The general solution is therefore

$$X = c_1 e^{-t}\left[\begin{pmatrix} 13 \\ 5 \end{pmatrix}\cos t - \begin{pmatrix} 0 \\ -1 \end{pmatrix}\sin t\right] + c_2 e^{-t}\left[\begin{pmatrix} 0 \\ -1 \end{pmatrix}\cos t + \begin{pmatrix} 13 \\ 5 \end{pmatrix}\sin t\right].$$

5. The characteristic equation is $\begin{vmatrix} 12-m & -17 \\ 4 & -4-m \end{vmatrix} = m^2 - 8m + 20 = (m-4)^2 + 4 = 0.$

If $m = 4+2i$, $\begin{pmatrix} 8-2i & -17 \\ 4 & -8-2i \end{pmatrix} \begin{pmatrix} b_1 \\ b_2 \end{pmatrix} = O$, so that $(8-2i)b_1 - 17b_2 = 0$. One such eigenvector is $\begin{pmatrix} 17 \\ 8-2i \end{pmatrix} = \begin{pmatrix} 17 \\ 8 \end{pmatrix} + i\begin{pmatrix} 0 \\ -2 \end{pmatrix}$. The general solution is therefore

$$X = c_1 e^{4t}\left[\begin{pmatrix} 17 \\ 8 \end{pmatrix}\cos 2t - \begin{pmatrix} 0 \\ -2 \end{pmatrix}\sin 2t\right] + c_2 e^{4t}\left[\begin{pmatrix} 0 \\ -2 \end{pmatrix}\cos 2t + \begin{pmatrix} 17 \\ 8 \end{pmatrix}\sin 2t\right].$$

7. The characteristic equation is $\begin{vmatrix} 1-m & 0 & 0 \\ 2 & 1-m & -2 \\ 3 & 2 & 1-m \end{vmatrix} = -(m-1)\left[(m-1)^2 + 4\right] = 0.$

If $m = 1$, $\begin{pmatrix} 0 & 0 & 0 \\ 2 & 0 & -2 \\ 3 & 2 & 0 \end{pmatrix} \begin{pmatrix} b_1 \\ b_2 \\ b_3 \end{pmatrix} = O$ so that $2b_1 - 2b_3 = 0$ and $3b_1 + 2b_2 = 0$. One such eigenvector is $\begin{pmatrix} 2 \\ -3 \\ 2 \end{pmatrix}$ with corresponding solution $\begin{pmatrix} 2 \\ -3 \\ 2 \end{pmatrix} e^t.$

If $m = 1 + 2i$, $\begin{pmatrix} -2i & 0 & 0 \\ 2 & -2i & -2 \\ 3 & 2 & -2i \end{pmatrix} \begin{pmatrix} b_1 \\ b_2 \\ b_3 \end{pmatrix} = O$, so that $b_1 = 0$ and $-2ib_2 - 2b_3 = 0$. One such

eigenvector is $\begin{pmatrix} 0 \\ 1 \\ -i \end{pmatrix} = \begin{pmatrix} 0 \\ 1 \\ 0 \end{pmatrix} + i \begin{pmatrix} 0 \\ 0 \\ -1 \end{pmatrix}$. The general solution is therefore

$$X = c_1 e^t \begin{pmatrix} 2 \\ -3 \\ 2 \end{pmatrix} + c_2 e^t \left[\begin{pmatrix} 0 \\ 1 \\ 0 \end{pmatrix} \cos 2t - \begin{pmatrix} 0 \\ 0 \\ -1 \end{pmatrix} \sin 2t \right] + c_3 e^t \left[\begin{pmatrix} 0 \\ 0 \\ -1 \end{pmatrix} \cos 2t + \begin{pmatrix} 0 \\ 1 \\ 0 \end{pmatrix} \sin 2t \right].$$

9. We let $u = y'$ so that $u' = y'' = -2y - 2u$. Thus $\begin{pmatrix} y \\ u \end{pmatrix}' = \begin{pmatrix} 0 & 1 \\ -2 & -2 \end{pmatrix} \begin{pmatrix} y \\ u \end{pmatrix}$. The characteristic

equation is $\begin{vmatrix} -m & 1 \\ -2 & -2-m \end{vmatrix} = m^2 + 2m + 2 = (m+1)^2 + 1 = 0$.

If $m = -1 + i$ then $\begin{pmatrix} 1-i & 1 \\ -2 & -1-i \end{pmatrix} \begin{pmatrix} b_1 \\ b_2 \end{pmatrix} = O$, so that $(1-i)b_1 + b_2 = 0$. One such eigenvector

is $\begin{pmatrix} 1 \\ -1+i \end{pmatrix} = \begin{pmatrix} 1 \\ -1 \end{pmatrix} + \begin{pmatrix} 0 \\ 1 \end{pmatrix} i$. The general solution is

$$\begin{pmatrix} y \\ u \end{pmatrix} = c_1 e^{-x} \left[\begin{pmatrix} 1 \\ -1 \end{pmatrix} \cos x - \begin{pmatrix} 0 \\ 1 \end{pmatrix} \sin x \right] + c_2 e^{-x} \left[\begin{pmatrix} 0 \\ 1 \end{pmatrix} \cos x + \begin{pmatrix} 1 \\ -1 \end{pmatrix} \sin x \right].$$

Note that $y = e^{-x}(c_1 \cos x + c_2 \sin x)$.

11. We let $u = y'$ so that $u' = y'' = -4y$. Thus $\begin{pmatrix} y \\ u \end{pmatrix}' = \begin{pmatrix} 0 & 1 \\ -4 & 0 \end{pmatrix} \begin{pmatrix} y \\ u \end{pmatrix}$. The characteristic

equation is $\begin{vmatrix} -m & 1 \\ -4 & -m \end{vmatrix} = m^2 + 4 = 0$.

If $m = 2i$ then $\begin{pmatrix} -2i & 1 \\ -4 & -2i \end{pmatrix} \begin{pmatrix} b_1 \\ b_2 \end{pmatrix} = O$, so that $-2ib_1 + b_2 = 0$. One such eigenvector is

$\begin{pmatrix} 1 \\ 2i \end{pmatrix} = \begin{pmatrix} 1 \\ 0 \end{pmatrix} + \begin{pmatrix} 0 \\ 2 \end{pmatrix} i$. The general solution is

$$\begin{pmatrix} y \\ u \end{pmatrix} = c_1 \left[\begin{pmatrix} 1 \\ 0 \end{pmatrix} \cos 2x - \begin{pmatrix} 0 \\ 2 \end{pmatrix} \sin 2x \right] + c_2 \left[\begin{pmatrix} 0 \\ 2 \end{pmatrix} \cos 2x + \begin{pmatrix} 1 \\ 0 \end{pmatrix} \sin 2x \right].$$

Note that $y = c_1 \cos 2x + c_2 \sin 2x$.

13.

For $m = p + qi$,

$$\begin{pmatrix} a - p - qi & b \\ c & d - p - qi \end{pmatrix} \begin{pmatrix} c_1 \\ c_2 \end{pmatrix} = O,$$

so that $(a - p - qi)c_1 + bc_2 = 0$.
One such eigenvector is

$$\begin{pmatrix} b \\ p - a + qi \end{pmatrix} = \begin{pmatrix} b \\ p - a \end{pmatrix} + \begin{pmatrix} 0 \\ q \end{pmatrix} i.$$

For $m = p - qi$,

$$\begin{pmatrix} a - p + qi & b \\ c & d - p + qi \end{pmatrix} \begin{pmatrix} c_1 \\ c_2 \end{pmatrix} = O,$$

so that $(a - p + qi)c_1 + bc_2 = 0$.
One such eigenvector is

$$\begin{pmatrix} b \\ p - a - qi \end{pmatrix} = \begin{pmatrix} b \\ p - a \end{pmatrix} - \begin{pmatrix} 0 \\ q \end{pmatrix} i.$$

These eigenvectors are complex conjugates of each other.

15. $W\left[X_1(0),\ X_2(0)\right] = \begin{vmatrix} b & 0 \\ p - a & q \end{vmatrix} = bq \neq 0.$

11.7 Repeated Eigenvalues

1. The characteristic equation is $\begin{vmatrix} 4 - m & 1 \\ -4 & 8 - m \end{vmatrix} = m^2 - 12m + 36 = (m - 6)^2 = 0.$

If $m = 6$, $\begin{pmatrix} -2 & 1 \\ -4 & 2 \end{pmatrix} \begin{pmatrix} c_1 \\ c_2 \end{pmatrix} = O$, so that $-2c_1 + c_2 = 0$. One solution of the system is

$X_1 = \begin{pmatrix} 1 \\ 2 \end{pmatrix} e^{6t}$. We now seek $X_2 = \begin{pmatrix} 1 \\ 2 \end{pmatrix} te^{6t} + \begin{pmatrix} c_3 \\ c_4 \end{pmatrix} e^{6t}$. The system of differential equa-

tions now requires that $\begin{pmatrix} 1 \\ 2 \end{pmatrix} e^{6t} + 6 \begin{pmatrix} c_3 \\ c_4 \end{pmatrix} e^{6t} = \begin{pmatrix} 4 & 1 \\ -4 & 8 \end{pmatrix} \begin{pmatrix} c_3 \\ c_4 \end{pmatrix} e^{6t}$ or $\begin{pmatrix} -2 & 1 \\ -4 & 2 \end{pmatrix} \begin{pmatrix} c_3 \\ c_4 \end{pmatrix} = \begin{pmatrix} 1 \\ 2 \end{pmatrix}$.

One solution of this system is $\begin{pmatrix} c_3 \\ c_4 \end{pmatrix} = \begin{pmatrix} 0 \\ 1 \end{pmatrix}$. Thus $X_2 = \begin{pmatrix} 1 \\ 2 \end{pmatrix} te^{6t} + \begin{pmatrix} 0 \\ 1 \end{pmatrix} e^{6t}$. Note that

$W\left[X_1(0),\ X_2(0)\right] = \begin{vmatrix} 1 & 0 \\ 2 & 1 \end{vmatrix} = 1 \neq 0.$

3. The characteristic equation is $\begin{vmatrix} 2 - m & -1 \\ 4 & 6 - m \end{vmatrix} = m^2 - 8m + 16 = (m - 4)^2 = 0.$

If $m = 4$, $\begin{pmatrix} -2 & -1 \\ 4 & 2 \end{pmatrix} \begin{pmatrix} c_1 \\ c_2 \end{pmatrix} = O$, so that $-2c_1 - c_2 = 0$. One solution of the system is

$X_1 = \begin{pmatrix} 1 \\ -2 \end{pmatrix} e^{4t}$. We now seek $X_2 = \begin{pmatrix} 1 \\ -2 \end{pmatrix} te^{4t} + \begin{pmatrix} c_3 \\ c_4 \end{pmatrix} e^{4t}$. The system of differential equations

now requires that $\begin{pmatrix} 1 \\ -2 \end{pmatrix} e^{4t} + 4 \begin{pmatrix} c_3 \\ c_4 \end{pmatrix} e^{4t} = \begin{pmatrix} 2 & -1 \\ 4 & 6 \end{pmatrix} \begin{pmatrix} c_3 \\ c_4 \end{pmatrix} e^{4t}$ or $\begin{pmatrix} -2 & -1 \\ 4 & 2 \end{pmatrix} \begin{pmatrix} c_3 \\ c_4 \end{pmatrix} = \begin{pmatrix} 1 \\ -2 \end{pmatrix}$.

One solution of this system is $\begin{pmatrix} c_3 \\ c_4 \end{pmatrix} = \begin{pmatrix} 0 \\ -1 \end{pmatrix}$. Thus $X_2 = \begin{pmatrix} 1 \\ -2 \end{pmatrix} te^{4t} + \begin{pmatrix} 0 \\ -1 \end{pmatrix} e^{4t}$. Note that

$W\left[X_1(0),\ X_2(0)\right] = \begin{vmatrix} 1 & 0 \\ -2 & -1 \end{vmatrix} = -1 \neq 0.$

5. The characteristic equation is $\begin{vmatrix} 1-m & 2 & -1 \\ 0 & 1-m & 1 \\ 0 & 0 & 2-m \end{vmatrix} = -(m-1)^2(m-2) = 0.$

If $m = 2$, $\begin{pmatrix} -1 & 2 & -1 \\ 0 & -1 & 1 \\ 0 & 0 & 0 \end{pmatrix} \begin{pmatrix} b_1 \\ b_2 \\ b_3 \end{pmatrix} = O$, so that $-b_1 + 2b_2 - b_3 = 0$ and $-b_2 + b_3 = 0$.

One solution of the system is $X_1 = \begin{pmatrix} 1 \\ 1 \\ 1 \end{pmatrix} e^{2t}$. If $m = 1$, $\begin{pmatrix} 0 & 2 & -1 \\ 0 & 0 & 1 \\ 0 & 0 & 1 \end{pmatrix} \begin{pmatrix} b_4 \\ b_5 \\ b_6 \end{pmatrix} = O$, so that

$2b_5 - b_6 = 0$ and $b_6 = 0$. Thus a second solution of the system is $X_2 = \begin{pmatrix} 1 \\ 0 \\ 0 \end{pmatrix} e^t$. We now seek

a third solution of the form $X_3 = \begin{pmatrix} 1 \\ 0 \\ 0 \end{pmatrix} te^t + \begin{pmatrix} b_7 \\ b_8 \\ b_9 \end{pmatrix} e^t$. The system of differential equations

now requires that $\begin{pmatrix} 1 \\ 0 \\ 0 \end{pmatrix} e^t + \begin{pmatrix} b_7 \\ b_8 \\ b_9 \end{pmatrix} e^t = \begin{pmatrix} 1 & 2 & -1 \\ 0 & 1 & 1 \\ 0 & 0 & 2 \end{pmatrix} \begin{pmatrix} b_7 \\ b_8 \\ b_9 \end{pmatrix} e^t$ or $\begin{pmatrix} 0 & 2 & -1 \\ 0 & 0 & 1 \\ 0 & 0 & 1 \end{pmatrix} \begin{pmatrix} b_7 \\ b_8 \\ b_9 \end{pmatrix} = \begin{pmatrix} 1 \\ 0 \\ 0 \end{pmatrix}.$

One solution of this system is $\begin{pmatrix} b_7 \\ b_8 \\ b_9 \end{pmatrix} = \begin{pmatrix} 0 \\ 1/2 \\ 0 \end{pmatrix}$. Thus $X_3 = \begin{pmatrix} 1 \\ 0 \\ 0 \end{pmatrix} te^t + \begin{pmatrix} 0 \\ 1/2 \\ 0 \end{pmatrix} e^t$. The general

solution can now be written

$$X(t) = c_1 \begin{pmatrix} 1 \\ 1 \\ 1 \end{pmatrix} e^{2t} + c_2 \begin{pmatrix} 1 \\ 0 \\ 0 \end{pmatrix} e^t + c_3 \left[\begin{pmatrix} 2 \\ 0 \\ 0 \end{pmatrix} t + \begin{pmatrix} 0 \\ 1 \\ 0 \end{pmatrix} \right] e^t.$$

7. The characteristic equation is $\begin{vmatrix} -m & 3 & 1 \\ 1 & 2-m & 1 \\ 1 & 3 & -m \end{vmatrix} = -(m+1)^2(m-4) = 0.$

If $m = 4$, $\begin{pmatrix} -4 & 3 & 1 \\ 1 & -2 & 1 \\ 1 & 3 & -4 \end{pmatrix} \begin{pmatrix} b_1 \\ b_2 \\ b_3 \end{pmatrix} = O$. One solution of the system is $X_1 = \begin{pmatrix} 1 \\ 1 \\ 1 \end{pmatrix} e^{4t}$. If

$m = -1$, $\begin{pmatrix} 1 & 3 & 1 \\ 1 & 3 & 1 \\ 1 & 3 & 1 \end{pmatrix} \begin{pmatrix} b_4 \\ b_5 \\ b_6 \end{pmatrix} = O$, so that $b_4 + 3b_5 + b_6 = 0$. One solution is $\begin{pmatrix} b_4 \\ b_5 \\ b_6 \end{pmatrix} = \begin{pmatrix} 1 \\ 0 \\ -1 \end{pmatrix}.$

A second solution is $\begin{pmatrix} b_4 \\ b_5 \\ b_6 \end{pmatrix} = \begin{pmatrix} 0 \\ 1 \\ -3 \end{pmatrix}$. Thus two solutions of the system of differential equations are $X_2 = \begin{pmatrix} 1 \\ 0 \\ -1 \end{pmatrix} e^{-t}$ and $X_2 = \begin{pmatrix} 0 \\ 1 \\ -3 \end{pmatrix} e^{-t}$. The general solution can now be written

$$X(t) = \left[c_1 \begin{pmatrix} 1 \\ 0 \\ -1 \end{pmatrix} + c_2 \begin{pmatrix} 0 \\ 1 \\ -3 \end{pmatrix} \right] e^{-t} + c_3 \begin{pmatrix} 1 \\ 1 \\ 1 \end{pmatrix} e^{4t}.$$

9. The characteristic equation is $\begin{vmatrix} -m & -1 & 3 \\ 2 & -3-m & 3 \\ 2 & -1 & 1-m \end{vmatrix} = -(m+2)^2(m-2) = 0.$

If $m = 2$, $\begin{pmatrix} -2 & -1 & 3 \\ 2 & -5 & 3 \\ 2 & -1 & -1 \end{pmatrix} \begin{pmatrix} b_1 \\ b_2 \\ b_3 \end{pmatrix} = O$. One solution of the system is $X_1 = \begin{pmatrix} 1 \\ 1 \\ 1 \end{pmatrix} e^{2t}$. If

$m = -2$, $\begin{pmatrix} 2 & -1 & 3 \\ 2 & -1 & 3 \\ 2 & -1 & 3 \end{pmatrix} \begin{pmatrix} b_4 \\ b_5 \\ b_6 \end{pmatrix} = O$, so that $2b_4 - b_5 + 3b_6 = 0$. One solution is $\begin{pmatrix} b_4 \\ b_5 \\ b_6 \end{pmatrix} = \begin{pmatrix} 3 \\ 0 \\ -2 \end{pmatrix}$.

A second solution is $\begin{pmatrix} b_4 \\ b_5 \\ b_6 \end{pmatrix} = \begin{pmatrix} 1 \\ 2 \\ 0 \end{pmatrix}$. Thus two solutions of the system of differential equations

are $X_2 = \begin{pmatrix} 3 \\ 0 \\ -2 \end{pmatrix} e^{-2t}$ and $X_3 = \begin{pmatrix} 1 \\ 2 \\ 0 \end{pmatrix} e^{-2t}$. The general solution can now be written

$$X(t) = c_1 \begin{pmatrix} 1 \\ 1 \\ 1 \end{pmatrix} e^{2t} + \left[c_2 \begin{pmatrix} 3 \\ 0 \\ -2 \end{pmatrix} + c_3 \begin{pmatrix} 1 \\ 2 \\ 0 \end{pmatrix} \right] e^{-2t}.$$

11. Each of the functions $X_1 = \begin{pmatrix} 1 \\ 0 \\ 0 \end{pmatrix} e^{at}$, $X_2 = \begin{pmatrix} 0 \\ 1 \\ 0 \end{pmatrix} e^{bt}$, $X_3 = \begin{pmatrix} 0 \\ 0 \\ 1 \end{pmatrix} e^{ct}$ is a solution of the system.

The solutions are linearly independent since $W[X_1(0), X_2(0), X_3(0)] = \begin{vmatrix} 1 & 0 & 0 \\ 0 & 1 & 0 \\ 0 & 0 & 1 \end{vmatrix} = 1 \neq 0.$

11.8 The Phase Plane

1. The general solution is $X = c_1 \begin{pmatrix} 3 \\ 4 \end{pmatrix} e^{4t} + c_2 \begin{pmatrix} 1 \\ 4 \end{pmatrix} e^{-4t}$. Here $m_1 = -4$ and $m_2 = 4$ so that almost all of the solutions approach infinity as $t \to \infty$. The exceptions being those for which $c_1 = 0$, in which case the solution approaches the origin along the line $y = 4x$. The origin is an unstable saddle.

3. The general solution is $X(t) = c_1 \left[\begin{pmatrix} 5 \\ -4 \end{pmatrix} \cos 2t - \begin{pmatrix} 0 \\ 2 \end{pmatrix} \sin 2t \right] + c_2 \left[\begin{pmatrix} 0 \\ 2 \end{pmatrix} \cos 2t + \begin{pmatrix} 5 \\ -4 \end{pmatrix} \sin 2t \right]$.
Here $a = 0$, and all trajectories move around the origin. The origin is a stable center.

5. The general solution is

$$X = c_1 e^{-t} \left[\begin{pmatrix} 13 \\ 5 \end{pmatrix} \cos t - \begin{pmatrix} 0 \\ -1 \end{pmatrix} \sin t \right] + c_2 e^{-t} \left[\begin{pmatrix} 0 \\ -1 \end{pmatrix} \cos t + \begin{pmatrix} 13 \\ 5 \end{pmatrix} \sin t \right].$$

Since $a = -1$ the origin is a stable spiral.

7. Here the characteristic equation is $\begin{vmatrix} 4 - m & -9 \\ 4 & -8 - m \end{vmatrix} = (m+2)^2 = 0.$ Since the repeated roots are negative the origin is a stable node.

15. The characteristic equation is $\begin{vmatrix} -1 - m & 0 \\ 0 & -1 - m \end{vmatrix} = (m + 1)^2 = 0.$ If $m = -1$ we have $\begin{pmatrix} 0 & 0 \\ 0 & 0 \end{pmatrix} \begin{pmatrix} b_1 \\ b_2 \end{pmatrix} = O$, so that there are two linearly independent eigenvectors $\begin{pmatrix} 1 \\ 0 \end{pmatrix}$ and $\begin{pmatrix} 0 \\ 1 \end{pmatrix}$. The general solution is $X = \left[c_1 \begin{pmatrix} 1 \\ 0 \end{pmatrix} + c_2 \begin{pmatrix} 0 \\ 1 \end{pmatrix} \right] e^{-t} = \begin{pmatrix} c_1 \\ c_2 \end{pmatrix} e^{-t}.$ For any choice of c_1 and c_2, not both zero, the corresponding trajectory is a half line approaching the origin as $t \to \infty$. The origin is a stable node.

Chapter 12

Nonhomogeneous Systems of Equations

12.1 Nonhomogeneous Systems

1. As in Example 12.1

$$a_1'(t) = (2e^t - 2)e^{-t} = 2 - 2e^{-t},$$
$$a_2'(t) = (2 - e^t)e^{-2t} = 2e^{-2t} - e^{-t},$$

so that $a_1(t) = 2t + 2e^{-t}$ and $a_2(t) = -e^{-2t} + e^{-t}$. The particular solution is

$$\begin{pmatrix} x \\ y \end{pmatrix}_p = (2t + 2e^{-t}) \begin{pmatrix} 1 \\ 1 \end{pmatrix} e^t + (-e^{-2t} + e^{-t}) \begin{pmatrix} 1 \\ 2 \end{pmatrix} e^{2t} = te^t \begin{pmatrix} 2 \\ 2 \end{pmatrix} + e^t \begin{pmatrix} 1 \\ 2 \end{pmatrix} + \begin{pmatrix} 1 \\ 0 \end{pmatrix}.$$

The general solution is $X = a_1 \begin{pmatrix} 1 \\ 1 \end{pmatrix} e^t + a_2 \begin{pmatrix} 1 \\ 2 \end{pmatrix} e^{2t} + te^t \begin{pmatrix} 2 \\ 2 \end{pmatrix} + e^t \begin{pmatrix} 1 \\ 2 \end{pmatrix} + \begin{pmatrix} 1 \\ 0 \end{pmatrix}.$

3. From Exercise 5 in Section 11.5 we have $X_h = a_1 \begin{pmatrix} 1 \\ -1 \end{pmatrix} + a_2 \begin{pmatrix} 3 \\ -1 \end{pmatrix} e^{2t}$. Substituting into the original system of equations forces us to take $a_1'(t) \begin{pmatrix} 1 \\ -1 \end{pmatrix} + a_2'(t) \begin{pmatrix} 3 \\ -1 \end{pmatrix} e^{2t} = \begin{pmatrix} t \\ 1 \end{pmatrix}$. Thus $a_1'(t) = -\dfrac{t}{2} - \dfrac{3}{2}$ and $a_2'(t) = \dfrac{(t+1)e^{-2t}}{2}$, so that $a_1(t) = -\dfrac{t^2}{4} - \dfrac{3t}{2}$ and $a_2(t) = -\dfrac{1}{4}te^{-2t} - \dfrac{3}{8}e^{-2t}$. One particular solution is

$$X_p = \left(-\frac{t^2}{4} - \frac{3t}{2} \right) \begin{pmatrix} 1 \\ -1 \end{pmatrix} + \left(-\frac{1}{4}t - \frac{3}{8} \right) \begin{pmatrix} 3 \\ -1 \end{pmatrix} = \frac{1}{8} \begin{pmatrix} -2t^2 - 18t - 9 \\ 2t^2 + 14t + 3 \end{pmatrix}.$$

A slightly simpler particular solution may be obtained by rewriting this particular solution as $X_p = \dfrac{1}{8} \begin{pmatrix} -2t^2 - 18t - 6 \\ 2t^2 + 14t \end{pmatrix} - \dfrac{3}{8} \begin{pmatrix} 1 \\ -1 \end{pmatrix}$ and incorporating the last term as part of the homogeneous solution. The general solution then becomes

$$X = c_1 \begin{pmatrix} 1 \\ -1 \end{pmatrix} + c_2 \begin{pmatrix} 3 \\ -1 \end{pmatrix} e^{2t} + \frac{1}{8} \begin{pmatrix} -2t^2 - 18t - 6 \\ 2t^2 + 14t \end{pmatrix}.$$

5. We follow Example 12.1 and have $a_1'(t) = (-3e^t)e^{-t} = -3$ and $a_2'(t) = (3e^t)e^{-2t} = 3e^{-t}$ so that $a_1(t) = -3t$ and $a_2(t) = -3e^{-t}$. Finally $X_p(t) = -3te^t \begin{pmatrix} 1 \\ 1 \end{pmatrix} - 3e^t \begin{pmatrix} 1 \\ 2 \end{pmatrix}.$

12.2   Arms Races

1. The system to be solved is $X' = \begin{pmatrix} -5 & 2 \\ 4 & -3 \end{pmatrix} X + \begin{pmatrix} 1 \\ 2 \end{pmatrix}$. The constant particular solution $\begin{pmatrix} e \\ f \end{pmatrix}$ must satisfy $\begin{pmatrix} -5 & 2 \\ 4 & -3 \end{pmatrix} \begin{pmatrix} e \\ f \end{pmatrix} + \begin{pmatrix} 1 \\ 2 \end{pmatrix} = \begin{pmatrix} 0 \\ 0 \end{pmatrix}$ so that $e = 1$ and $f = 2$.

   The characteristic equation of A is

   $$\begin{vmatrix} -5 - m & 2 \\ 4 & -3 - m \end{vmatrix} = m^2 + 8m + 7 = (m + 1)(m + 7) = 0$$

   giving eigenvalues -1 and -7. These eigenvalues correspond to eigenvectors $\begin{pmatrix} 1 \\ 2 \end{pmatrix}$ and $\begin{pmatrix} 1 \\ -1 \end{pmatrix}$. Thus the general solution is

   $$X(t) = c_1 \begin{pmatrix} 1 \\ 2 \end{pmatrix} e^{-t} + c_2 \begin{pmatrix} 1 \\ -1 \end{pmatrix} e^{-7t} + \begin{pmatrix} 1 \\ 2 \end{pmatrix}.$$

   The initial conditions $x_0 = 8$, $y_0 = 7$ require

   $$c_1 \begin{pmatrix} 1 \\ 2 \end{pmatrix} + c_2 \begin{pmatrix} 1 \\ -1 \end{pmatrix} + \begin{pmatrix} 1 \\ 2 \end{pmatrix} = \begin{pmatrix} 8 \\ 7 \end{pmatrix},$$

   an equation that has solution $c_1 = 4$, $c_2 = 3$. Finally,

   $$X(t) = 4 \begin{pmatrix} 1 \\ 2 \end{pmatrix} e^{-t} + 3 \begin{pmatrix} 1 \\ -1 \end{pmatrix} e^{-7t} + \begin{pmatrix} 1 \\ 2 \end{pmatrix}.$$

   The solution will approach a stable solution $x = 1$ and $y = 2$.

3. The system to be solved is $X' = \begin{pmatrix} -2 & 4 \\ 4 & -2 \end{pmatrix} X + \begin{pmatrix} 8 \\ 2 \end{pmatrix}$, $X(0) = \begin{pmatrix} 5 \\ 2 \end{pmatrix}$. The constant particular solution $\begin{pmatrix} e \\ f \end{pmatrix}$ must satisfy $\begin{pmatrix} -2 & 4 \\ 4 & -2 \end{pmatrix} \begin{pmatrix} e \\ f \end{pmatrix} + \begin{pmatrix} 8 \\ 2 \end{pmatrix} = \begin{pmatrix} 0 \\ 0 \end{pmatrix}$ so that $e = -2$ and $f = -3$.

   The characteristic equation of A is

   $$\begin{vmatrix} -2 - m & 4 \\ 4 & -2 - m \end{vmatrix} = m^2 + 4m - 12 = (m + 6)(m - 2) = 0$$

   giving eigenvalues 2 and -6. These eigenvalues correspond to eigenvectors $\begin{pmatrix} 1 \\ 1 \end{pmatrix}$ and $\begin{pmatrix} 1 \\ -1 \end{pmatrix}$. Thus the general solution is

   $$X(t) = c_1 \begin{pmatrix} 1 \\ 1 \end{pmatrix} e^{2t} + c_2 \begin{pmatrix} 1 \\ -1 \end{pmatrix} e^{-6t} + \begin{pmatrix} -2 \\ -3 \end{pmatrix}.$$

   The initial conditions $x_0 = 5$, $y_0 = 2$ require

   $$c_1 \begin{pmatrix} 1 \\ 1 \end{pmatrix} + c_2 \begin{pmatrix} 1 \\ -1 \end{pmatrix} - \begin{pmatrix} 2 \\ 3 \end{pmatrix} = \begin{pmatrix} 5 \\ 2 \end{pmatrix},$$

an equation that has solution $c_1 = 6$, $c_2 = 1$. Finally,

$$X(t) = 6 \begin{pmatrix} 1 \\ 1 \end{pmatrix} e^{2t} + \begin{pmatrix} 1 \\ -1 \end{pmatrix} e^{-6t} - \begin{pmatrix} 2 \\ 3 \end{pmatrix}.$$

There will be a runaway arms race.

5. The system to be solved is $X' = \begin{pmatrix} -2 & 4 \\ 4 & -2 \end{pmatrix} X + \begin{pmatrix} -2 \\ -2 \end{pmatrix}$. The constant particular solution $\begin{pmatrix} e \\ f \end{pmatrix}$ must satisfy $\begin{pmatrix} -2 & 4 \\ 4 & -2 \end{pmatrix} \begin{pmatrix} e \\ f \end{pmatrix} + \begin{pmatrix} -2 \\ -2 \end{pmatrix} = \begin{pmatrix} 0 \\ 0 \end{pmatrix}$ so that $e = 1$ and $f = 1$. The homogeneous solution is the same as in Exercise 3. Thus the general solution is

$$X(t) = c_1 \begin{pmatrix} 1 \\ 1 \end{pmatrix} e^{2t} + c_2 \begin{pmatrix} 1 \\ -1 \end{pmatrix} e^{-6t} + \begin{pmatrix} 1 \\ 1 \end{pmatrix}.$$

The initial conditions require that

$$\begin{pmatrix} 1 & 1 \\ 1 & -1 \end{pmatrix} \begin{pmatrix} c_1 \\ c_2 \end{pmatrix} = \begin{pmatrix} x_0 - 1 \\ y_0 - 1 \end{pmatrix},$$

an equation that has solution $c_1 = \frac{1}{2}(x_0 + y_0 - 2)$, $c_2 = \frac{1}{2}(x_0 - y_0)$. Finally,

$$X(t) = \frac{1}{2}(x_0 + y_0 - 2)e^{2t} \begin{pmatrix} 1 \\ 1 \end{pmatrix} + \frac{1}{2}(x_0 - y_0)e^{-6t} \begin{pmatrix} 1 \\ -1 \end{pmatrix} + \begin{pmatrix} 1 \\ 1 \end{pmatrix}.$$

If $x_0 + y_0 < 2$ the arms expenditures for both countries will become and remain negative, a condition of disarmament. If $x_0 + y_0 > 2$ there will be a runaway arms race.

7. The system to be solved is $X' = \begin{pmatrix} -2 & 4 \\ 4 & -2 \end{pmatrix} X + \begin{pmatrix} r \\ s \end{pmatrix}$. The constant particular solution $\begin{pmatrix} e \\ f \end{pmatrix}$ must satisfy $\begin{pmatrix} -2 & 4 \\ 4 & -2 \end{pmatrix} \begin{pmatrix} e \\ f \end{pmatrix} = \begin{pmatrix} -r \\ -s \end{pmatrix}$ so that $e = -\frac{1}{6}(r + 2s)$ and $f = -\frac{1}{6}(2r + s)$. The homogeneous solution is the same as in Exercise 3. Thus the general solution is

$$X(t) = c_1 \begin{pmatrix} 1 \\ 1 \end{pmatrix} e^{2t} + c_2 \begin{pmatrix} 1 \\ -1 \end{pmatrix} e^{-6t} - \frac{1}{6} \begin{pmatrix} r + 2s \\ 2r + s \end{pmatrix}.$$

The initial conditions require that

$$\begin{pmatrix} 1 & 1 \\ 1 & -1 \end{pmatrix} \begin{pmatrix} c_1 \\ c_2 \end{pmatrix} = \begin{pmatrix} x_0 \\ y_0 \end{pmatrix} + \frac{1}{6} \begin{pmatrix} r + 2s \\ 2r + s \end{pmatrix},$$

an equation for which $c_1 = \frac{1}{2} \left(x_0 + y_0 + \frac{r + s}{2} \right)$. Because of the factor e^{2t} in the first term of the general solution, that term will dictate the behavior of the solution as $t \to \infty$. If $c_1 > 0$ there will be a runaway arms race. This will occur if $x_0 + y_0 > \frac{-r - s}{2}$.

9. Since $pq - ab < 0$ the two eigenvalues are real and have opposite signs. We set

$$u = \frac{-(p+q) + \sqrt{(p+q)^2 - 4(pq - ab)}}{2} \quad \text{and} \quad v = \frac{-(p+q) - \sqrt{(p+q)^2 - 4(pq - ab)}}{2}.$$

We also note in passing that since $pq - ab < 0$ and $q > 0$,

$$p + v < p - \frac{p + q + \sqrt{(p+q)^2}}{2} = -q < 0. \tag{12.1}$$

The constant particular solution $\begin{pmatrix} e \\ f \end{pmatrix}$ must satisfy $\begin{pmatrix} -p & a \\ b & -q \end{pmatrix} \begin{pmatrix} e \\ f \end{pmatrix} = \begin{pmatrix} -r \\ -s \end{pmatrix}$ so that we have $e = \dfrac{rq + sa}{pq - ab}$ and $f = \dfrac{ps + rb}{pq - ab}$. Note that e and f are both negative. If we use the eigenvalue $m = u$ we must satisfy $\begin{pmatrix} -p - u & a \\ b & -q - u \end{pmatrix} \begin{pmatrix} c_1 \\ c_2 \end{pmatrix} = O$ so that one eigenvector is $\begin{pmatrix} a \\ p + u \end{pmatrix}$. If we use the other eigenvalue $m = v$ we get in the same manner an eigenvector $\begin{pmatrix} a \\ p + v \end{pmatrix}$. Thus the general solution of the system is

$$X(t) = c_1 e^{ut} \begin{pmatrix} a \\ p + u \end{pmatrix} + c_2 e^{vt} \begin{pmatrix} a \\ p + v \end{pmatrix} + \begin{pmatrix} e \\ f \end{pmatrix}.$$

Since u, a, and $p + u$ are all positive and v is negative, the character of these solutions will depend on whether c_1 is positive or negative. If it is positive we will have a runaway arms race. We now apply the initial conditions $X(0) = \begin{pmatrix} x_0 \\ y_0 \end{pmatrix}$ and get

$$\begin{pmatrix} a & a \\ p + u & p + v \end{pmatrix} \begin{pmatrix} c_1 \\ c_2 \end{pmatrix} = \begin{pmatrix} x_0 - e \\ y_0 - f \end{pmatrix}.$$

Therefore $c_1 = \dfrac{(x_0 - e)(p + v) - a(y_0 - f)}{a(v - u)}$. Remembering that e, f, $p + v$, and v are negative while x_0, y_0, u, and a are positive establishes that c_1 is positive. We therefore have a runaway arms race.

12.4   Simple Networks

1. This is Example 12.6.

3. The initial value problem is

$$I'(t) + \frac{R}{L} I(t) = \frac{E}{L} \sin \omega t, \quad I(0) = 0.$$

The linear differential equation has as an integrating factor $\exp\left(\dfrac{R}{L}t\right)$ so that

$$\exp\left(\frac{R}{L}\right)I(t) = \frac{E}{L}\int \sin\omega t \exp\left(\frac{R}{L}t\right)\,dt$$

and $I(t) = \dfrac{E}{Z^2}(-\omega L\cos\omega t + R\sin\omega t) + c_1\exp\left(-\dfrac{R}{L}t\right)$. The initial condition forces us to choose $c_1 = \dfrac{E\omega L}{Z^2}$ so that finally

$$I(t) = \frac{E}{Z^2}\left[-\omega L\cos\omega t + R\sin\omega t + \omega L\exp\left(-\frac{R}{L}t\right)\right].$$

5. The initial value problem is $\dfrac{dQ}{dt} + 250Q = 5$, $Q(0) = 0.015$. The general solution is $Q(t)\exp(250t) = 0.02\exp(250t) + c_1$. The initial condition requires that $c_1 = -0.005$, so that $Q(t) = 0.02 - 0.005\exp(-250t)$. Therefore $I(t) = \dfrac{dQ}{dt} = 1.25\exp(-250t)$(amp) and $I(0) = 1.25$(amp).

7. Here $I(t) = 3000t\exp(-1000t)$. Thus $I'(t) = (-3\cdot 10^6 t + 3000)\exp(-1000t)$ so that the maximum current occurs at $t = 0.001$(sec). We have $I_{\max} = 3e^{-1}$(amp).

9. We have the problem $RI + L\dfrac{dI}{dt} + \dfrac{1}{C}Q = E\sin\omega t$, $Q(0) = 0$, $I(0) = 0$ which may be written $LC\dfrac{d^2Q}{dt^2} + RC\dfrac{dQ}{dt} + Q = EC\sin\omega t$, $Q(0) = 0$, $Q'(0) = 0$. Here $LCm^2 + RCm + 1 = 0$ so that

$$m == \frac{-RC \pm \sqrt{R^2C^2 - 4LC}}{2LC} = -\frac{R}{2L} \pm \sqrt{\frac{R^2}{4L^2} - \frac{1}{LC}} = -a \pm \beta i.$$

Thus $Q_h = e^{-at}(c_1\cos\beta t + c_2\sin\beta t)$. We seek a particular solution that has the form $Q_p = A\sin\omega t + B\cos\omega t$. Substituting into the differential equation and determining the undetermined coefficients yields $A = -\dfrac{\gamma E}{\omega Z^2}$ and $B = -\dfrac{ER}{\omega Z^2}$. Thus $Q_p = -\dfrac{\gamma E}{\omega Z^2}\sin\omega t - \dfrac{ER}{\omega Z^2}\cos\omega t$ and

$$Q(t) = e^{-at}(c_1\cos\beta t + c_2\sin\beta t) - \frac{\gamma E}{\omega Z^2}\sin\omega t - \frac{ER}{\omega Z^2}\cos\omega t,$$

$$Q'(t) = e^{-at}(-c_1\beta\sin\beta t + c_2\beta\cos\beta t) - ae^{-at}(c_1\cos\beta t + c_2\sin\beta t) - \frac{\gamma E}{Z^2}\cos\omega t + \frac{ER}{Z^2}\sin\omega t.$$

But $Q(0) = 0$ and $Q'(0) = 0$ so that $c_1 - \dfrac{ER}{\omega Z^2} = 0$ and $c_2\beta - ac_1 - \dfrac{\gamma E}{Z^2} = 0$. Thus $c_1 = \dfrac{ER}{\omega Z^2}$ and $c_2 = \dfrac{E}{\beta Z^2}\left(\dfrac{aR}{\omega} + \gamma\right)$. Substituting c_1 and c_2 into $I(t) = Q'(t)$ gives us

$$I(t) = e^{-at}\left[\left\{-\frac{\beta ER}{\omega Z^2} - \frac{aE}{\beta Z^2}\left(\frac{aR}{\omega} + \gamma\right)\right\}\sin\beta t + \left\{-\frac{aER}{\omega Z^2} + \frac{E}{Z^2}\left(\frac{aR}{\omega} + \gamma\right)\right\}\cos\beta t\right]$$

$$\qquad - \frac{\gamma E}{Z^2}\cos\omega t + \frac{ER}{Z^2}\sin\omega t$$

$$= e^{-at}\left[\frac{E}{Z^2}\left(-\frac{\beta R}{\omega} - \frac{a^2 R}{\beta\omega} - \frac{\gamma a}{\beta}\right)\sin\beta t + \frac{\gamma E}{Z^2}\cos\beta t\right] + \frac{E}{Z^2}(R\sin\omega t - \gamma\cos\omega t)$$

$$= \frac{E}{Z^2}(R\sin\omega t - \gamma\cos\omega t) + \frac{Ee^{-at}}{\beta Z^2}\left[\beta\gamma\cos\beta t - a\left(\gamma + \frac{2}{\omega C}\right)\sin\beta t\right].$$

11. We know that $C = \dfrac{4L}{R^2}$ so that $\gamma = \omega L - \dfrac{1}{\omega C} = \omega L - \dfrac{R^2}{4L\omega}$. Thus $\dfrac{\gamma}{L} = \omega - \dfrac{R^2}{4L^2\omega} = \omega - \dfrac{a^2}{\omega}$. From this we get $-\dfrac{a^2 R}{\omega} = \dfrac{R\gamma}{L} - R\omega = 2a\gamma - R\omega$. Finally $-a\gamma - \dfrac{a^2 R}{\omega} = a\gamma - R\omega$. Substituting this expression for the coefficient of t in the last term in the answer of Exercise 10 yields the desired form.

13. We have from Kirchhoff's laws

$$10I_1 + 2000Q_2 = 60, \quad Q_2(0) = 0.03,$$
$$10I_1 + 20I_3 = 60,$$
$$I_1 = I_2 + I_3.$$

From the second equation we have $I_3 = 3 - \frac{1}{2}I_1$. Substituting into the last equation yields $I_1 = \frac{2}{3}I_2 + 2$. Thus the differential equation may be written $\dfrac{dQ_2}{dt} + 300Q_2 = 6$. The general solution of this equation is $Q_2 = \frac{1}{50} + c_1 e^{-300t}$. The initial condition yields $c_1 = 1/100$ so that

$$Q_2 = \frac{1}{50} + \frac{1}{100}e^{-300t},$$
$$I_2 = -3e^{-300t},$$
$$I_1 = 2 - 2e^{-300t},$$
$$I_3 = 2 + e^{-300t}.$$

15. Application of Kirchhoff's laws yields the system

$$R_1 I_1 + R_2 I_2 + L_2\frac{dI_2}{dt} = E,$$
$$R_1 I_1 + R_3 I_3 + \frac{1}{C_3}Q_3 = E,$$
$$I_1 = I_2 + I_3.$$

The first of these equations can be written $\dfrac{dI_2}{dt} = -\dfrac{R_1}{L_2}I_1 - \dfrac{R_2}{L_2}I_2 + \dfrac{E}{L_2}$. Differentiating the second equation and using the third equation to eliminate I_1 yields

$$R_1 \frac{dI_1}{dt} + R_3 \frac{dI_3}{dt} + \frac{1}{C_3}I_3 = 0,$$

$$R_1 \frac{dI_1}{dt} + R_3 \left(\frac{dI_1}{dt} - \frac{dI_2}{dt} \right) + \frac{1}{C_3}(I_1 - I_2) = 0.$$

Replacing $\dfrac{dI_2}{dt}$ by its equivalent gives us

$$(R_1 + R_3)\frac{dI_1}{dt} - \frac{R_3}{L_2}(-R_1 I_1 - R_2 I_2 + E) + \frac{1}{C_3}(I_1 - I_2) = 0,$$

$$(R_1 + R_3)\frac{dI_1}{dt} = \left(-\frac{R_1 R_3}{L_2} - \frac{1}{C_3} \right) I_1 + \left(-\frac{R_2 R_3}{L_2} + \frac{1}{C_3} \right) I_2 + \frac{R_3 E}{L_2}.$$

The system in I_1 and I_2 can now be written

$$\begin{pmatrix} I_1 \\ I_2 \end{pmatrix}' = \begin{pmatrix} -\frac{C_3 R_1 R_3 + L_2}{C_3 L_2 (R_1 + R_3)} & -\frac{C_3 R_2 R_3 - L_2}{C_3 L_2 (R_1 + R_3)} \\ -\frac{R_1}{L_2} & -\frac{R_2}{L_2} \end{pmatrix} \begin{pmatrix} I_1 \\ I_2 \end{pmatrix} + \begin{pmatrix} \frac{R_3 E}{L_2 (R_1 + R_3)} \\ \frac{E}{L_2} \end{pmatrix}.$$

The nature of the solutions of this system depend upon the roots of the characteristic equation

$$\begin{vmatrix} -\frac{C_3 R_1 R_3 + L_2}{C_3 L_2 (R_1 + R_3)} - m & -\frac{C_3 R_2 R_3 - L_2}{C_3 L_2 (R_1 + R_3)} \\ -\frac{R_1}{L_2} & -\frac{R_2}{L_2} - m \end{vmatrix} = 0,$$

which may be written

$$m^2 + \left[\frac{C_3 R_1 R_3 + L_2}{C_3 L_2 (R_1 + R_3)} + \frac{R_2}{L_2} \right] m + \frac{R_2 (C_3 R_1 R_3 + L_2)}{C_3 L_2^2 (R_1 + R_3)} - \frac{R_1 (C_3 R_2 R_3 - L_2)}{C_3 L_2^2 (R_1 + R_3)} = 0,$$

or

$$C_3 L_2 (R_1 + R_3)m^2 + \left[C_3 (R_1 R_2 + R_2 R_3 + R_3 R_1) + L_2 \right] m + R_1 + R_2 = 0.$$

Note that the answer given in the book has a typographic error. The last term should be R_2, not R_3.

Chapter 13

The Existence and Uniqueness of Solutions

13.2   An Existence and Uniqueness Theorem

1. Here $f(x, y) = x$, $x_0 = 2$, and $y_0(x) = 1$. The sequence defined by (2) becomes

$$y_0(x) = 1,$$

$$y_1(x) = 1 + \int_2^x t \, dt = 1 + \left[\frac{t^2}{2}\right]_2^x = \frac{x^2}{2} - 1,$$

$$\vdots$$

$$y_n(x) = 1 + \int_2^x t \, dt = \frac{x^2}{2} - 1.$$

It follows that $\lim_{x \to \infty} y_n(x) = \dfrac{x^2}{2} - 1.$

3. Here $f(x, y) = 2y$, $x_0 = 0$, and $y_0(x) = 1$. The sequence defined by (2) becomes

$$y_0(x) = 1,$$

$$y_1(x) = 1 + \int_0^x 2 \, dt = 1 + [2t]_0^x = 1 + (2x),$$

$$y_2(x) = 1 + \int_0^x [2 + 2(2t)] \, dt = 1 + \left[(2t) + \frac{(2t)^2}{2!}\right]_0^x = 1 + (2x) + \frac{(2x)^2}{2!}.$$

Let us suppose that $y_k(x) = 1 + (2x) + \dfrac{(2x)^2}{2!} + \cdots + \dfrac{(2x)^k}{k!}$. Then

$$y_{k+1}(x) = 1 + \int_0^x \left[2 + 2(2t) + \cdots + \frac{2(2t)^k}{k!}\right] dt = 1 + \left[(2t) + \frac{(2t)^2}{2!} + \cdots + \frac{(2t)^{k+1}}{k!}\right]_0^x$$

$$= 1 + (2x) + \frac{(2x)^2}{2!} + \cdots + \frac{(2x)^{k+1}}{(k+1)!}.$$

By induction we have $y_n(x) = \displaystyle\sum_{k=0}^n \frac{(2x)^k}{k!}$. It follows that $\displaystyle\lim_{n \to \infty} y_n(x) = \sum_{k=0}^\infty \frac{(2k)^k}{k!} = e^{2x}.$

Chapter 14

The Laplace Transform

14.3 Transforms of Elementary Functions

1. From elementary calculus we obtain

$$L\{\cos kt\} = \int_0^\infty e^{-st} \cos kt \; dt = \left[\frac{e^{-st}(-s\cos kt + k\sin kt)}{s^2 + k^2} \right]_0^\infty .$$

For positive s, $e^{-st} \to 0$ as $t \to \infty$. Furthermore $\sin kt$ and $\cos kt$ are bounded as $t \to \infty$. Therefore we have

$$L\{\cos kt\} = \frac{s}{s^2 + k^2}, \quad s > 0.$$

3. From Euler's formula we get $\sin kt = \dfrac{e^{ikt} - e^{-ikt}}{2i}$. Formally using the result of Example 14.1 we have

$$L\{\sin kt\} = \frac{1}{2i}L\{e^{ikt}\} - \frac{1}{2i}L\{e^{-ikt}\} = \frac{1}{2i}\left[\frac{1}{s - ik} - \frac{1}{s + ik} \right] = \frac{k}{s^2 + k^2}.$$

5. $L\{t^3 - t^2 + 4t\} = L\{t^3\} - L\{t^2\} + 4L\{t\} = \dfrac{6}{s^4} - \dfrac{2}{s^3} + \dfrac{4}{s^2}, \quad s > 0.$

7. Using the result of Example 14.1 we have

$$L\{3e^{4t} - e^{-2t}\} = 3L\{e^{4t}\} - L\{e^{-2t}\} = \frac{3}{s - 4} - \frac{1}{s + 2} = \frac{2s + 10}{(s - 4)(s + 2)}, \quad s > 4.$$

9. $L\{\sinh kt\} = \dfrac{1}{2}L\{e^{kt} - e^{-kt}\} = \dfrac{1}{2}\left[\dfrac{1}{s - k} - \dfrac{1}{s + k} \right] = \dfrac{k}{s^2 - k^2}, \quad s > |k|.$

11. $L\{\sin^2 kt\} = L\left\{ \dfrac{1 - \cos 2kt}{2} \right\} = \dfrac{1}{2}\left[\dfrac{1}{s} - \dfrac{s}{s^2 + 4k^2} \right] = \dfrac{2k^2}{s(s^2 + 4k^2)}, \quad s > 0.$

13. $L\{\sin^2 kt\} = L\{1 - \cos^2 kt\} = \dfrac{1}{s} - \dfrac{s^2 + 2k^2}{s(s^2 + 4k^2)} = \dfrac{2k^2}{s(s^2 + 4k^2)}, \quad s > 0.$

15. $L\{e^{-at} - e^{-bt}\} = \dfrac{1}{s + a} - \dfrac{1}{s + b} = \dfrac{b - a}{(s + a)(s + b)}, \quad s > \max(-a, -b).$

17. From the definition of the Laplace transform we have

$$L\{\phi(t)\} = \int_0^2 e^{-st}\, dt + \int_2^\infty te^{-st}\, dt = \left[\frac{e^{-st}}{-s}\right]_0^2 + \left[\frac{-te^{-st}}{s} - \frac{e^{-st}}{s^2}\right]_2^\infty$$

$$= \frac{e^{-2s}}{s} + \frac{1}{s} + \frac{e^{-2s}}{s^2}, \quad s > 0.$$

19. From the definition of the Laplace transform we have

$$L\{B(t)\} = \int_0^\pi e^{-st} \sin 2t\, dt = \left[\frac{e^{-st}(-s\sin 2t - 2\cos 2t)}{s^2 + 4}\right]_0^\pi$$

$$= \frac{2(1 - e^{-\pi s})}{s^2 + 4}, \quad s > 0.$$

14.6   Functions of Class A

1. If $F_1(t)$ and $F_2(t)$ are of exponential order as $t \to \infty$ then there exist constants M_1, M_2, and b_1, b_2 and fixed numbers t_1, t_2 such that

$$|F_1(t)| < M_1 e^{b_1 t}, \text{ for } t \geq t_1,$$

and

$$|F_2(t)| < M_2 e^{b_2 t}, \text{ for } t \geq t_2.$$

If we take $t_3 = \max(t_1, t_2)$, we have for $t \geq t_3$

$$|F_1(t)| \cdot |F_2(t)| < M_1 M_2 \exp\left[(b_1 + b_2)t\right]$$

or

$$|F_1(t) \cdot F_2(t)| < M_1 M_2 \exp\left[(b_1 + b_2)t\right].$$

Thus $F_1(t) \cdot F_2(t)$ is of exponential order as $t \to \infty$.

Similarly, if we take $M_3 = \max(M_1, M_2)$, $b_3 = \max(b_1, b_2)$, and $t_3 = \max(t_1, t_2)$ we may write

$$|F_1(t)| < M_1 e^{b_1 t} \leq M_3 e^{b_3 t}, \text{ for } t \geq t_3,$$

and

$$|F_2(t)| < M_2 e^{b_2 t} \leq M_3 e^{b_3 t}, \text{ for } t \geq t_3.$$

Thus

$$|F_1(t) + F_2(t)| \leq |F_1(t)| + |F_2(t)| < 2M_3 e^{b_3 t}, \text{ for } t \geq t_3.$$

That is, $F_1(t) + F_2(t)$ is of exponential order as $t \to \infty$.

3. Consider $\lim\limits_{t\to\infty} e^{-t}t^x = \lim\limits_{t\to\infty} \dfrac{t^x}{e^t}$. If $x \le 0$ this limit is zero and $t^x = O(e^t)$ as $t \to \infty$. If $0 < x < 1$, we use the tools of elementary calculus to evaluate the indeterminate form

$$\lim_{t\to\infty} \frac{t^x}{e^t} = \lim_{t\to\infty} \frac{xt^{x-1}}{e^t} = 0.$$

Again we see that $t^x = O(e^t)$. Finally, if $0 < n \le x < n+1$, this argument may be repeated $n+1$ times to obtain

$$\lim_{t\to\infty} \frac{t^x}{e^t} = \lim_{t\to\infty} \frac{x(x-1)\cdots(x-n+1)t^{x-n}}{e^t}$$

$$= \lim_{t\to\infty} \frac{x(x-1)\cdots(x-n)t^{x-n-1}}{e^t} = 0.$$

Thus

$$t^x = O(e^t).$$

5. The function $\cos kt$ is continuous and bounded by unity for all t, hence $|\cos kt| < 2e^{0t}$ for all t. Thus $\cos kt$ is of exponential order as $t \to \infty$. Hence $\cos kt$ is of class A.

7. The function $\sinh kt$ is continuous for all t and if $k > 0$ then

$$|\sinh kt| = \left| \frac{e^{kt} - e^{-kt}}{2} \right| < \frac{e^{kt}}{2}, \text{ for all } t \ge 0.$$

If $k < 0$ then

$$|\sinh kt| = |-\sinh[(-k)t]| = |\sinh[(-k)t]| < \frac{1}{2}e^{-kt}, \text{ for } t > 0.$$

Hence $\sinh kt$ is of exponential order as $t \to \infty$. Therefore $\sinh kt$ is of class A.

9. For $t > 1$, $\left| \dfrac{\sin kt}{t} \right| < |\sin kt| < e^{0t}$. Hence $\dfrac{\sin kt}{t}$ is of exponential order as $t \to \infty$. Moreover, $\dfrac{\sin kt}{t}$ is continuous for $t > 0$. It is also true that $\lim\limits_{t\to 0} \dfrac{\sin kt}{t} = \lim\limits_{t\to 0} k\cos kt = k$. Therefore $\dfrac{\sin kt}{t}$ is sectionally continuous over every finite interval in the range $t \ge 0$. Thus $\dfrac{\sin kt}{t}$ is of class A.

11. By Exercise 8, t^n is of class A. Also e^{kt} is of class A. Hence by Exercise 2, $t^n e^{kt}$ is of class A.

13. By Exercise 8 and Exercise 5, t^n and $\cos kt$ are of class A. Hence by Exercise 2 their product is of class A.

15. By Exercise 8, t^n is of class A, and by Exercise 6, $\cosh kt$ is also of class A. It follows from Exercise 2 that $t^n \cosh kt$ is of class A.

17. We let $F(t) = \dfrac{\cos t - \cosh t}{t}$. Then for $t > 0$, $F(t)$ is a continuous function. Moreover, $\lim\limits_{t \to 0} F(t) = \lim\limits_{t \to 0}(-\sin t - \cosh t) = -1$. Hence, $F(t)$ is sectionally continuous for $t \geq 0$. For

$t > 1$, $\left| \dfrac{\cos t - \cosh t}{t} \right| < |\cos t - \cosh t|$. But by Exercises 5, 6, and 2, $\cos t - \cosh t$ is of

exponential order as $t \to \infty$; that is, there exist constants M and b and a fixed t_0 such that $|\cos t - \cosh t| < M e^{bt}$ for $t > t_0$. It follows from the above inequalities that for $t > \max(1, t_0)$,

$\left| \dfrac{\cos t - \cosh t}{t} \right| < M e^{bt}$. That is, $F(t)$ is of exponential order as $t \to \infty$. Therefore, $F(t)$ is of class A.

14.10   Periodic Functions

1. By equation (6), page 269,

$$
\begin{aligned}
L\{t^{1/2}\} &= \frac{\Gamma(3/2)}{s^{3/2}}, \\
&= \frac{1}{2}\frac{\Gamma(1/2)}{s^{3/2}}, \quad \text{by Theorem 14.10,} \\
&= \frac{1}{2}\frac{\sqrt{\pi}}{s^{3/2}}, \quad \text{by equation (7), page 269,} \\
&= \frac{1}{2s}\left(\frac{\pi}{s}\right)^{1/2}, \quad \text{for } s > 0.
\end{aligned}
$$

3. $\sin kt$ is the unique solution of the initial value problem

$$y'' + k^2 y = 0, \ y(0) = 0, \ y'(0) = k.$$

Using the Laplace transform on this problem gives

$$s^2 L\{\sin kt\} - k + k^2 L\{\sin kt\} = 0.$$

Thus $L\{\sin kt\} = \dfrac{k}{s^2 + k^2}$.

5. If $F(t) = \sin kt$ then $F'(t) = k \cos kt$. For this $F(t)$, Theorem 14.5 states that

$$L\{k \cos kt\} = s L\{\sin kt\} - 0$$

or

$$k L\{\cos kt\} = s L\{\sin kt\}.$$

7. Using Theorem 14.9 we have

$$L\{t^2 \sin kt\} = L\{(-t)^2 \sin kt\} = \frac{d^2}{ds^2}\left[\frac{k}{s^2+k^2}\right]$$

$$= \frac{-2k(s^2+k^2)^2 + 2ks \cdot 2(s^2+k^2) \cdot 2s}{(s^2+k^2)^4} = \frac{2k(3s^2-k^2)}{(s^2+k^2)^3}, \quad s > 0.$$

9. From the definition of the Laplace transform we obtain

$$L\{F(t)\} = \int_0^2 e^{-st}(t+1)\,dt + \int_2^\infty 3e^{-st}\,dt$$

$$= \left[\frac{-e^{-st}(t+1)}{s} - \frac{e^{-st}}{s^2}\right]_0^2 + \left[\frac{-3e^{-st}}{s}\right]_2^\infty = -\frac{e^{-2s}}{s^2} + \frac{1}{s} + \frac{1}{s^2}.$$

From Theorem 14.5 we obtain

$$L\{F'(t)\} = s\left[-\frac{e^{-2s}}{s^2} + \frac{1}{s} + \frac{1}{s^2}\right] - 1 = \frac{1 - e^{-2s}}{s}, \quad s > 0.$$

A second derivation of this formula can be obtained by using the definition of the Laplace transform directly on the function $F'(t) = 1$, $0 \leq t < 2$; $F'(t) = 0$, $t > 2$.

11. Using Theorem 14.12 we obtain

$$L\{T(t,\ c)\} = \frac{\int_0^{2c} e^{-s\beta} T(\beta,\ c)\,d\beta}{1 - e^{-2cs}} = \frac{\int_0^c \beta e^{-s\beta}\,d\beta + \int_c^{2c}(2c - \beta)e^{-s\beta}\,d\beta}{1 - e^{-2cs}}$$

$$= \frac{1}{1 - e^{-2cs}} \cdot \left\{\left[\frac{-\beta}{s}e^{-s\beta} - \frac{1}{s^2}e^{-s\beta}\right]_0^c + \left[\frac{-2c}{s}e^{-s\beta} + \frac{\beta}{s}e^{-s\beta} + \frac{1}{s^2}e^{-s\beta}\right]_c^{2c}\right\}$$

$$= \frac{1}{s^2} \cdot \frac{1 - 2e^{-cs} + e^{-2cs}}{1 - e^{-2cs}} = \frac{1}{s^2} \cdot \frac{(1 - e^{-cs})^2}{1 - e^{-2cs}}$$

$$= \frac{1}{s^2} \cdot \frac{1 - e^{-cs}}{1 + e^{-cs}} = \frac{1}{s^2} \cdot \frac{e^{cs/2} - e^{-cs/2}}{e^{cs/2} + e^{-cs/2}}$$

$$= \frac{1}{s^2} \tanh\frac{cs}{2}.$$

13. Using Theorem 14.12 we have

$$L\{|\sin kt|\} = \frac{\int_0^{\pi/k} e^{-s\beta}\sin k\beta\,d\beta}{1 - e^{-s\pi/k}} = \frac{\left[e^{-s\beta}(-s\sin k\beta - k\cos k\beta)\right]_0^{\pi/k}}{(1 - e^{-s\pi/k})(s^2 + k^2)}$$

$$= \frac{e^{-s\pi/k}(k) + k}{(1 - e^{-s\pi/k})(s^2 + k^2)} = \frac{k}{s^2 + k^2} \cdot \frac{1 + e^{-s\pi/k}}{1 - e^{-s\pi/k}}$$

$$= \frac{k}{s^2 + k^2} \cdot \coth\frac{s\pi}{2k}.$$

15. From Theorem 14.12 we have

$$L\{G(t)\} = \frac{\int_0^c e^{-s\beta} e^{\beta} \, d\beta}{1 - e^{-cs}} = \frac{\int_0^c e^{(-s+1)\beta} \, d\beta}{1 - e^{-cs}}$$

$$= \left[\frac{e^{(-s+1)\beta}}{(1-s)(1-e^{-cs})}\right]_0^c = -\frac{1}{s-1} \cdot \frac{1 - \exp\left[c(1-s)\right]}{1 - \exp\left(-cs\right)}, \quad s > 1.$$

17. From Theorem 14.12 we obtain

$$L\{F(t)\} = \frac{1}{1 - e^{-2\pi s/\omega}} \cdot \int_0^{2\pi/\omega} e^{-s\beta} F(\beta) \, d\beta = \frac{1}{1 - e^{-2\pi s/\omega}} \cdot \int_0^{\pi/\omega} e^{-s\beta} \sin\omega\beta \, d\beta$$

$$= \frac{1}{1 - e^{-2\pi s/\omega}} \cdot \left[\frac{e^{-s\beta}(-s\sin\omega\beta - \omega\cos\omega\beta)}{s^2 + \omega^2}\right]_0^{\pi/\omega} = \frac{\omega}{s^2 + \omega^2} \cdot \frac{1}{1 - e^{-\pi s/\omega}}.$$

19. Using an argument similar to the proof given in the text for Theorem 14.8, we have

$$\int_s^{\infty} f(\beta) \, d\beta = \int_s^{\infty} \int_0^{\infty} e^{-\beta t} F(t) \, dt \, d\beta = \int_0^{\infty} \int_s^{\infty} e^{-\beta t} F(t) \, d\beta \, dt$$

$$= \int_0^{\infty} \left[\frac{e^{-\beta t} F(t)}{-t}\right]_s^{\infty} \, dt = \int_0^{\infty} \frac{e^{-st} F(t)}{t} \, dt$$

$$= L\left\{\frac{F(t)}{t}\right\}.$$

Chapter 15

Inverse Transforms

15.1 Definition of an Inverse Transform

1. Using Theorem 15.2 we obtain

$$L^{-1}\left\{\frac{1}{s^2+2s+10}\right\} = L^{-1}\left\{\frac{1}{(s+1)^2+9}\right\} = e^{-t}L^{-1}\left\{\frac{1}{s^2+9}\right\} = \frac{1}{3}e^{-t}\sin 3t.$$

3. Using Theorem 15.2 we obtain

$$L^{-1}\left\{\frac{3s}{s^2+4s+13}\right\} = L^{-1}\left\{\frac{3(s+2)-6}{(s+2)^2+9}\right\} = e^{-2t}L^{-1}\left\{\frac{3s-6}{s^2+9}\right\}$$

$$= e^{-2t}\left[3L^{-1}\left\{\frac{s}{s^2+9}\right\} - 2L^{-1}\left\{\frac{3}{s^2+9}\right\}\right]$$

$$= e^{-2t}(3\cos 3t - 2\sin 3t).$$

5. Using Theorem 15.2 we obtain

$$L^{-1}\left\{\frac{1}{s^2+4s+4}\right\} = L^{-1}\left\{\frac{1}{(s+2)^2}\right\} = e^{-2t}L^{-1}\left\{\frac{1}{s^2}\right\} = te^{-2t}.$$

7. Using Theorem 15.2 we obtain

$$L^{-1}\left\{\frac{2s-3}{s^2-4s+8}\right\} = L^{-1}\left\{\frac{2s-3}{(s-2)^2+4}\right\} = e^{2t}L^{-1}\left\{\frac{2s+1}{s^2+4}\right\} = e^{2t}\left(2\cos 2t + \frac{1}{2}\sin 2t\right).$$

9. Using Theorem 15.2 we obtain

$$L^{-1}\left\{\frac{2s+3}{(s+4)^3}\right\} = L^{-1}\left\{\frac{2(s+4)-5}{(s+4)^3}\right\} = e^{-4t}L^{-1}\left\{\frac{2s-5}{s^3}\right\} = e^{-4t}(2t - \tfrac{5}{2}t^2).$$

11. Using Theorem 15.2 we obtain

$$L^{-1}\left\{\frac{1}{(s+a)^{n+1}}\right\} = e^{-at}L^{-1}\left\{\frac{1}{s^{n+1}}\right\} = \frac{t^n e^{-at}}{n!}.$$

13. Using Theorem 15.2 we obtain

$$L^{-1}\left\{\frac{1}{(s+a)^2+b^2}\right\} = e^{-at}L^{-1}\left\{\frac{1}{s^2+b^2}\right\} = \frac{1}{b}e^{-at}\sin bt.$$

15. Consider $L\left\{F\left(\frac{t}{a}\right)\right\} = \int_0^\infty e^{-s\beta}F\left(\frac{\beta}{a}\right)\,d\beta$. We change the variable of integration by setting $\beta = at$ and $d\beta = a\,dt$ to get

$$L\left\{F\left(\frac{t}{a}\right)\right\} = a\int_0^\infty e^{-sat}F(t)\,dt = af(as).$$

That is $L^{-1}\{f(as)\} = \frac{1}{a}F\left(\frac{t}{a}\right)$.

15.2   Partial Fractions

1. $L^{-1}\left\{\frac{1}{s^2+as}\right\} = \frac{1}{a}L^{-1}\left\{\frac{1}{s}-\frac{1}{s+a}\right\} = \frac{1}{a}(1-e^{-at})$.

3. $L^{-1}\left\{\frac{2s^2+5s-4}{s^3+s^2-2s}\right\} = L^{-1}\left\{\frac{2}{s}+\frac{1}{s-1}-\frac{1}{s+2}\right\} = 2+e^t-e^{-2t}$.

5. $L^{-1}\left\{\frac{4s+4}{s^2(s-2)}\right\} = L^{-1}\left\{\frac{3}{s-2}-\frac{3}{s}-\frac{2}{s^2}\right\} = 3e^{2t}-3-2t$.

7. $L^{-1}\left\{\frac{5s-2}{s^2(s+2)(s-1)}\right\} = L^{-1}\left\{-\frac{2}{s}+\frac{1}{s+2}+\frac{1}{s-1}+\frac{1}{s^2}\right\} = -2+t+e^t+e^{-2t}$.

9. $L^{-1}\left\{\frac{s}{(s^2+a^2)(s^2+b^2)}\right\} = \frac{1}{a^2-b^2}L^{-1}\left\{\frac{s}{s^2+b^2}-\frac{s}{s^2+a^2}\right\} = \frac{1}{a^2-b^2}(\cos bt - \cos at)$.

15.3   Initial Value Problems

1. $y' = e^t$; $y(0) = 2$. Applying the Laplace transform to both sides of the differential equation and using Theorem 14.6 allows us to write

$$su(s) - 2 = \frac{1}{s-1},$$
$$u(s) = \frac{2s-1}{s(s-1)} = \frac{1}{s}+\frac{1}{s-1}.$$

The inverse transform now gives $y(t) = 1 + e^t$. Verification of this solution is straightforward.

3. $y' + y = e^{2t}$; $y(0) = 0$. Applying the Laplace transform to both sides of the differential equation and using Theorem 14.6 allows us to write

$$su(s) + u(s) = \frac{1}{s-2},$$

$$u(s) = \frac{1}{(s-2)(s+1)} = \frac{\frac{1}{3}}{s-2} - \frac{\frac{1}{3}}{s+1}.$$

The inverse transform now gives $y(t) = \frac{1}{3}e^{2t} - \frac{1}{3}e^{-t}$. Verification of this solution is straightforward.

5. $y'' + a^2 y = 0$; $y(0) = 1$, $y'(0) = 0$. Applying the Laplace transform to both sides of the differential equation and using Theorem 14.6 allows us to write

$$s^2 u(s) - s + a^2 u(s) = 0,$$

$$u(s) = \frac{s}{s^2 + a^2}.$$

The inverse transform now gives $y(t) = \cos at$. Verification of this solution is straightforward.

7. $y'' - 3y' + 2y = e^{3t}$; $y(0) = 0$, $y'(0) = 0$. Applying the Laplace transform to both sides of the differential equation and using Theorem 14.6 allows us to write

$$s^2 u(s) - 3su(s) + 2u(s) = \frac{1}{s-3},$$

$$u(s) = \frac{1}{(s-1)(s-2)(s-3)} = \frac{1}{2}\left[\frac{1}{s-1} - \frac{2}{s-2} + \frac{1}{s-3}\right].$$

The inverse transform now gives $y(t) = \frac{1}{2}\left[e^t - 2e^{2t} + e^{3t}\right]$. Verification of this solution is straightforward.

9. $y'' - 2y' = -4$; $y(0) = 0$, $y'(0) = 4$. Applying the Laplace transform to both sides of the differential equation and using Theorem 14.6 allows us to write

$$s^2 u(s) - 4 - 2su(s) = -\frac{4}{s},$$

$$u(s) = \frac{-4(1-s)}{s^2(s-2)} = \frac{1}{s-2} - \frac{1}{s} + \frac{2}{s^2}.$$

The inverse transform now gives $y(t) = e^{2t} - 1 + 2t$. Verification of this solution is straightforward.

11. $x'' - 4x' + 4x = 4e^{2t}$; $x(0) = -1$, $x'(0) = -4$. Applying the Laplace transform to both sides of the differential equation and using Theorem 14.6 allows us to write

$$s^2 u(s) + s + 4 - 4su(s) - 4 + 4u(s) = \frac{4}{s-2},$$

$$u(s) = \frac{-s^2 + 2s + 4}{(s-2)^3} = \frac{4}{(s-2)^3} - \frac{2}{(s-2)^2} - \frac{1}{s-2}.$$

The inverse transform now gives $x(t) = e^{2t}(2t^2 - 2t - 1)$. Verification of this solution is straightforward.

13. $y'' - y = 4\cos t$; $y(0) = 0$, $y'(0) = 1$. Applying the Laplace transform to both sides of the differential equation and using Theorem 14.6 allows us to write

$$s^2 u(s) - 1 - u(s) = \frac{4s}{s^2 + 1},$$

$$u(s) = \frac{4s}{(s^2 - 1)(s^2 + 1)} + \frac{1}{s^2 - 1} = \frac{2s}{s^2 - 1} - \frac{2s}{s^2 + 1} + \frac{1}{s^2 - 1}.$$

The inverse transform now gives $y(t) = 2\cosh t + \sinh t - 2\cos t$. Verification of this solution is straightforward.

15. $x'' + 4x = t + 4$; $x(0) = 1$, $x'(0) = 0$. Applying the Laplace transform to both sides of the differential equation and using Theorem 14.6 allows us to write

$$s^2 u(s) - s + 4u(s) = \frac{1}{s^2} + \frac{4}{s},$$

$$u(s) = \frac{1}{s^2(s^2 + 4)} + \frac{4}{s(s^2 + 4)} + \frac{s}{s^2 + 4} = \frac{1}{4s^2} - \frac{1}{4(s^2 + 4)} + \frac{1}{s}.$$

The inverse transform now gives $x(t) = \frac{1}{4}t - \frac{1}{8}\sin 2t + 1$. Verification of this solution is straightforward.

17. $x'' + x = 4e^t$; $x(0) = 1$, $x'(0) = 3$. Applying the Laplace transform to both sides of the differential equation and using Theorem 14.6 allows us to write

$$s^2 u(s) - s - 3 + u(s) = \frac{4}{s - 1},$$

$$u(s) = \frac{4}{(s - 1)(s^2 + 1)} + \frac{s}{(s^2 + 1)} + \frac{3}{s^2 + 1} = \frac{2}{s - 1} + \frac{1}{s^2 + 1} - \frac{s}{s^2 + 1}.$$

The inverse transform now gives $x(t) = 2e^t + \sin t - \cos t$. Verification of this solution is straightforward.

19. $y'' + 9y = 40e^x$; $y(0) = 5$, $y'(0) = -2$. Applying the Laplace transform to both sides of the differential equation and using Theorem 14.6 allows us to write

$$s^2 u(s) - 5s + 2 + 9u(s) = \frac{40}{s - 1},$$

$$u(s) = \frac{40}{(s - 1)(s^2 + 9)} + \frac{5s - 2}{s^2 + 9} = \frac{4}{s - 1} + \frac{s}{s^2 + 9} - \frac{6}{s^2 + 9}.$$

The inverse transform now gives $y(x) = 4e^x + \cos 3x - 2\sin 3x$. Verification of this solution is straightforward.

21. $x'' + 3x' + 2x = 4t^2$; $x(0) = 0$, $x'(0) = 0$. Applying the Laplace transform to both sides of the differential equation and using Theorem 14.6 allows us to write

$$s^2 u(s) + 3su(s) + 2u(s) = \frac{8}{s^3},$$

$$u(s) = \frac{8}{s^3(s+1)(s+2)} = \frac{4}{s^3} - \frac{6}{s^2} + \frac{7}{s} - \frac{8}{s+1} + \frac{1}{s+2}.$$

The inverse transform now gives $x(t) = 2t^2 - 6t + 7 - 8e^{-t} + e^{-2t}$. Verification of this solution is straightforward.

23. $y'' + y' = -\cos x$; $y(0) = a$, $y'(0) = b$. Applying the Laplace transform to both sides of the differential equation and using Theorem 14.6 allows us to write

$$s^2 u(s) - sa - b + su(s) - a = -\frac{s}{s^2 + 1},$$

$$u(s) = \frac{-s}{s(s+1)(s^2+1)} + \frac{as + a + b}{s(s+1)}$$

$$= \frac{a+b}{s} - \frac{b + \frac{1}{2}}{s+1} + \frac{\frac{1}{2}s}{s^2+1} - \frac{\frac{1}{2}}{s^2+1}.$$

The inverse transform now gives $y(x) = c_1 + c_2 e^{-x} + \frac{1}{2}\cos x - \frac{1}{2}\sin x$. Verification of this solution is straightforward.

25. $y'' + 3y' + 2y = 12x^2$; $y(0) = a$, $y'(0) = b$. Applying the Laplace transform to both sides of the differential equation and using Theorem 14.6 allows us to write

$$s^2 u(s) - sa - b + 3su(s) - a + 2u(s) = \frac{24}{s^3},$$

$$u(s) = \frac{24}{s^3(s+1)(s+2)} + \frac{as + a + b}{(s+1)(s+2)}$$

$$= \frac{b - 24}{s+1} + \frac{a - b + 3}{s+2} + \frac{12}{s^3} - \frac{18}{s^2} + \frac{21}{s}.$$

The inverse transform now gives $y(x) = c_1 e^{-x} + c_2 e^{-2x} + 6x^2 - 18x + 21$. Verification of this solution is straightforward.

27. $y'' + 4y' + 5y = 50x + 13e^{3x}$; $y(0) = a$, $y'(0) = b$. Applying the Laplace transform to both sides of the differential equation and using Theorem 14.6 allows us to write

$$s^2 u(s) - sa - b + 4su(s) - a + 5u(s) = \frac{50}{s^2} + \frac{13}{s-3},$$

$$u(s) = \frac{50}{s^2(s^2 + 4s + 5)} + \frac{13}{(s-3)(s^2 + 4s + 5)} + \frac{as + a + b}{s^2 + 4s + 5}$$

$$= \frac{(a + \frac{15}{2})(s+2)}{(s+2)^2 + 1} + \frac{b - a + \frac{8}{3}}{(s+2)^2 + 1} + \frac{10}{s^2} - \frac{8}{s} + \frac{\frac{1}{2}}{s-3}.$$

The inverse transform now gives $y(x) = e^{-2x}(c_1 \cos x + c_2 \sin x) + 10x - 8 + \frac{1}{2}e^{3x}$. Verification of this solution is straightforward.

29. $y''' + y'' - 4y' - 4y = 3e^{-x} - 4x - 6$; $y(0) = a$, $y'(0) = b$, $y''(0) = c$. Applying the Laplace transform to both sides of the differential equation and using Theorem 14.6 allows us to write

$$s^3 u(s) - s^2 a - sb - c + s^2 u(s) - sa - b - 4su(s) + 4a - 4u(s) = \frac{3}{s+1} - \frac{4}{s^2} - \frac{6}{s},$$

$$u(s) = \frac{3s^2 - 4(s+1) - 6s(s+1)}{s^2(s+1)^2(s-2)(s+2)} + \frac{s^2 a + s(a+b) + b + c - 4a}{(s+1)(s-2)(s+2)}$$

$$= \frac{\frac{1}{12}(-3 + 2a + 3b + c)}{s-2} - \frac{\frac{1}{4}(1 + 2a + b - c)}{s+2} + \frac{\frac{1}{6}(-9 + 8a - 2c)}{s+1} - \frac{1}{(s+1)^2} + \frac{1}{s^2} + \frac{\frac{1}{2}}{s}.$$

The inverse transform now gives $y(x) = c_1 e^{2x} + c_2 e^{-2x} + c_3 e^{-x} - xe^{-x} + x + \frac{1}{2}$. Verification of this solution is straightforward.

31. $y'''' - y = e^{-x}$; $y(0) = a$, $y'(0) = b$, $y''(0) = c$, $y'''(0) = d$. Applying the Laplace transform to both sides of the differential equation and using Theorem 14.6 allows us to write

$$s^4 u(s) - s^3 a - s^2 b - sc - d - u(s) = \frac{1}{s+1},$$

$$u(s) = \frac{1}{(s-1)(s+1)^2(s^2+1)} + \frac{s^3 a + s^2 b + sc + d}{(s-1)(s+1)(s^2+1)}$$

$$= \frac{c_1}{s-1} + \frac{c_2}{s+1} - \frac{\frac{1}{4}}{(s+1)^2} + \frac{c_3 s}{s^2+1} + \frac{c_4}{s^2+1}.$$

The inverse transform now gives $y(x) = c_1 e^x + c_2 e^{-x} + c_3 \cos x + c_4 \sin x - \frac{1}{4} e^{-x}$, where $c_1 = \frac{1}{8}(1 + 2a + 2b + 2c + 2d)$, $c_2 = \frac{1}{8}(-3 + 2a - 2b + 2c + 6d)$, $c_3 = \frac{1}{4}(-1 + 2a + 4b - 2c - 4d)$, and $c_4 = \frac{1}{4}(-1 + 2b - 2d)$. Verification of this solution is straightforward.

33. $y'' - 4y = 2 - 8x$; $y(0) = 0$, $y'(0) = 5$. Applying the Laplace transform to both sides of the differential equation and using Theorem 14.6 allows us to write

$$s^2 u(s) - 5 - 4u(s) = \frac{2}{s} - \frac{8}{s^2},$$

$$u(s) = \frac{2}{s(s^2-4)} - \frac{8}{s^2(s^2-4)} + \frac{5}{s^2-4} = \frac{1}{s-2} - \frac{\frac{1}{2}}{s+2} + \frac{2}{s^2} - \frac{\frac{1}{2}}{s}.$$

The inverse transform now gives $y(x) = e^{2x} - \frac{1}{2} e^{-2x} + 2x - \frac{1}{2}$. Verification of this solution is straightforward.

35. $y'' + 4y' + 5y = 10e^{-3x}$; $y(0) = 4$, $y'(0) = 0$. Applying the Laplace transform to both sides of the differential equation and using Theorem 14.6 allows us to write

$$s^2 u(s) - 4s + 4su(s) - 16 + 5u(s) = \frac{10}{s+3},$$

$$u(s) = \frac{10}{(s+3)(s^2+4s+5)} + \frac{4s+16}{s^2+4s+5}$$

$$= \frac{13}{(s+2)^2+1} - \frac{s+2}{(s+2)^2+1} + \frac{5}{s+3}.$$

The inverse transform now gives $y(x) = e^{-2x}(13\sin x - \cos x) + 5e^{-3x}$. Verification of this solution is straightforward.

37. $\ddot{x} + 4\dot{x} + 5x = 8\sin t$; $x(0) = 0$, $\dot{x}(0) = 0$. Applying the Laplace transform to both sides of the differential equation and using Theorem 14.6 allows us to write

$$s^2 u(s) + 4su(s) + 5u(s) = \frac{8}{s^2 + 1},$$

$$u(s) = \frac{8}{(s^2 + 4s + 5)(s^2 + 1)}$$

$$= \frac{s + 2}{(s + 2)^2 + 1} + \frac{1}{(s + 2)^2 + 1} - \frac{s}{s^2 + 1} + \frac{1}{s^2 + 1}.$$

The inverse transform now gives $x(t) = e^{-2t}(\cos t - \sin t) - \cos t + \sin t$. Verification of this solution is straightforward.

39. $y''' + 4y'' + 9y' + 10y = -24e^x$; $y(0) = 0$, $y'(0) = -4$, $y''(0) = 10$. Applying the Laplace transform to both sides of the differential equation and using Theorem 14.6 allows us to write

$$s^3 u(s) + 4s - 10 + 4s^2 u(s) + 16 + 9su(s) + 10u(s) = \frac{-24}{s - 1},$$

$$u(s) = \frac{-4s^2 - 2s - 18}{(s - 1)(s^3 + 4s^2 + 9s + 10)}$$

$$= \frac{2}{s + 2} - \frac{s + 1}{(s + 1)^2 + 4} - \frac{1}{s - 1}.$$

The inverse transform now gives $y(x) = 2e^{-2x} - e^{-x}\cos 2x - e^x$. Verification of this solution is straightforward.

41. $y'' + 2y' + y = x$; $y(0) = -3$, $y'(0) = a$. Applying the Laplace transform to both sides of the differential equation and using Theorem 14.6 allows us to write

$$s^2 u(s) + 3s - a + 2su(s) - 6 + u(s) = \frac{1}{s^2},$$

$$u(s) = \frac{-3s^3 + (a + 6)s^2}{s^2(s + 1)^2}$$

$$= \frac{-2}{s} + \frac{1}{s^2} - \frac{1}{s + 1} + \frac{a + 10}{(s + 1)^2}.$$

The inverse transform now gives $y(x) = -2 + x - e^{-x} + (a + 10)xe^{-x}$. From this we get $y(1) = -1 - e^{-1} + (a + 10)e^{-1} = -1$, so that $a = -9$ and

$$y(x) = -2 + x - e^{-x} + xe^{-x},$$
$$y'(x) = 1 + 2e^{-x} - xe^{-x}.$$

It follows that $y(2) = e^{-2}$ and $y'(2) = 1$.

43. $x'' - 4x' + 4x = e^{2t}$; $x(0) = b$, $x'(0) = 0$. Applying the Laplace transform to both sides of the differential equation and using Theorem 14.6 allows us to write

$$s^2 u(s) - bs - 4su(s) + 4b + 4u(s) = \frac{1}{s-2},$$

$$u(s) = \frac{1}{(s-2)^3} + \frac{b}{s-2} - \frac{2b}{(s-2)^2}.$$

The inverse transform now gives the function $x(t) = \frac{1}{2}t^2 e^{2t} + be^{2t} - 2bte^{2t}$. From this we get $x(1) = \frac{1}{2}e^2 + be^2 - 2be^2 = 0$, so that $b = \frac{1}{2}$, and $x(t) = \frac{1}{2}(1-t)^2 e^{2t}$. Verification of this solution is straightforward.

15.4   A Step Function

9. We first write the function $F(t)$ in terms of the α function and then use Theorem 15.3 to obtain the Laplace transform.

$$F(t) = 4 - 4\alpha(t-2) + (2t-1)\alpha(t-2) = 4 - \alpha(t-2) + 2(t-2)\alpha(t-2),$$

$$L\{F(t)\} = \frac{4}{s} - \frac{e^{-2s}}{s} + \frac{2e^{-2s}}{s^2}.$$

11. We first write the function $F(t)$ in terms of the α function and then use Theorem 15.3 to obtain the Laplace transform.

$$F(t) = t^2 - t^2\alpha(t-1) + 3\alpha(t-1) - 3\alpha(t-2)$$
$$= t^2 - (t-1)^2\alpha(t-1) - 2(t-1)\alpha(t-1) + 2\alpha(t-1) - 3\alpha(t-2),$$

$$L\{F(t)\} = \frac{2}{s^3} - \frac{2e^{-s}}{s^3} - \frac{2e^{-s}}{s^2} + \frac{2e^{-s}}{s} - \frac{3e^{-2s}}{s} = \frac{2}{s^3} + e^{-s}\left(\frac{2}{s} - \frac{2}{s^2} - \frac{2}{s^3}\right) - \frac{3e^{-2s}}{s}.$$

13. We first write the function $F(t)$ in terms of the α function and then use Theorem 15.3 to obtain the Laplace transform.

$$F(t) = e^{-t} - e^{-t}\alpha(t-2) = e^{-t} - e^{-2}e^{-(t-2)}\alpha(t-2),$$

$$L\{F(t)\} = \frac{1}{s+1} - \frac{e^{-2}e^{-2s}}{s+1} = \frac{1}{s+1} - \frac{e^{-2s-2}}{s+1}.$$

15. We first write the function $F(t)$ in terms of the α function and then use Theorem 15.3 to obtain the Laplace transform.

$$F(t) = \sin 3t - \sin 3t \, \alpha(t-\pi) = \sin 3t + \sin 3(t-\pi) \, \alpha(t-\pi),$$

$$L\{F(t)\} = \frac{3}{s^2+9} + \frac{3e^{-\pi s}}{s^2+9}.$$

17. We have

$$L^{-1}\left\{\frac{1}{s^3}\right\} = \frac{t^2}{2},$$

$$L^{-1}\left\{\frac{1}{(s+2)^3}\right\} = \frac{1}{2}t^2 e^{-2t},$$

$$L^{-1}\left\{\frac{e^{-4s}}{(s+2)^3}\right\} = \frac{1}{2}(t-4)^2 \exp\left[-2(t-4)\right] \alpha(t-4).$$

19. We have

$$F(t) = L^{-1}\left\{\frac{(1-e^{-2s})(1-3e^{-2s})}{s^2}\right\} = L^{-1}\left\{\frac{1}{s^2} - \frac{4e^{-2s}}{s^2} + \frac{3e^{-4s}}{s^2}\right\}$$

$$= t - 4(t-2)\,\alpha(t-2) + 3(t-4)\,\alpha(t-4).$$

Therefore $F(1) = 1$, $F(3) = 3 - 4 = -1$, and $F(5) = 5 - 4 \cdot 3 + 3 = -4$.

21. We have

$$F(t) = \sin t\,\psi(t,\ \pi) = \sum_{n=0}^{\infty}(-1)^n \sin t\,\alpha(t-n\pi) = \sum_{n=0}^{\infty}\sin(t-n\pi)\,\alpha(t-n\pi),$$

$$L\{F(t)\} = \sum_{n=0}^{\infty}\frac{e^{-n\pi s}}{s^2+1} = \sum_{n=0}^{\infty}\frac{(e^{-s\pi})^n}{s^2+1} = \frac{1}{s^2+1}\cdot\frac{1}{1-e^{-s\pi}}.$$

23. We first rewrite $H(t)$ and find its transform.

$$H(t) = 3 - 3\,\alpha(t-4) + (2t-5)\,\alpha(t-4)$$
$$= 3 - 3\,\alpha(t-4) + 2(t-4)\,\alpha(t-4) + 3\,\alpha(t-4)$$
$$= 3 + 2(t-4)\,\alpha(t-4)$$
$$L\{H(t)\} = \frac{3}{s} + \frac{2e^{-4s}}{s^2}.$$

Now applying the Laplace transform to the differential equation we get

$$s^2 L\{x(t)\} - s + L\{x(t)\} = \frac{3}{s} + \frac{2e^{-4s}}{s^2},$$

$$L\{x(t)\} = \frac{s}{s^2+1} + \frac{3}{s(s^2+1)} + \frac{2e^{-4s}}{s^2(s^2+1)}$$

$$= \frac{-2s}{s^2+1} + \frac{3}{s} + \left(\frac{1}{s^2} - \frac{1}{s^2+1}\right)2e^{-4s}.$$

The inverse transform now yields

$$x(t) = 3 - 2\cos t + 2\left[(t-4) - \sin(t-4)\right]\alpha(t-4).$$

25. $x'' + 4x = M(t)$; $x(0) = 0$, $x'(0) = 0$. The Laplace transform of $M(t)$ can be determined to be $L\{M(t)\} = \dfrac{1}{s^2 + 1} - \dfrac{e^{-2\pi s}}{s^2 + 1}$. We now apply the transform to the differential equation to obtain

$$s^2 u(s) + 4u(s) = \frac{1}{s^2 + 1} - \frac{e^{-2\pi s}}{s^2 + 1},$$

$$u(s) = \frac{1}{(s^2 + 1)(s^2 + 4)} - \frac{e^{-2\pi s}}{(s^2 + 1)(s^2 + 4)}$$

$$= \frac{\frac{1}{3}}{s^2 + 1} - \frac{\frac{1}{3}}{s^2 + 4} - \frac{\frac{1}{3}e^{-2\pi s}}{s^2 + 1} + \frac{\frac{1}{3}e^{-2\pi s}}{s^2 + 4},$$

$$x(t) = \tfrac{1}{3}\sin t - \tfrac{1}{6}\sin 2t - \tfrac{1}{3}\sin(t - 2\pi)\,\alpha(t - 2\pi) + \tfrac{1}{6}\sin 2(t - 2\pi)\,\alpha(t - 2\pi)$$

$$= \tfrac{1}{3}\sin t - \tfrac{1}{6}\sin 2t - \tfrac{1}{3}\sin t\,\alpha(t - 2\pi) + \tfrac{1}{6}\sin 2t\,\alpha(t - 2\pi)$$

$$= \tfrac{1}{6}\big[1 - \alpha(t - 2\pi)\big](2\sin t - \sin 2t).$$

27. $x''(t) + 2x'(t) + x(t) = 2 + (t - 3)\,\alpha(t - 3)$; $x(0) = 2$, $x'(0) = 1$. We use the Laplace transform on both sides of the differential equation and get

$$s^2 u(s) - 2s - 1 + 2su(s) - 4 + u(s) = \frac{2}{s} + \frac{e^{-3s}}{s^2}.$$

$$u(s) = \frac{e^{-3s}}{s^2(s + 1)^2} + \frac{2s^2 + 5s + 2}{s(s + 1)^2}$$

$$= e^{-3s}\left[-\frac{2}{s} + \frac{1}{s^2} + \frac{2}{s + 1} + \frac{1}{(s + 1)^2}\right] + \frac{2}{s} + \frac{1}{(s + 1)^2}.$$

We obtain $x(t) = \left[-2 + (t - 3) + 2e^{-(t-3)} + (t - 3)e^{-(t-3)}\right]\alpha(t - 3) + 2 + te^{-t}$. We now have $x(1) = 2 + e^{-1}$ and $x(4) = \left[-2 + 1 + 2e^{-1} + e^{-1}\right] + 2 + 4e^{-4} = 1 + 3e^{-1} + 4e^{-4}$.

15.5   A Convolution Theorem

1. $L\left\{\displaystyle\int_0^t (t - \beta)\sin 3\beta\, d\beta\right\} = L\{t\} \cdot L\{\sin 3t\} = \dfrac{3}{s^2(s^2 + 9)}.$

3. $L\left\{\displaystyle\int_0^t (t - \beta)^3 e^\beta\, d\beta\right\} = L\left\{t^3\right\} \cdot L\left\{e^t\right\} = \dfrac{6}{s^4} \cdot \dfrac{1}{s - 1} = \dfrac{6}{(s - 1)s^4}.$

5. $L^{-1}\left\{\dfrac{1}{s(s + 2)}\right\} = \displaystyle\int_0^t e^{-2\beta}\, d\beta = \left[-\tfrac{1}{2}e^{-2\beta}\right]_0^t = \dfrac{1}{2} - \dfrac{1}{2}e^{-2t}.$

7. We use the convolution theorem to obtain

$$L^{-1}\left\{\frac{1}{(s^2+1)^2}\right\} = \int_0^t \sin\beta \sin(t-\beta)\,d\beta = -\frac{1}{2}\int_0^t \left[\cos t - \cos(2\beta - t)\right]\,d\beta$$

$$= -\frac{1}{2}\cos t\,[\beta]_0^t - \frac{1}{4}\left[\sin(2\beta - t)\right]_0^t = -\frac{1}{2}t\cos t - \frac{1}{4}(\sin t + \sin t)$$

$$= -\frac{1}{2}(t\cos t - \sin t).$$

9. We use the Laplace transform to solve the initial value problem

$$y'' - k^2 y = H(t); \ \ y(0) = 0, \ y'(0) = 0.$$
$$s^2 u(s) - k^2 u(s) = L\{H(t)\},$$
$$u(s) = \frac{L\{H(t)\}}{s^2 - k^2} = \frac{1}{k}L\{H(t)\}\cdot L\{\sinh kt\}.$$

An inverse transform yields $y(t) = \dfrac{1}{k}\displaystyle\int_0^t H(t-\beta)\sinh k\beta\,d\beta.$

11. We use the Laplace transform to solve the initial problem

$$x'' + 6x' + 9x = F(t); \ \ x(0) = A, \ x'(0) = B.$$
$$s^2 u(s) - sA - B + 6su(s) - 6A + 9u(s) = L\{F(t)\},$$
$$u(s) = \frac{L\{F(t)\} + sA + 6A + B}{(s+3)^2} = \frac{L\{F(t)\}}{(s+3)^2} + \frac{A(s+3)}{(s+3)^2} + \frac{3A+B}{(s+3)^2}.$$

An inverse transform yields $x(t) = e^{-3t}\left[A + (3A+B)t\right] + \displaystyle\int_0^t \beta e^{-3\beta}F(t-\beta)\,d\beta..$

15.6   Special Integral Equations

1. $F(t) = 1 + 2\displaystyle\int_0^t F(t-\beta)e^{-2\beta}\,d\beta.$ We apply the Laplace transform to the integral equation, solve for $u(s)$, and find the inverse transform.

$$u(s) = \frac{1}{s} + \frac{2u(s)}{s+2},$$
$$u(s) = \frac{1}{s} + \frac{2}{s^2},$$
$$F(t) = 1 + 2t.$$

Verification of this solution is straightforward.

3. $F(t) = 1 + \int_0^t F(t-\beta)e^{-\beta}\,d\beta$. We apply the Laplace transform to the integral equation, solve for $u(s)$, and find the inverse transform.

$$u(s) = \frac{1}{s^2} + \frac{u(s)}{s+1},$$
$$u(s) = \frac{1}{s^3} + \frac{1}{s^2},$$
$$F(t) = \frac{1}{2}t^2 + t.$$

Verification of this solution is straightforward and tedious.

5. $F(t) = t^3 + \int_0^t F(\beta)\sin(t-\beta)\,d\beta$. We apply the Laplace transform to the integral equation, solve for $u(s)$, and find the inverse transform.

$$u(s) = \frac{6}{s^4} + \frac{u(s)}{s^2+1},$$
$$u(s) = \frac{6}{s^4} + \frac{6}{s^6},$$
$$F(t) = t^3 + \frac{t^5}{20}.$$

Verification of this solution is straightforward and tedious.

7. $F(t) = t^2 - 2\int_0^t F(t-\beta)\sinh 2\beta\,d\beta$. We apply the Laplace transform to the integral equation, solve for $u(s)$, and find the inverse transform.

$$u(s) = \frac{2}{s^3} - \frac{4u(s)}{s^2-4},$$
$$u(s) = \frac{2}{s^3} - \frac{8}{s^5},$$
$$F(t) = t^2 - \frac{t^4}{3}.$$

Verification of this solution is straightforward and tedious.

9. $H(t) = 9e^{2t} - 2\int_0^t H(t-\beta)\cos\beta\,d\beta$. We apply the Laplace transform to the integral equation, solve for $u(s)$, and find the inverse transform.

$$u(s) = \frac{9}{s-2} - \frac{2su(s)}{s^2+1},$$
$$u(s) = \frac{9s^2+9}{(s-2)(s+1)^2} = \frac{5}{s-2} + \frac{4}{s+1} - \frac{6}{(s+1)^2},$$
$$H(t) = 5e^{2t} + 4e^{-t} - 6te^{-t}.$$

Verification of this solution is straightforward and tedious.

11. $g(x) = e^{-x} - 2 \int_0^x g(\beta) \cos(x - \beta) \, d\beta$. We apply the Laplace transform to the integral equation, solve for $u(s)$, and find the inverse transform.

$$u(s) = \frac{1}{s+1} - \frac{2su(s)}{s^2+1},$$

$$u(s) = \frac{s^2+1}{(s+1)^3} = \frac{1}{s+1} - \frac{2}{(s+1)^2} + \frac{2}{(s+1)^3},$$

$$g(x) = e^{-x} - 2xe^{-x} + x^2e^{-x} = e^{-x}(1-x)^2.$$

Verification of this solution is straightforward and tedious.

13. $F'(t) = t + \int_0^t F(t-\beta) \cos\beta \, d\beta$; $F(0) = 4$. We apply the Laplace transform to the equation, solve for $u(s)$, and find the inverse transform.

$$su(s) - 4 = \frac{1}{s^2} + \frac{su(s)}{s^2+1},$$

$$u(s) = \frac{(s^2+1)(4s^2+1)}{s^5} = \frac{4}{s} + \frac{5}{s^3} + \frac{1}{s^5},$$

$$F(t) = 4 + \frac{5t^2}{2} + \frac{t^4}{24}.$$

Verification of this solution is straightforward and tedious.

15. Given $F(t) = t + \int_0^t F(t-\beta)e^{-\beta} \, d\beta$, we multiply both sides of the equation by e^t to obtain

$$e^t F(t) = te^t + \int_0^t F(t-\beta)e^{t-\beta} \, d\beta.$$

In the integral we let $t - \beta = \mu$ to get

$$e^t F(t) = te^t - \int_t^0 F(\mu)e^{\mu} \, d\mu = te^t + \int_0^t e^{\beta} F(\beta) \, d\beta.$$

Differentiating both sides of this equation with respect to t yields

$$e^t F'(t) + e^t F(t) = te^t + e^t + e^t F(t),$$
$$F'(t) = t + 1,$$
$$F(t) = \frac{1}{2}t^2 + t + c.$$

But $F(0) = 0$, so that $c = 0$, and finally $F(t) = t + \frac{1}{2}t^2$.

15.8   The Deflection of Beams

1. This problem is the same as the problem in Example 15.22 except for the second pair of boundary conditions. Here $y(2c) = 0$ and $y'(2c) = 0$. Equations (11) and (12), page 309, are still valid here. We substitute $x = 2c$ into each of these equations to obtain

$$0 = \frac{1}{2}A(4c^2) + \frac{1}{6}B(8c^3) + \frac{w_0}{120c}\left[5c(16c^4) - 32c^5 + c^5\right],$$

$$0 = A(2c) + \frac{1}{2}B(4c^2) + \frac{w_0}{24c}\left[4c(8c^3) - 16c^4 + c^4\right].$$

Solving these equations for A and B yields $A = \frac{23}{240}w_0c^2$ and $B = -\frac{9}{20}w_0c$. Equation (11), page 309, now becomes

$$EIy(x) = \frac{23}{480}w_0c^2x^2 - \frac{3}{40}w_0cx^3 + \frac{w_0}{120c}\left[5cx^4 - x^5 + (x - c)^5\,\alpha(x - c)\right].$$

3. The boundary value problem to be solved is

$$EI\frac{d^4y}{dx^4} = w_0\left[1 - \alpha(x - c)\right];\ y(0) = 0,\ y'(0) = 0,\ y(2c) = 0,\ y''(2c) = 0.$$

To solve this problem we use the same notations as in Example 15.22. Applying the Laplace transform to the differential equation and solving for $u(s)$ yields

$$s^4u(s) - s^3 \cdot 0 - s^2 \cdot 0 - s \cdot A - B = w_0\left(\frac{1}{s} - \frac{e^{-cs}}{s}\right),$$

$$u(s) = \frac{a}{s^3} + \frac{B}{s^4} + w_0\left(\frac{1}{s^5} - \frac{e^{-cs}}{s^5}\right).$$

It follows that

$$EIy(x) = \frac{1}{2}Ax^2 + \frac{1}{6}Bx^3 + \frac{w_0}{24}\left[x^4 - (x - c)^4\,\alpha(x - c)\right],$$

$$EIy'(x) = Ax + \frac{1}{2}Bx^2 + \frac{w_0}{6}\left[x^3 - (x - c)^3\,\alpha(x - c)\right],$$

$$EIy''(x) = A + Bx + \frac{w_0}{2}\left[x^2 - (x - c)^2\,\alpha(x - c)\right].$$

Using the boundary conditions at $x = 2c$ yields

$$0 = 2Ac^2 + \frac{4}{3}Bc^3 + w_0\left[\frac{2c^4}{3} - \frac{c^4}{24}\right],$$

$$0 = A + 2Bc + w_0\left[2c^2 - \frac{c^2}{2}\right].$$

Solving this system for A and B gives us $A = \frac{9w_0c^2}{32}$ and $B = -\frac{57w_0c}{64}$. Finally we have

$$EIy(x) = \frac{9}{64}w_0c^2x^2 - \frac{19}{128}w_0cx^3 + \frac{1}{24}w_0\left[x^4 - (x - c)^4\,\alpha(x - c)\right].$$

15.9 Systems of Equations

1. Applying the Laplace transform directly gives

$$s^2 u(s) - 3su(s) - sv(s) + 6.5 + 2v(s) = \frac{14}{s^2} + \frac{3}{s},$$

$$su(s) - 3u(s) + sv(s) - 6.5 = \frac{1}{s},$$

or

$$(s^2 - 3s)u(s) - (s - 2)v(s) = \frac{14}{s^2} + \frac{3}{s} - 6.5,$$

$$(s - 3)u(s) + sv(s) = \frac{1}{s} + 6.5.$$

Solving for $u(s)$ and $v(s)$ gives us

$$u(s) = \frac{-9s + 12}{s(s-1)(s-3)(s+2)} = \frac{2}{s} - \frac{1/2}{s-1} - \frac{1}{s+2} - \frac{1/2}{s-3},$$

$$v(s) = \frac{6.5s^3 + 7.5s^2 - 3s - 14}{s^2(s-1)(s+2)} = \frac{5}{s} + \frac{7}{s^2} - \frac{1}{s-1} + \frac{5/2}{s+2}.$$

The inverse transform now yields

$$x(t) = 2 - \frac{1}{2}e^t - e^{-2t} - \frac{1}{2}e^{3t},$$

$$y(t) = 5 + 7t - e^t + \frac{5}{2}e^{-2t}.$$

3. Applying the Laplace transform directly gives

$$su(s) - 3 - 2u(s) - sv(s) - v(s) = \frac{6}{s-3},$$

$$2su(s) - 6 - 3u(s) + sv(s) - 3v(s) = \frac{6}{s-3},$$

or

$$(s - 2)u(s) - (s + 1)v(s) = \frac{3(s-1)}{s-3},$$

$$(2s - 3)u(s) + (s - 3)v(s) = \frac{6(s-2)}{s-3}.$$

Solving for $u(s)$ and $v(s)$ gives us

$$u(s) = \frac{3s^2 - 6s - 1}{(s-1)^2(s-3)} = \frac{1}{s-1} + \frac{2}{(s-1)^2} + \frac{2}{s-3},$$

$$v(s) = \frac{-3s + 5}{(s-1)^2(s-3)} = \frac{1}{s-1} - \frac{1}{(s-1)^2} - \frac{1}{s-3}.$$

The inverse transform now yields

$$x(t) = e^t + 2te^t + 2e^{3t},$$
$$y(t) = e^t - te^t - e^{3t}.$$

5. Applying the Laplace transform directly gives

$$s^2 u(s) - u(s) + 5sv(s) - 5 = \frac{1}{s^2},$$
$$s^2 v(s) - s - 4v(s) - 2su(s) = -\frac{2}{s},$$

or

$$(s^2 - 1)u(s) + 5sv(s) = \frac{5s^2 + 1}{s^2},$$
$$-2su(s) + (s^2 - 4)v(s) = \frac{s^2 - 2}{s}.$$

Solving for $u(s)$ and $v(s)$ gives us

$$u(s) = \frac{-9s^2 - 4}{s^2(s^2 + 1)(s^2 + 4)} = -\frac{1}{s^2} - \frac{5/3}{s^2 + 1} + \frac{8/3}{s^2 + 4},$$
$$v(s) = \frac{s^4 + 7s^2 + 4}{s(s^2 + 1)(s^2 + 4)} = \frac{1}{s} + \frac{\frac{2}{3}s}{s^2 + 1} - \frac{\frac{2}{3}s}{s^2 + 4}.$$

The inverse transform now yields

$$x(t) = -t - \frac{5}{3}\sin t + \frac{4}{3}\sin 2t,$$
$$y(t) = 1 + \frac{2}{3}\cos t - \frac{2}{3}\cos 2t.$$

7. Applying the Laplace transform directly gives

$$s^2 u(s) - 1 + sv(s) - v(s) = 0,$$
$$2su(s) - u(s) + sw(s) - w(s) = 0,$$
$$su(s) + 3u(s) + sv(s) - 4v(s) + 3w(s) = 0,$$

or

$$s^2 u(s) + (s - 1)v(s) = 1,$$
$$(2s - 1)u(s) + (s - 1)w(s) = 0,$$
$$(s + 3)u(s) + (s - 4)v(s) + 3w(s) = 0.$$

Solving for $u(s)$, $v(s)$, and $w(s)$ gives us

$$u(s) = \frac{1}{s(s-1)} = -\frac{1}{s} + \frac{1}{s-1},$$

$$v(s) = -\frac{1}{(s-1)^2},$$

$$w(s) = \frac{1-2s}{s(s-1)^2} = \frac{1}{s} - \frac{1}{s-1} - \frac{1}{(s-1)^2}.$$

The inverse transform now yields

$$x(t) = -1 + e^t,$$

$$y(t) = -te^t,$$

$$z(t) = 1 - e^t - te^t.$$

9. We have

$$u(s) = \frac{c_1 s + c_2}{s^2 - 1} + \frac{sf(s) + g(s)}{s^2 - 1},$$

$$v(s) = \frac{c_2 s + c_1}{s^2 - 1} + \frac{sg(s) + f(s)}{s^2 - 1}.$$

The inverse transform now yields

$$x(t) = c_1 \cosh t + c_2 \sinh t + \int_0^t \left[\cosh \beta F(t - \beta) + \sinh \beta G(t - \beta) \right] d\beta,$$

$$y(t) = c_1 \sinh t + c_2 \cosh t + \int_0^t \left[\cosh \beta G(t - \beta) + \sinh \beta F(t - \beta) \right] d\beta.$$

11. Applying the Laplace transform directly gives

$$su(s) - 1 - 2v(s) = f(s),$$
$$sv(s) + 2u(s) = g(s),$$

or

$$su(s) - 2v(s) = f(s) + 1,$$
$$2u(s) + sv(s) = g(s).$$

Solving for $u(s)$ and $v(s)$ gives us

$$u(s) = \frac{sf(s) + s + 2g(s)}{s^2 + 4} = \frac{s}{s^2 + 4} + \frac{sf(s)}{s^2 + 4} + \frac{2g(s)}{s^2 + 4},$$

$$v(s) = \frac{sg(s) - 2 - 2f(s)}{s^2 + 4} = -\frac{2}{s^2 + 4} + \frac{sg(s)}{s^2 + 4} - \frac{2f(s)}{s^2 + 4}.$$

The inverse transform now yields

$$x(t) = \cos 2t + \int_0^t \left[\cos 2\beta F(t - \beta) + \sin 2\beta G(t - \beta)\right] d\beta,$$

$$y(t) = -\sin 2t + \int_0^t \left[\cos 2\beta G(t - \beta) - \sin 2\beta F(t - \beta)\right] d\beta.$$

13. Applying the Laplace transform directly gives

$$su(s) - c_1 = au(s) + bv(s) + L\{f(t)\},$$
$$sv(s) - c_2 = cu(s) + dv(s) + L\{g(t)\},$$

or

$$(s - a)u(s) - bv(s) = c_1 + L\{f(t)\},$$
$$-cu(s) + (s - d)v(s) = c_2 + L\{g(t)\}.$$

Solving for $u(s)$ and $v(s)$ gives us

$$u(s) = \frac{\begin{vmatrix} c_1 + L\{f(t)\} & -b \\ c_2 + L\{g(t)\} & s - d \end{vmatrix}}{s^2 - (a+d)s + ad - bc} = \frac{\begin{vmatrix} c_1 & -b \\ c_2 & s - d \end{vmatrix}}{s^2 - (a+d)s + ad - bc} + \frac{\begin{vmatrix} L\{f(t)\} & -b \\ L\{g(t)\} & s - d \end{vmatrix}}{s^2 - (a+d)s + ad - bc},$$

$$v(s) = \frac{\begin{vmatrix} s - a & c_1 + L\{f(t)\} \\ -c & c_2 + L\{g(t)\} \end{vmatrix}}{s^2 - (a+d)s + ad - bc} = \frac{\begin{vmatrix} s - a & c_1 \\ -c & c_2 \end{vmatrix}}{s^2 - (a+d)s + ad - bc} + \frac{\begin{vmatrix} s - a & L\{f(t)\} \\ -c & L\{g(t)\} \end{vmatrix}}{s^2 - (a+d)s + ad - bc}.$$

We note that the first term on the right side of each equation depends on c_1 and c_2, but is independent of $L\{f(t)\}$ and $L\{g(t)\}$. Also the second term on the right side of each equation is independent of c_1 and c_2, but depends on $L\{f(t)\}$ and $L\{g(t)\}$. It follows that

$$x(t) = x_c(t) + x_p(t),$$
$$y(t) = y_c(t) + y_p(t),$$

where $x_c(t)$ and $y_c(t)$ depend on c_1 and c_2 whereas $x_p(t)$ and $y_p(t)$ depend on $f(t)$ and $g(t)$.

15. Kirchhoff's laws require that the following system of equations be satisfied:

$$R_1 I_1 + R_2 I_2 + L_2 \frac{dI_2}{dt} = E,$$

$$R_1 I_1 + R_3 I_3 + \frac{1}{C_3} Q_3 = E,$$

$$I_1 = I_2 + I_3,$$

$$\frac{dQ_3}{dt} = I_3.$$

An application of the Laplace transform gives

$$R_1 i_1 + R_2 i_2 + L_2 s i_2 = E/s,$$

$$R_1 i_1 + R_3 i_3 + \frac{1}{C_3} q_3 = E/s,$$

$$i_1 = i_2 + i_3,$$

$$s q_3 = i_3.$$

Multiplying the second equation by sC_3 and substituting for sq_3 from the fourth equation allows us to write

$$R_1 i_1 + (R_2 + L_2 s)i_2 = E/s,$$

$$C_3 R_1 s i_1 + (C_3 R_3 s + 1)i_3 = C_3 E,$$

$$i_1 - i_2 - i_3 = 0.$$

The nature of the solutions of this system will depend on the value of the determinant

$$\begin{vmatrix} R_1 & R_2 + L_2 s & 0 \\ C_3 R_1 s & 0 & C_3 R_3 s + 1 \\ 1 & -1 & -1 \end{vmatrix} = \begin{vmatrix} R_1 & R_1 + R_2 + L_2 s & R_1 \\ C_3 R_1 s & C_3 R_1 s & C_3 (R_1 + R_3)s + 1 \\ 1 & 0 & 0 \end{vmatrix}$$

$$= \begin{vmatrix} L_2 s + R_1 + R_2 & R_1 \\ C_3 R_1 s & C_3 (R_1 + R_3)s + 1 \end{vmatrix}$$

$$= C_3 L_2 (R_1 + R_3)s^2 + [C_3 (R_1 R_2 + R_2 R_3 + R_3 R_1) + L_2]s + R_1 + R_2.$$

Chapter 16

Nonlinear Equations

16.2   Factoring the Left Member

1. $x^2p^2 - y^2 = (xp - y)(xp + y) = 0$. It follows that

$$xp = y \qquad\qquad \text{or} \qquad\qquad xp = -y,$$

$$\frac{p}{y} = \frac{1}{x} \qquad\qquad \text{or} \qquad\qquad \frac{p}{y} = -\frac{1}{x},$$

$$y = c_1 x \qquad\qquad \text{or} \qquad\qquad y = c_2/x.$$

3. $x^2p^2 - 5xyp + 6y^2 = (xp - 2y)(xp - 3y) = 0$. It follows that

$$xp = 2y \qquad\qquad \text{or} \qquad\qquad xp = 3y,$$

$$\frac{p}{y} = \frac{2}{x} \qquad\qquad \text{or} \qquad\qquad \frac{p}{y} = \frac{3}{x},$$

$$y = c_1 x^2 \qquad\qquad \text{or} \qquad\qquad y = c_2 x^3.$$

5. $xp^2 + (1 - x^2y)p - xy = (xp + 1)(p - xy) = 0$. It follows that

$$xp = -1 \qquad\qquad \text{or} \qquad\qquad p = xy,$$

$$p = -\frac{1}{x} \qquad\qquad \text{or} \qquad\qquad \frac{p}{y} = x,$$

$$y = -\ln|c_2 x| \qquad\qquad \text{or} \qquad\qquad y = c_1 \exp\left(\frac{1}{2}x^2\right).$$

7. $xp^2 - (1 + xy)p + y = (xp - 1)(p - y) = 0$. It follows that

$$p = \frac{1}{x} \qquad\qquad \text{or} \qquad\qquad \frac{p}{y} = 1,$$

$$y = \ln|c_1 x| \qquad\qquad \text{or} \qquad\qquad x = \ln|c_1 y|.$$

9. $(x + y)^2p^2 - y^2 = \left[(x + y)p - y\right]\left[(x + y)p + y\right] = 0$. It follows that

$$(x + y)p - y = 0 \qquad\qquad \text{or} \qquad\qquad (x + y)p + y = 0.$$

Each of these equations is homogeneous, so we substitute $y = vx$ in order to separate the variables. The resulting equations are

$$\frac{dx}{x} + \frac{(v+1)dv}{v^2} = 0 \qquad \text{or} \qquad \frac{dx}{x} + \frac{(v+1)dv}{v^2 + 2v} = 0,$$

$$v \ln|c_1 xv| = 1 \qquad \text{or} \qquad \ln|x^2(v^2 + 2v)| = 2\ln|a|,$$

$$x = y \ln|c_1 y| \qquad \text{or} \qquad y(2x + y) = c_2.$$

11. $p^2 - xy(x + y)p + x^3 y^3 = (p - x^2 y)(p - xy^2) = 0$. It follows that

$$p = x^2 y \qquad \text{or} \qquad p = xy^2,$$

$$\frac{p}{y} = x^2 \qquad \text{or} \qquad \frac{p}{y^2} = x,$$

$$x^3 = 3\ln|c_2 y| \qquad \text{or} \qquad y(x^2 + c_1) = -2.$$

13. $(x - y)^2 p^2 - y^2 = \big[(x - y)p - y\big]\big[(x - y)p + y\big] = 0$. It follows that

$$(x - y)p - y = 0 \qquad \text{or} \qquad (x - y)p + y = 0.$$

Each of these equations is homogeneous, so we substitute $y = vx$ in order to separate the variables. The resulting equations are

$$-\frac{dx}{x} + \frac{(1 - v)dv}{v^2} = 0 \qquad \text{or} \qquad \frac{dx}{x} + \frac{(1 - v)dv}{2v - v^2} = 0,$$

$$-\frac{1}{v} = \ln|c_1 xv| \qquad \text{or} \qquad x^2(2v - v^2) = c_2,$$

$$x = -y \ln|c_1 y| \qquad \text{or} \qquad y(2x - y) = c_2.$$

15. $(x^2 + y^2)^2 p^2 - 4x^2 y^2 = \big[(x^2 + y^2)p - 2xy\big]\big[(x^2 + y^2)p + 2xy\big] = 0$. It follows that

$$(x^2 + y^2)p - 2xy = 0 \qquad \text{or} \qquad (x^2 + y^2)p + 2xy = 0.$$

Each of these equations is homogeneous, so we substitute $y = vx$ in order to separate the variables. The resulting equations are

$$\frac{dx}{x} + \frac{(1 + v^2)dv}{v^3 - v} = 0 \qquad \text{or} \qquad \frac{dx}{x} + \frac{(1 + v^2)dv}{v^3 + 3v} = 0,$$

$$\ln|x(v^2 - 1)| = \ln|av| \qquad \text{or} \qquad \ln|x^3(v^3 + 3v)| = \ln|b|,$$

$$x(v^2 - 1) = c_1 v \qquad \text{or} \qquad x^3(v^3 + 3v) = c_2,$$

$$y^2 - x^2 = c_1 y \qquad \text{or} \qquad y(3x^2 + y^2) = c_2.$$

17. $xy(x^2 + y^2)p^2 - (x^4 + x^2 y^2 + y^4)p - xy(x^2 + y^2) = \big[(xyp - (x^2 + y^2)\big]\big[(x^2 + y^2)p + xy\big] = 0$. It follows that

$$xyp - (x^2 + y^2) = 0 \qquad \text{or} \qquad (x^2 + y^2)p + xy = 0.$$

Each of these equations is homogeneous, so we substitute $y = vx$ in order to separate the variables. The resulting equations are

$$-\frac{dx}{x} + v\,dv = 0 \qquad \text{or} \qquad \frac{dx}{x} + \frac{(1+v^2)dv}{v^3 + 2v} = 0,$$

$$-\ln|c_2 x| + \frac{1}{2}v^2 = 0 \qquad \text{or} \qquad x^4 v^2(v^2 + 2) = c_1,$$

$$y^2 = 2x^2 \ln|c_2 x| \qquad \text{or} \qquad y^2(y^2 + 2x^2) = c_1.$$

19. In order for the solution to pass through the point $(1, 1)$ the solution must have the form $y = \sqrt{-2(x - c_1)}$ and $1 = \sqrt{-2(1 - c_1)}$. Therefore $c_1 = 3/2$ and $y = \sqrt{3 - 2x}$, valid for $x < 3/2$.

21. Let

$$F(x) = \sqrt{2 - 2x}, \text{ for } x \le 1, \qquad \text{and} \qquad F'(x) = \frac{-1}{\sqrt{2 - 2x}}, \text{ for } x < 1,$$

$$F(x) = -\ln x, \text{ for } x \ge 1, \qquad \text{and} \qquad F'(x) = \frac{-1}{x}, \text{ for } x > 1.$$

Note that $F(x)$ is defined for all x, but $F'(x)$ is not defined for $x = 1$. We have for $x < 1$,

$$xyp^2 + (x + y)p + 1 = \frac{x\sqrt{2 - 2x}}{2 - 2x} - \frac{x + \sqrt{2 - 2x}}{\sqrt{2 - 2x}} + 1 = 0.$$

And for $x > 1$

$$xyp^2 + (x + y)p + 1 = -\frac{x \ln x}{x} - \frac{x - \ln x}{x} + 1 = 0.$$

Therefore $F(x)$ is a solution of the differential equation so long as $x \ne 1$.

23. Let

$$G(x) = \sqrt{3 - 2x}, \text{ for } x \le 1, \qquad \text{and} \qquad G'(x) = \frac{-1}{\sqrt{3 - 2x}}, \text{ for } x < 1,$$

$$G(x) = 1 - \ln x, \text{ for } x \ge 1, \qquad \text{and} \qquad G'(x) = \frac{-1}{x}, \text{ for } x > 1.$$

Note that $G(x)$ and $G'(x)$ are defined for all x. Substituting $y = G$ and $p = G'$ into the differential equation leads us to

$$xyp^2 + (x + y)p + 1 = \frac{x\sqrt{3 - 2x}}{3 - 2x} - \frac{x + \sqrt{3 - 2x}}{\sqrt{3 - 2x}} + 1 = 0, \quad \text{for } x \le 1,$$

$$xyp^2 + (x + y)p + 1 = \frac{x(1 - \ln x)}{x^2} - \frac{x + 1 - \ln x}{x} + 1 = 0, \quad \text{for } x \ge 1.$$

Therefore $G(x)$ is a solution of the differential equation for all x.

25. From equation (5) we seek a solution $y = \sqrt{-2(x - c_1)}$ that passes through $(-\frac{1}{2}, 2)$. We must have $2 = \sqrt{-2(-\frac{1}{2} - c_1)}$ or $c_1 = 3/2$. Thus $y = \sqrt{3 - 2x}$, valid for $x < 3/2$. From the equation (6) we seek a solution $y = -\ln|c_2 x|$ that passes through $(-\frac{1}{2}, 2)$. This requires that $2 = -\ln|-\frac{1}{2}c_2|$ or $-\frac{1}{2}c_2 = e^{-2}$, so that $c_2 = -2e^{-2}$ and $y = -\ln|-2e^{-2}x| = 2-\ln 2-\ln(-x)$, valid for $x < 0$. Each of these solutions is valid on the interval $-1 < x < -\frac{1}{4}$.

27. The function $y = \sqrt{3 - 2x}$ is the only solution of the equation that is valid for $x < 3/2$ and passes through the point $(-\frac{1}{2}, 2)$. In order to define a function that is a solution on the interval $-1 < x < 2$ we must find a function of the form $y = -\ln|c_2 x|$ that passes through the point $(1, 1)$. We require that $1 = -\ln|c_2|$; that is, $c_2 = e^{-1}$ and $y = -\ln|e^{-1}x| = 1 - \ln x$ for $x \geq 1$. We now define the function

$$G(x) = \sqrt{3 - 2x}, \quad \text{for } x \leq 1,$$
$$= 1 - \ln x, \quad \text{for } x \geq 1.$$

This is the function of Figure 16.1

16.5 The p-Discriminant Equation

1. $f = Ap^2 + Bp + C = 0$, $\dfrac{\partial f}{\partial p} = 2Ap + B = 0$. If we multiply the first equation by 2, the second equation by p, and subtract the second from the first, we obtain the pair of equations $2Ap + B = 0$, $Bp + 2C = 0$. Multiplying the first of these equations by B, the second by $2A$, and subtracting yields $B^2 - 4AC = 0$.

3. $f = p^3 + Ap^2 + B = 0$, $\dfrac{\partial f}{\partial p} = 3p^2 + 2Ap = 0$. If we multiply the first equation by 3, the second by p, and subtract we obtain the pair of equations $3p^2 + 2Ap = 0$, $Ap^2 + 3B = 0$. Elimination of p^2 from these equations yields $2A^2 p = 9B$, giving us the pair of equations $4A^2 p^2 - 81B^2 = 0$, $Ap^2 + 3B = 0$. Elimination of p^2 from these equations gives the desired result $B(4A^3 + 27B) = 0$.

5. $xyp^2 + (x + y)p + 1 = 0$. Solving for p we have

$$p = \frac{-(x + y) \pm \sqrt{(x + y)^2 - 4xy}}{2xy} = \frac{-(x + y) \pm \sqrt{(x - y)^2}}{2xy}.$$

Thus the condition that the equation have equal roots in p is $(x - y)^2 = 0$. But for $y = x$ we have $p = 1$ so that $xyp^2 + (x + y)p + 1 = x^2 + 2x + 1 \neq 0$. Hence $y = x$ is not a solution of the differential equation. There are no singular solutions.

7. The function has as its graph the left half of the semicircle $y = \left[a^2 - (x+2a)^2\right]^{1/2}$ and the right half of the semicircle $y = \left[a^2 - (x - 2a)^2\right]^{1/2}$, connected by the line segment $y = a$ between $x = -2a$ and $x = 2a$. Note that at the points where these arcs are joined, the slopes of the arcs are both zero, so that the given function is a solution for the entire interval $-3a \leq x \leq 3a$.

9. $3x^4p^2 - xp - y = 0$. From Exercise 1 the p-discriminant equation is

$$B^2 - 4AC = x^2 + 12x^4y = x^2(1 + 12x^2y) = 0.$$

Consider the function $y = \dfrac{-1}{12x^2}$ with $p = \dfrac{1}{6x^3}$. Substituting this function into the differential equation gives $3x^4p^2 - xp - y = \dfrac{3x^4}{36x^6} - \dfrac{x}{6x^3} + \dfrac{1}{12x^2} = 0$. That is, one solution of the differential equation is given by $12x^2y = -1$.

11. $p^2 - xp + y = 0$. The p-discriminant equation is $B^2 - 4AC = x^2 - 4y = 0$. For $y = x^2/4$, $p = x/2$ and $p^2 - xp + y = \dfrac{x^2}{4} - \dfrac{x^2}{2} + \dfrac{x^2}{4} = 0$. Thus $y = x^2/4$ is a solution of the differential equation.

13. $4y^3p^2 - 4xp + y = 0$. The p-discriminant equation is $B^2 - 4AC = 16(x - y^2)(x + y^2) = 0$. Rather than testing each of the four parabolic arcs $y = \sqrt{x}$, $y = -\sqrt{x}$, $y = \sqrt{-x}$, and $y = -\sqrt{-x}$ individually to see if they represent solutions of the differential equation, we note that $y^4 = x^2$ implies that $2y^3p = x$. Multiplying the differential equation by y^3 and checking $4y^6p^2 - 4xy^3p + y^4 = x^2 - 2x^2 + x^2 = 0$, we see that each of the four functions represents a solution of the differential equation.

15. $y^2p^3 - 6xp + 2y = 0$. In order to use the formula given in Exercise 2, we need to divide first by y^4. We must therefore check to see if $y = 0$ is itself a solution, and indeed it is. Then the differential equation becomes $p^3 - \dfrac{6xp}{y^4} + \dfrac{2}{y^3} = 0$. Now the p-discriminant equation is given by

$$4A^3 + 27B^2 = \frac{108}{y^{12}}(y^2 - 2x)(y^4 + 2xy^2 + 4x^2) = 0.$$

For the parabola $y^2 = 2x$ we have $yp = 1$. If we multiply the differential equation by y we may then write for the parabola $y^5p^3 - 6xyp + 2y^2 = 2x - 6x + 4x = 0$. Thus both $y = \sqrt{2x}$ and $y = -\sqrt{2x}$ are solutions of the differential equation.

17. The quadratic equation $c^2 + cxy + 4x = 0$ has equal roots when $x^2y^2 - 16x = 0$. That is, when $xy^2 = 16$. This the same as the singular solution of the differential equation.

16.7   Clairaut's Equation

1. Here we have

$$x = -f'(\alpha) \qquad\qquad \text{and} \qquad\qquad y = f(\alpha) - \alpha f'(\alpha),$$
$$dx = -f''(\alpha)\, d\alpha \qquad\qquad \text{and} \qquad\qquad dy = -\alpha f''(\alpha)\, d\alpha.$$

It follows that $p = \dfrac{dy}{dx} = \alpha$ and

$$y - px - f(p) = f(\alpha) - \alpha f'(\alpha) + \alpha f'(\alpha) - f(\alpha) = 0.$$

That is, the original equations are the parametric form of a solution of the differential equation $y = px + f(p)$.

3. $p^2 + x^3 p - 2x^2 y = 0$. Dividing by x^2 and differentiating with respect to x eliminates y and yields $x^2(2p + x^3)\dfrac{dp}{dx} = px(2p + x^3)$. Thus $x\dfrac{dp}{dx} = p$ or $p = -x^3/2$. We solve this differential equation and obtain $p = cx$. Substituting $p = cx$ into the original differential equation yields $2y = c^2 + cx^2$. Substituting $p = -x^3/2$ into the original differential equation yields $8y = -x^4$.

5. $2xp^3 - 6yp^2 + x^4 = 0$. Dividing by p^2 and differentiating with respect to x eliminates y and yields $x(p^3 - x^3)\dfrac{dp}{dx} = 2p(p^3 - x^3)$. Thus $x\dfrac{dp}{dx} = 2p$ or $p = x$. We solve this differential equation and obtain $p = cx^2$. Substituting $p = cx^2$ into the original differential equation yields $2c^3 x^3 = 1 - 6c^2 y$. Substituting $p = x$ into the original differential equation yields $2y = x^2$.

7. $y = px + kp^2$. This is a Clairaut equation with $f(p) = kp^2$ and $f'(p) = 2kp$. The general solution is $y = cx + kc^2$, and the singular solution is given parametrically by $x = -2xp$, $y = -kp^2$. Eliminating p from these equations gives us $x^2 = -4ky$.

9. $x^4 p^2 + 2x^3 yp - 4 = 0$. Dividing by $x^3 y$ and differentiating with respect to x eliminates y and yields $-x(x^4 p^2 + 4)\dfrac{dp}{dx} = 3p(x^4 p^2 + 4)$. Since $x^4 p^2 + 4$ cannot be zero we have $x\dfrac{dp}{dx} + 3p = 0$. We solve this differential equation and obtain $p = k/x^3$. Substituting $p = k/x^3$ into the original differential equation yields $x^2(1 + cy) = c^2$, where $c = -k/2$. There is no singular solution.

11. $3x^4 p^2 - xp - y = 0$. Differentiating both sides with respect to x eliminates y and results in $x(6x^3 p - 1)\dfrac{dp}{dx} + 2p(6x^3 p - 1) = 0$. Thus $x\dfrac{dp}{dx} + 2p = 0$ or $p = 1/(6x^3)$. We solve this differential equation and obtain $p = c/x^2$. Substituting $p = c/x^2$ into the original differential equation yields $xy = c(3cx - 1)$. Substituting $p = 1/(6x^3)$ into the original differential equation yields $12x^2 y = -1$.

13. $p(xp - y + k) + a = 0$. We can write this equation in the Clairaut form $y = xp + k + a/p$. The general solution is $y = cx + k + a/c$ or $c(cx - y + k) + a = 0$. Here $f(p) = k + a/p$ and $f'(p) = -a/p^2$. Parametric equations for the singular solution are $x = a/p^2$, $y = k + 2a/p$. It follows that $(y - k)^2 = 4ax$.

15. $y = x^6 p^3 - xp$. Differentiating with respect to x eliminates y and gives us the differential equation $x(3x^5 p^2 - 1)\dfrac{dp}{dx} + 2p(3x^5 p^2 - 1) = 0$. Thus $x\dfrac{dp}{dx} + 2p = 0$ or $p^2 = 1/(3x^5)$. We solve this differential equation and obtain $p = c/x^2$. Substituting $p = c/x^2$ into the original differential equation gives us $xy = c(c^2 x - 1)$. The original differential equation may be replaced by $y^2 = x^{12} p^6 - 2x^7 p^4 + x^2 p^2$. Substituting $p = 1/(3x^5)$ into this differential equation yields $27x^3 y^2 = 4$.

17. $xp^3 - yp^2 + 1 = 0$. We can write this equation in the Clairaut form $y = px + 1/p^2$. The general solution is $y = cx + 1/c^2$ or $c^2 y = c^3 x + 1$. Parametric equations for the singular solution are $x = \dfrac{2}{p^3}$, $y = \dfrac{1}{p^2} + \dfrac{2}{p^2} = \dfrac{3}{p^2}$. It follows that $4y^3 = 27x^2$.

19. $p^2 - xp - y = 0$. Differentiating both sides with respect to x eliminates y and gives us the differential equation $(2p - x)\dfrac{dp}{dx} = 2p$. We solve this homogeneous differential equation and

obtain $3x = 2p + kp^{(-1/2)}$. Substituting this value for $3x$ into the original differential equation yields a second equation $3y = p^2 - kp^{1/2}$. The last two equations give the general solution in terms of the parameter p.

21. $2p^2 + xp - 2y = 0$. Differentiating both sides with respect to x eliminates y and gives us the differential equation $\dfrac{dx}{dp} - \dfrac{1}{p}x = 4$. We solve this linear differential equation and obtain $x = 4p \ln|pc|$. Substituting this value for x into the original differential equation yields a second equation $y = p^2[1 + 2\ln|pc|]$. The last two equations give the general solution in terms of the parameter p.

23. $4xp^2 - 3yp + 3 = 0$. Dividing by p and differentiating both sides with respect to x eliminates y and gives us the differential equation $\dfrac{dx}{dp} + \dfrac{4}{p}x = \dfrac{3}{p^3}$. We solve this linear differential equation and obtain $2x = 3p^{-2} + cp^{-4}$. Substituting this value for $2x$ into the original differential equation yields a second equation $3y = 9p^{-1} + 2cp^{-3}$. The last two equations give the general solution in terms of the parameter p.

25. $5p^2 + 6xp - 2y = 0$. Differentiating both sides with respect to x eliminates y and gives us the differential equation $\dfrac{dx}{dp} + \dfrac{3}{2p}x = -\dfrac{5}{2}$. We solve this linear differential equation and obtain $x = -p + kp^{(-3/2)}$. The form of this solution may be changed to get $p^3(x + p)^2 = c$, which, when taken with the original differential equation, gives parametric equations for the solution.

27. $5p^2 + 3xp - y = 0$. Differentiating both sides with respect to x eliminates y and gives us the differential equation $\dfrac{dx}{dp} + \dfrac{3}{2p}x = -5$. We solve this linear differential equation and obtain $x = -2p + kp^{(-3/2)}$. The form of this solution may be changed to get $p^3(x + 2p)^2 = c$, which, when taken with the original differential equation, gives parametric equations for the solution.

29. $y = xp + x^3p^2$. Differentiating both sides of this differential equation eliminates y and yields $3xp^2\dfrac{dx}{dp} + 2x^2p + 1 = 0$. Substituting $v = x^2$ changes this equation into $\dfrac{dv}{dp} + \dfrac{4v}{3p} = -\dfrac{2}{3}p^{-2}$. We solve this linear differential equation and obtain $v = x^2 = cp^{(-4/3)} - 2p^{-1}$, which, when taken with the original differential equation, gives parametric equations for the solution.

16.9   Independent Variable Missing

1. $y'' = x(y')^3$. Since y does not appear explicitly we set $p = y'$ and get a differential equation we must solve for p. Then an integration yields y. We get

$$\frac{dp}{p^3} = x\,dx, \qquad \text{and} \qquad p = \frac{dy}{dx} = \frac{\pm 1}{\sqrt{c_3^2 - x^2}},$$

$$y + c_2 = \pm \arcsin\left(\frac{x}{c_3}\right), \qquad \text{and} \qquad x = c_1 \sin(y + c_2).$$

3. $y^2 y'' + (y')^3 = 0$. Since x does not appear explicitly we set $p = y'$ and $y'' = p\dfrac{dp}{dx}$ to get a differential equation we must solve for p. Then an integration yields y. We get

$$\frac{dp}{p^2} + \frac{dy}{y^2} = 0, \qquad \text{and} \qquad p = \frac{dy}{dx} = \frac{y}{c_1 y - 1},$$

$$x = c_1 y - \ln|c_2 y|.$$

5. $2a y'' + (y')^3 = 0$. Since y does not appear explicitly we set $p = y'$ and get a differential equation we must solve for p. Then an integration yields y. We get

$$2a\frac{dp}{p^3} + dx = 0, \qquad \text{and} \qquad p = \frac{dy}{dx} = \frac{\pm\sqrt{a}}{\sqrt{x - c_1}},$$

$$y = \pm 2\sqrt{a}\sqrt{x - c_1} + c_2, \qquad \text{and} \qquad (y - c_2)^2 = 4a(x - c_1).$$

7. $y'' = 2y(y')^3$. Since x does not appear explicitly we set $p = y'$ and $y'' = p\dfrac{dp}{dx}$ to get a differential equation we must solve for p. Then an integration yields y. We get

$$\frac{dp}{p^2} = 2y\, dy, \qquad \text{and} \qquad p = \frac{dy}{dx} = -\frac{1}{y^2 + c_1},$$

$$y^3 = 3(c_2 - x - c_1 y).$$

9. $y y'' + (y')^3 = 0$. Since x does not appear explicitly we set $p = y'$ and $y'' = p\dfrac{dp}{dx}$ to get a differential equation we must solve for p. Then an integration yields y. We get

$$\frac{dp}{p^2} + \frac{dy}{y} = 0, \qquad \text{and} \qquad p = \frac{dy}{dx} = \frac{1}{\ln|c_2 y|},$$

$$x = c_1 + y\ln|c_2 y| - y.$$

11. $x^3 y'' - x^2 y' = 3 - x^2$. Since y does not appear explicitly we set $p = y'$ and get a differential equation we must solve for p. Then an integration yields y. We get

$$p' - \frac{p}{x} = \frac{3}{x^3} - \frac{1}{x} \qquad \text{and} \qquad p = \frac{dy}{dx} = -x^{-2} + 1 + c_3 x,$$

$$y = x^{-1} + x + c_1 x^2 + c_2.$$

13. $y'' = e^x (y')^2$. Since y does not appear explicitly we set $p = y'$ and get a differential equation we must solve for p. Then an integration yields y. We get

$$\frac{dp}{p^2} = e^x dx \qquad \text{and} \qquad p = \frac{dy}{dx} = \frac{1}{c_1 - e^x} = \frac{e^{-x}}{c_1 e^{-x} - 1},$$

$$y = -\frac{1}{c_1}\ln|c_1 e^{-x} - 1| - c_3, \qquad \text{and} \qquad c_1 y + c_2 = -\ln|c_1 e^{-x} - 1|.$$

15. $y'' = 1 + (y')^2$. Since y does not appear explicitly we set $p = y'$ and get a differential equation we must solve for p. Then an integration yields y. We get

$$\frac{dp}{1+p^2} = dx, \qquad \text{and} \qquad p = \frac{dy}{dx} = \tan(x + c_1),$$

$$y = -\ln|\cos(x + c_1)| + \ln c_3, \qquad \text{and} \qquad e^y \cos(x + c_1) = c_2.$$

17. $(1 + y^2)y'' + (y')^3 + y' = 0$. Since x does not appear explicitly we set $p = y'$ and $y'' = p\dfrac{dp}{dy}$ to get a differential equation we must solve for p. Then an integration yields y. We get

$$\frac{dp}{p^2 + 1} + \frac{dy}{y^2 + 1} = 0, \qquad \text{and} \qquad p = \frac{dy}{dx} = \frac{y + c_1}{c_1 y - 1},$$

$$x = c_2 + c_1 y - (1 + c_1^2)\ln|y + c_1|.$$

19. $xy'' = y'(2 - 3xy')$. Since y does not appear explicitly we set $p = y'$ and get a differential equation we must solve for p. Then an integration yields y. We get

$$\frac{dp}{dx} - \frac{2p}{x} = -3p^2 \quad \text{(Bernoulli)} \qquad \text{and} \qquad p = \frac{dy}{dx} = \frac{x^2}{x^3 + c_1},$$

$$3y = \ln|x^3 + c_1| + c_2.$$

21. $(y'')^2 - xy'' + y' = 0$. We set $w = y'$, $w' = p$, and get $w = px - p^2$, a Clairaut equation. The general solution is $w = c_1 x - c_1^2$, so that $y = \frac{1}{2}c_1 x^2 - c_1^2 x + c_3$ or $2y = c_1 x^2 - 2c_1^2 x + c_2$. In the Clairaut equation $f(p) = -p^2$ and $f'(p) = -2p$. The parametric equations of the singular solution are $x = 2p$ and $w = -p^2 + 2p^2 = p^2$. Eliminating the parameter p gives us $w = \frac{1}{4}x^2$. It follows that $y = \frac{1}{12}x^3 + c_4$ or $12y = x^3 + k$ is a family of singular solutions.

23. $3yy'y'' = (y')^3 - 1$. Since x does not appear explicitly we set $p = y'$ and $y'' = p\dfrac{dp}{dx}$ to get a differential equation we must solve for p. Then an integration yields y. We get

$$\frac{3p^2\,dp}{p^3 - 1} = \frac{dy}{y}, \qquad \text{and} \qquad p = \frac{dy}{dx} = \frac{(y + c_1)^{1/3}}{c_1^{1/3}},$$

$$27c_1(y + c_1)^2 = 8(x + c_2)^3.$$

25. $x^2 y'' + (y')^2 - 2xy' = 0$; $y(2) = 5$, $y'(2) = -4$. Since y does not appear explicitly we set $p = y'$ to get a differential equation we must solve for p. We then use an initial condition to obtain c_1. Then an integration yields y. Finally, we determine the value of c_2. We have

$$\frac{dp}{dx} - \frac{2p}{x} = -\frac{p^2}{x^2} \quad \text{(Bernoulli)}, \qquad \text{and} \qquad p = \frac{x^2}{x + c_1};$$

$$p(2) = -4 \text{ gives us } c_1 = -3, \qquad \text{so that} \qquad p = \frac{dy}{dx} = \frac{x^2}{x - 3} = x + 3 + \frac{9}{x - 3};$$

$$y = \frac{x^2}{2} + 3x + 9\ln|x - 3| + c_2, \qquad \text{but} \qquad y(2) = 5, \text{ so that } c_2 = -3;$$

$$y = \frac{x^2}{2} + 3x + 9\ln|x - 3| - 3.$$

27. $xy'' = y' + x^5$; $y(1) = \frac{1}{2}$, $y'(1) = 1$. Since y does not appear explicitly we set $p = y'$ to get a differential equation we must solve for p. We then use an initial condition to obtain c_1. Then an integration yields y. Finally, we determine the value of c_2. We have

$$\frac{dp}{dx} - \frac{p}{x} = x^4, \qquad \text{and} \qquad p = \frac{x^5}{4} + c_1 x;$$

$$p(1) = 1 \text{ gives us } c_1 = \frac{3}{4}, \qquad \text{so that} \qquad p = \frac{dy}{dx} = \frac{x^5}{4} + \frac{3x}{4};$$

$$y = \frac{x^6}{24} + \frac{3x^2}{8} + c_2, \qquad \text{but} \qquad y(1) = \frac{1}{2}, \text{ so that } c_2 = \frac{1}{12};$$

$$y = \frac{x^6}{24} + \frac{3x^2}{8} + \frac{1}{12}, \qquad 24y = x^6 + 9x^2 + 2.$$

29. $y'' + \beta^2 y = 0$. Since x does not appear explicitly we set $p = y'$ and $y'' = p\dfrac{dp}{dx}$ to get a differential equation we must solve for p. Then an integration yields y. We get

$$p\,dp + \beta^2 y\,dy = 0, \qquad \text{and} \qquad p = \frac{dy}{dx} = \pm\sqrt{c_3^2 - \beta^2 y^2},$$

$$\arcsin \frac{\beta y}{c_3} = \beta(x + c_2), \qquad \text{and} \qquad y = c_1 \sin \beta(x + c_2).$$

31. $y'' = x(y')^2$; $y(0) = 1$, $y'(0) = \frac{1}{2}$. Since y does not appear explicitly we set $p = y'$ to get a differential equation we must solve for p. We then use an initial condition to obtain c_1. Then an integration yields y. Finally, we determine the value of c_2. We have

$$\frac{dp}{p^2} = x\,dx, \qquad \text{and} \qquad p = -\frac{2}{x^2 + c_1};$$

$$p(0) = \frac{1}{2} \text{ gives us } c_1 = -4, \qquad \text{so that} \qquad p = -\frac{2}{x^2 - 4} = \frac{1/2}{x+2} - \frac{1/2}{x-2};$$

$$y = \frac{1}{2}\ln\left|\frac{x+2}{x-2}\right| + c_2, \qquad \text{but} \qquad y(0) = 1, \text{ so that } c_2 = 1;$$

$$y = 1 + \frac{1}{2}\ln\left|\frac{x+2}{x-2}\right|.$$

33. $y'' = -e^{-2y}$; $y(3) = 0$, $y'(3) = -1$. Since x does not appear explicitly we set $p = y'$ and $y'' = p\dfrac{dp}{dy}$ to get a differential equation we must solve for p. We then use an initial condition to obtain c_1. Then an integration yields y. Finally, we determine the value of c_2. We have

$$p\,dp = -e^{-2y}\,dy, \qquad \text{and} \qquad p^2 = e^{-2y} + c_1;$$

$$p(3) = -1 \text{ gives us } c_1 = 0, \qquad \text{so that} \qquad p = -e^{-y};$$

$$e^y = -x + c_2, \qquad \text{but} \qquad y(3) = -1, \text{ so that } c_2 = 4;$$

$$e^y = 4 - x, \qquad \text{or} \qquad y = \ln(4 - x).$$

35. $2y'' = \sin 2y$; $y(0) = -\pi/2$, $y'(0) = 1$. Since x does not appear explicitly we set $p = y'$ and $y'' = p\dfrac{dp}{dy}$ to get a differential equation we must solve for p. We then use an initial condition to obtain c_1. Then an integration yields y. Finally, we determine the value of c_2. We have

$4p\,dp = 2\sin 2y\,dy,$	and	$2p^2 = -\cos 2y + c_1;$
$p(0) = 1$ and $y(0) = -\pi/2$ gives us $c_1 = 1,$	so that	$p = -\sin y;$
$-\ln|\csc y + \cot y| = -x + c_2,$	but	$y(0) = -\pi/2,$ so that $c_2 = 0;$
$x = \ln(-\csc y - \cot y).$		

37. $2y'' = (y')^3 \sin 2x$; $y(0) = 1$, $y'(0) = 1$. Since y does not appear explicitly we set $p = y'$ to get a differential equation we must solve for p. We then use an initial condition to obtain c_1. Then an integration yields y. Finally, we determine the value of c_2. We have

$\dfrac{2\,dp}{p^3} = \sin 2x\,dx,$	and	$-p^{-2} = -\dfrac{\cos 2x}{2} + c_1;$
$p(0) = 1$ gives us $c_1 = -\dfrac{1}{2},$	so that	$p^2 = \dfrac{2}{1+\cos 2x} = \sec^2 x;$
$y = \ln|\sec x + \tan x| + c_2,$	but	$y(0) = 1,$ so that $c_2 = 1;$
$y = 1 + \ln(\sec x + \tan x).$		

39. $yy'' = (y')^2[1 - y'\sin y - yy'\cos y]$. Since x does not appear explicitly we set $p = y'$ and $y'' = p\dfrac{dp}{dx}$ to get a differential equation we must solve for p. Then an integration yields y. We get

$\dfrac{dp}{dx} - \dfrac{p}{y} = p^2\left[\dfrac{-\sin y - y\cos y}{y}\right]$ (Bernoulli),	and	$p = \dfrac{dy}{dx} = \dfrac{y}{y\sin y + c_1},$
$x = c_1 \ln|c_2 y| - \cos y.$		

41. $x^2 y'' = y'(2x - y')$; $y(-1) = 5$, $y'(-1) = 1$. Since y does not appear explicitly we set $p = y'$ to get a differential equation we must solve for p. We then use an initial condition to obtain c_1. Then an integration yields y. Finally, we determine the value of c_2. We have

$\dfrac{dp}{dx} - \dfrac{2p}{x} = -\dfrac{p^2}{x^2},$ (Bernoulli)	and	$p = \dfrac{x^2}{x + c_1};$
$p(-1) = 1$ gives us $c_1 = 2,$	so that	$p = \dfrac{x^2}{x+2} = x - 2 + \dfrac{4}{x+2};$
$y = \dfrac{x^2}{2} - 2x + 4\ln|x+2| + c_2,$	but	$y(-1) = 5,$ so that $c_2 = 5/2;$
$y = \dfrac{x^2}{2} - 2x + 4\ln|x+2| + \dfrac{5}{2},$		$(2y - 1) = (x-2)^2 + 8\ln|x+2|.$

43. $(y'')^2 - 2y'' + (y')^2 - 2xy' + x^2 = 0$; $y(0) = \frac{1}{2}$, $y'(0) = 1$. Since y does not appear explicitly we set $p = y'$ and get $(p')^2 - 2p' + p^2 - 2xp + x^2 = 0$. This equation may be rewritten as

$(p'-1)^2 + (p-x)^2 = 1$. Here we observe that $(p-x)' = p'-1$. This observation suggests that we let $q = p - x$. This substitution gives us $q' = \pm\sqrt{1-q^2}$ and $q = \sin(x+c_1)$. Thus $p = x + \sin(x+c_1)$. The condition $p(0) = 1$ now forces us to choose $c_1 = \pi/2$, so that $p = x + \cos x$, and $y = \dfrac{x^2}{2} + \sin x + c_2$. The condition $y(0) = \frac{1}{2}$ now requires that $c_2 = \frac{1}{2}$, and finally $2y = 1 + x^2 + 2\sin x$.

Miscellaneous Exercises

1. $x^3p^2 + x^2yp + 4 = 0$. Dividing both sides of this equation by x^2p and differentiating both sides with respect to x eliminates y and gives $x(x^3p^2 - 4)\dfrac{dp}{dx} + 2p(x^3p^2 - 4) = 0$. Thus $x\dfrac{dp}{dx} + 2p = 0$ or $x^3p^2 = 4$. We solve this differential equation and obtain $p = c/x^2$. Substituting $p = c/x^2$ into the original differential equation gives $c^2 + cxy + 4x = 0$. Substituting $x^3p^2 = 4$ into the original differential equation gives $p = -\dfrac{8}{x^2y}$. Eliminating p from the last two equations gives us the singular solution $xy^2 = 16$.

3. $9p^2 + 3xy^4p + y^5 = 0$. Dividing both sides of this equation by $3y^4p$ and differentiating both sides with respect to x eliminates x and gives $3y^4(y^5 - 9p^2)\dfrac{dp}{dy} - 12y^3p(y^5 - 9p^2) = 0$. Remember that $\dfrac{dp}{dx} = p\dfrac{dp}{dy}$. Thus $3y^4\dfrac{dp}{dy} - 12y^3p = 0$ or $9p^2 = y^5$. We solve this differential equation and obtain $p = ky^4$. Substituting $p = ky^4$ into the original differential equation gives $cy^3(x-c) = 1$, where $c = -3k$. Substituting $9p^2 = y^5$ into the original differential equation gives $p = -\dfrac{2y}{3x}$. Eliminating p from the last two equations gives us the singular solution $x^2y^3 = 4$.

5. $x^6p^2 - 2xp - 4y = 0$. Differentiating both sides of this equation with respect to x eliminates y and gives $x(x^5p - 1)\dfrac{dp}{dx} + 3p(x^5p - 1) = 0$. Thus $x\dfrac{dp}{dx} + 3p = 0$ or $p = x^{-5}$. We solve this differential equation and obtain $p = -\dfrac{2c}{x^3}$. Substituting $p = -\dfrac{2c}{x^3}$ into the original differential equation yields $x^2(y - c^2) = c$. Substituting $p = x^{-5}$ into the original differential equation yields $4x^4y = -1$.

7. $5p^2 + 6xp - 2y = 0$. Differentiating both sides of this equation with respect to x eliminates y and gives $(10p + 6x)\dfrac{dp}{dx} + 4p = 0$. We solve this homogeneous differential equation and obtain $x = cp^{-3/2} - p$. Substituting into the original differential equation yields $2y = 6cp^{-1/2} - p^2$. The last two equations are parametric equations of the solution of the original differential equation.

9. $4x^5p^2 + 12x^4yp + 9 = 0$. Dividing both sides of this equation by x^4p and differentiating both sides with respect to x eliminates y and gives $x(4x^5p^2 - 9)\dfrac{dp}{dx} + 4p(4x^5p^2 - 9) = 0$. Thus $x\dfrac{dp}{dx} + 4p = 0$ or $4x^5p^2 = 9$. We solve this differential equation and obtain $p = \dfrac{k}{x^4}$.

Substituting $p = \dfrac{k}{x^4}$ into the original differential equation yields $x^3(2cy - 1) = c^2$, where $c = -2k/3$. Substituting $4x^5p^2 = 9$ into the original differential equation gives us $p = -\dfrac{3}{2x^4y}$. Eliminating p from the last two equations yields $x^3y^2 = 1$.

11. $p^4 + xp - 3y = 0$. Differentiating both sides of this equation with respect to x eliminates y and gives $\dfrac{dx}{dp} - \dfrac{x}{2p} = 2p^2$. We solve this linear differential equation and obtain $5x = 4p^3 + cp^{1/2}$. Substituting into the original differential equation yields $15y = 9p^4 + cp^{3/2}$. The last two equations are parametric equations of the solution of the original differential equation.

13. $x^2p^3 - 2xyp^2 + y^2p + 1 = 0$. Dividing by p and completing the square allows us to rewrite this equation as $(xp - y)^2 + \dfrac{1}{p} = 0$. We note that p must be negative, so that we can factor the left hand side to get $\left(xp - y + \dfrac{1}{\sqrt{-p}}\right)\left(xp - y - \dfrac{1}{\sqrt{-p}}\right) = 0$. Thus $y = px + \dfrac{1}{\sqrt{-p}}$ or $y = px - \dfrac{1}{\sqrt{-p}}$. Each of these equations is of Clairaut type so that their general solutions are $y = cx + \dfrac{1}{\sqrt{-c}}$ and $y = cx - \dfrac{1}{\sqrt{-c}}$. These two solutions can be combined to give a single equation in the form $x^2c^3 - 2xyc^2 + y^2c + 1 = 0$. For the first Clairaut differential equation $f(p) = (-p)^{-1/2}$ and $f'(p) = \dfrac{1}{2}(-p)^{-3/2}$, so that equations for the singular solution are given parametrically by $x = -\dfrac{1}{2}(-p)^{-3/2}$ and $y = \dfrac{3}{2}(-p)^{-1/2}$. Eliminating the parameter p from these equations gives $27x = -4y^3$.

15. $xp^2 - (x^2 + 1)p + x = 0$. Here it is possible to factor the equation into $(xp - 1)(p - x) = 0$. The first factor yields the solution $y = \ln|c_2x|$, the second $y = \dfrac{1}{2}x^2 + c_1$.

17. $p^3 - 2xp - y = 0$. Differentiating both sides of this equation with respect to x eliminates y and gives $\dfrac{dx}{dp} + \dfrac{2x}{3p} = p$. We solve this linear differential equation and obtain $8x = 3p^2 + cp^{-2/3}$. Substituting into the original differential equation yields $4y = p^3 - cp^{1/3}$. The last two equations are parametric equations of the solution of the original differential equation.

19. $x^2p^2 - (2xy + 1)p + y^2 + 1 = 0$. Completing the square allows us to rewrite this equation as $(xp - y)^2 = p - 1$. We note that $p - 1$ must be positive, so that $y = px + \sqrt{p - 1}$ or $y = px - \sqrt{p - 1}$. Each of these equations is of Clairaut type so that their general solutions are $y = cx + \sqrt{c - 1}$ and $y = cx - \sqrt{c - 1}$. These two solutions can be combined to give a single equation in the form $x^2c^2 - (2xy + 1)c + y^2 + 1 = 0$. For the first Clairaut differential equation $f(p) = (p - 1)^{1/2}$ and $f'(p) = \dfrac{1}{2}(p - 1)^{-1/2}$, so that equations for the singular solution are given parametrically by $x = -\dfrac{1}{2}(p - 1)^{-1/2}$ and $y = (p - 1)^{1/2} - \dfrac{1}{2}p(p - 1)^{-1/2}$. Eliminating the parameter p from these equations gives $4x^2 - 4xy - 1 = 0$.

21. $x^2 p^2 = (x - y)^2$. The equation factors and yields two differential equations that may be written $\dfrac{dy}{dx} + \dfrac{1}{x} y = 1$ and $\dfrac{dy}{dx} - \dfrac{1}{x} = -1$. The solution of the first of these linear equations is $xy = \dfrac{x^2}{2} + c$ which may be rewritten as $x(x - 2y) = c_1$. The second differential equation has the solution $y = -x \ln |c_2 x|$.

23. $p^3 - p^2 + xp - y = 0$. This equation may be rewritten $y = px + p^3 - p^2$ and recognized as a Clairaut equation. The general solution is $y = cx + c^3 - c^2$. Here $f(p) = p^3 - p^2$ and $f'(p) = 3p^2 - 2p$, so that we obtain $x = -3p^2 + 2p$ and $y = -2p^3 + p^2$ as parametric equations for the singular solution.

25. $yp^2 - (x + y)p + y = 0$. Dividing both sides of this equation by p and differentiating both sides with respect to x eliminates y and gives $y(p - 1)(p + 1)\dfrac{dp}{dy} + p^2(p - 1) = 0$. Remember that $\dfrac{dp}{dx} = p\dfrac{dp}{dy}$. Thus $y(p + 1)\dfrac{dp}{dy} + p^2 = 0$ or $p = 1$. We solve this differential equation and obtain $py = c \exp(-1/p)$. This equation taken with the original differential equation $px = y(p^2 - p + 1)$ gives the solution of the original equation in parametric form. Substituting $p = 1$ into the original differential equation gives the singular solution $y = x$.

27. $xp^3 - 2yp^2 + 4x^2 = 0$. Dividing both sides of this equation by p^2 and differentiating both sides with respect to x eliminates y and gives $x(p^3 - 8x)\dfrac{dp}{dx} = p(p^3 - 8x)$. Thus $x\dfrac{dp}{dx} = p$ or $p = 2x^{1/3}$. We solve this differential equation and obtain $p = kx$. Substituting $p = kx$ into the original differential equation yields $x^2 = 4c(y - 8c^2)$, where $c = 1/(2k)$. Substituting $p = 2x^{1/3}$ into the original differential equation yields $8y^3 = 27x^4$.

Chapter 17

Power Series Solutions

17.5   Solutions Near an Ordinary Point

1. $y'' + y = 0$. The elementary solution is $y = c_1 \cos x + \sin x$. Substituting $y = \sum_{n=0}^{\infty} a_n x^n$, we

obtain $\sum_{n=2}^{\infty} n(n-1)a_n x^{n-2} + \sum_{n=0}^{\infty} a_n x^n = 0$. Replacing n by $n+2$ in the first series gives us

$\sum_{n=0}^{\infty} (n+2)(n+1)a_{n+2} x^n + \sum_{n=0}^{\infty} a_n x^n = 0$. Combining these series and equating each coefficient

to zero yields the recurrence relation $a_{n+2} = \dfrac{-a_n}{(n+2)(n+1)}$, for $n \geq 0$. For $n = 0, 2, \cdots, 2k$

and $n = 1, 3, \cdots, 2k-1$ we have

$$a_2 = \frac{-a_0}{2 \cdot 1} \qquad\qquad a_3 = \frac{-a_1}{3 \cdot 2}$$

$$a_4 = \frac{-a_2}{4 \cdot 3} \qquad\qquad a_5 = \frac{-a_3}{5 \cdot 4}$$

$$\vdots \qquad\qquad\qquad \vdots$$

$$a_{2k} = \frac{-a_{2k-2}}{2k(2k-1)} \qquad\qquad a_{2k+1} = \frac{-a_{2k-1}}{(2k+1)(2k)}.$$

It follows that $a_{2k} = \dfrac{(-1)^k a_0}{(2k)!}$ and $a_{2k+1} = \dfrac{(-1)^k a_1}{(2k+1)!}$ for $k \geq 1$ with a_0 and a_1 arbitrary

constants. The original series becomes

$$y = a_0 \left[1 + \sum_{k=1}^{\infty} \frac{(-1)^k x^{2k}}{(2k)!} \right] + a_1 \left[x + \sum_{k=1}^{\infty} \frac{(-1)^k x^{2k+1}}{(2k+1)!} \right] = a_0 \cos x + a_1 \sin x.$$

3. $y'' + 3xy' + 3y = 0$. $\displaystyle\sum_{n=2}^{\infty} n(n-1)a_n x^{n-2} + 3\sum_{n=0}^{\infty} na_n x^n + 3\sum_{n=0}^{\infty} a_n x^n = 0$,

$$\sum_{n=0}^{\infty} [(n+2)(n+1)a_{n+2} + 3(n+1)a_n] x^n = 0.$$

The recurrence relation is $a_{n+2} = \dfrac{-3}{n+2} a_n$ for $n \geq 0$.

$$a_2 = \frac{-3a_0}{2} \qquad\qquad a_3 = \frac{-3a_1}{3}$$

$$a_4 = \frac{-3a_2}{4} \qquad\qquad a_5 = \frac{-3a_3}{5}$$

$$\vdots \qquad\qquad\qquad\qquad \vdots$$

$$a_{2k} = \frac{-3a_{2k-2}}{2k} \qquad\qquad a_{2k+1} = \frac{-3a_{2k-1}}{2k+1}$$

$$a_{2k} = \frac{(-1)^k 3^k a_0}{2^k k!} \qquad\qquad a_{2k+1} = \frac{(-1)^k 3^k a_1}{3 \cdot 5 \cdots (2k+1)}.$$

$$y = a_0 \left[1 + \sum_{k=1}^{\infty} \frac{(-3)^k x^{2k}}{2^k k!} \right] + a_1 \left[x + \sum_{k=1}^{\infty} \frac{(-3)^k x^{2k+1}}{3 \cdot 5 \cdots (2k+1)} \right] ; \text{ valid for all finite } x.$$

5. $(1 - 4x^2)y'' + 8y = 0.$ $\displaystyle\sum_{n=2}^{\infty} n(n-1)a_n x^{n-2} - \sum_{n=0}^{\infty} 4n(n-1)a_n x^n + \sum_{n=0}^{\infty} 8a_n x^n = 0,$

$\displaystyle\sum_{n=0}^{\infty} \left[(n+2)(n+1)a_{n+2} - 4(n-2)(n+1)a_n \right] x^n = 0.$ The recurrence relation is

$a_{n+2} = \dfrac{4(n-2)a_n}{n+2}$ for $n \geq 0$.

$$a_2 = -4a_0 \qquad\qquad a_3 = \frac{4 \cdot (-1)a_1}{3}$$

$$a_4 = 0 \qquad\qquad a_5 = \frac{4 \cdot 1 a_3}{5}$$

$$\vdots \qquad\qquad\qquad\qquad \vdots$$

$$a_{2k} = 0 \text{ for } k \geq 2 \qquad\qquad a_{2k+1} = \frac{4(2k-3)a_{2k-1}}{2k+1}$$

$$a_{2k} = 0 \text{ for } k \geq 2 \qquad\qquad a_{2k+1} = \frac{4^k(-1) \cdot 1 \cdot 3 \cdots (2k-3)a_1}{3 \cdot 5 \cdots (2k+1)}$$

$$a_{2k} = 0 \text{ for } k \geq 2 \qquad\qquad a_{2k+1} = \frac{-4^k a_1}{4k^2 - 1} \text{ for } k \geq 1.$$

$$y = a_0(1 - 4x^2) + a_1 \left[x - \sum_{k=1}^{\infty} \frac{2^{2k} x^{2k+1}}{4k^2 - 1} \right]. \text{ Replace } a_1 \text{ with } -b_1 \text{ to get}$$

$$y = a_0(1 - 4x^2) + b_1 \sum_{k=0}^{\infty} \frac{2^{2k} x^{2k+1}}{4k^2 - 1}; \text{ valid for } |x| < \frac{1}{2}.$$

7. $(1 + x^2)y'' + 10xy' + 20y = 0.$ $\displaystyle\sum_{n=2}^{\infty} n(n-1)a_n x^{n-2} + \sum_{n=0}^{\infty} [n(n-1) + 10n + 20]a_n x^n = 0,$

$\displaystyle\sum_{n=0}^{\infty} [(n+2)(n+1)a_{n+2} + (n+5)(n+4)a_n]x^n = 0.$ The recurrence relation is

$a_{n+2} = \dfrac{-(n+5)(n+4)a_n}{(n+2)(n+1)}$ for $n \geq 0.$

$$a_2 = \frac{-5 \cdot 4a_0}{2 \cdot 1} \qquad\qquad\qquad a_3 = \frac{-6 \cdot 5a_1}{3 \cdot 2}$$

$$a_4 = \frac{-7 \cdot 6a_2}{4 \cdot 3} \qquad\qquad\qquad a_5 = \frac{-8 \cdot 7a_3}{5 \cdot 4}$$

$$\vdots \qquad\qquad\qquad\qquad\qquad \vdots$$

$$a_{2k} = \frac{-(2k+3)(2k+2)a_{2k-2}}{2k(2k-1)} \qquad a_{2k+1} = \frac{-(2k+4)(2k+3)a_{2k-1}}{(2k+1)(2k)}$$

$$a_{2k} = \frac{(-1)^k 4 \cdot 5 \cdots (2k+3)a_0}{1 \cdot 2 \cdots (2k)} \qquad a_{2k+1} = \frac{(-1)^k 5 \cdot 6 \cdots (2k+4)a_1}{2 \cdot 3 \cdots (2k+1)}$$

$$a_{2k} = \frac{(-1)^k(k+1)(2k+1)(2k+3)a_0}{3} \qquad a_{2k+1} = \frac{(-1)^k(k+1)(k+2)(2k+3)a_1}{6}.$$

$$y_1 = 1 + \frac{1}{3}\sum_{k=1}^{\infty}(-1)^k(k+1)(2k+1)(2k+3)x^{2k} = \frac{1}{3}\sum_{k=0}^{\infty}(-1)^k(k+1)(2k+1)(2k+3)x^{2k};$$

$$y_2 = x + \frac{1}{6}\sum_{k=1}^{\infty}(-1)^k(k+1)(k+2)(2k+3)x^{2k+1} = \frac{1}{6}\sum_{k=0}^{\infty}(-1)^k(k+1)(k+2)(2k+3)x^{2k+1};$$ valid

for $|x| < 1.$

9. $(x^2 - 9)y'' + 3xy' - 3y = 0.$ $-9\displaystyle\sum_{n=2}^{\infty} n(n-1)a_n x^{n-2} + \sum_{n=0}^{\infty} [n(n-1) + 3n - 3]a_n x^n = 0,$

$\displaystyle\sum_{n=0}^{\infty} [-9(n+2)(n+1)a_{n+2} + (n+3)(n-1)a_n]x^n = 0.$ The recurrence relation is

$a_{n+2} = \dfrac{(n+3)(n-1)a_n}{9(n+2)(n+1)}$ for $n \geq 0.$

$$a_2 = \frac{3 \cdot (-1)a_0}{9 \cdot 2 \cdot 1} \qquad\qquad\qquad a_3 = 0$$

$$a_4 = \frac{5 \cdot 1a_2}{9 \cdot 4 \cdot 3} \qquad\qquad\qquad a_5 = 0$$

$$\vdots \qquad\qquad\qquad\qquad\qquad \vdots$$

$$a_{2k} = \frac{(2k+1)(2k-3)a_{2k-2}}{9(2k)(2k-1)} \qquad\qquad a_{2k+1} = 0 \text{ for } k > 1.$$

$$a_{2k} = \frac{3 \cdot 5 \cdots (2k+1) \cdot (-1) \cdot 1 \cdots (2k-3)a_0}{9^k 2^k k! \cdot 1 \cdot 3 \cdots (2k-1)} = \frac{-[3 \cdot 5 \cdots (2k+1)]a_0}{(18)^k(2k-1)k!}.$$

$$y = a_0 \left[1 - \sum_{k=1}^{\infty} \frac{[3 \cdot 5 \cdots (2k+1)]x^{2k}}{(18)^k(2k-1)k!}\right] + a_1 x; \text{ valid for } |x| < 3.$$

11. $(x^2 + 4)y'' + 6xy' + 4y = 0.$ $\quad 4\sum_{n=2}^{\infty} n(n-1)a_n x^{n-2} + \sum_{n=0}^{\infty} \left[n(n-1) + 6n + 4\right]a_n x^n = 0,$

$\sum_{n=0}^{\infty} \left[4(n+2)(n+1)a_{n+2} + (n+4)(n+1)a_n\right]x^n = 0.$ The recurrence relation is

$a_{n+2} = \dfrac{-(n+4)a_n}{4(n+2)}$ for $n \geq 0.$

$$a_2 = \frac{-4a_0}{4 \cdot 2} \qquad\qquad\qquad a_3 = \frac{-5a_1}{4 \cdot 3}$$

$$a_4 = \frac{-6a_2}{4 \cdot 4} \qquad\qquad\qquad a_5 = \frac{-7a_3}{4 \cdot 5}$$

$$\vdots \qquad\qquad\qquad\qquad\qquad \vdots$$

$$a_{2k} = \frac{-(2k+2)a_{2k-2}}{4(2k)} \qquad\qquad a_{2k+1} = \frac{-(2k+3)a_{2k-1}}{4(2k+1)}$$

$$a_{2k} = \frac{(-1)^k\left[4 \cdot 6 \cdots (2k+2)\right]a_0}{4^k\left[2 \cdot 4 \cdots (2k)\right]} \qquad a_{2k+1} = \frac{(-1)^k\left[5 \cdot 7 \cdots (2k+3)\right]a_1}{4^k\left[3 \cdot 5 \cdots (2k+1)\right]}$$

$$a_{2k} = \frac{(-1)^k(k+1)a_0}{2^{2k}} \qquad\qquad a_{2k+1} = \frac{(-1)^k(2k+3)a_1}{3 \cdot 2^{2k}}$$

$$y = a_0\left[1 + \sum_{k=1}^{\infty} \frac{(-1)^k(k+1)x^{2k}}{2^{2k}}\right] + a_1\left[x + \sum_{k=1}^{\infty} \frac{(-1)^k(2k+3)x^{2k+1}}{3 \cdot 2^{2k}}\right]; \text{ valid for } |x| < 2.$$

13. $y'' + x^2 y = 0.$ $\quad \sum_{n=2}^{\infty} n(n-1)a_n x^{n-2} + \sum_{n=0}^{\infty} a_n x^{n+2} = 0,$

$\sum_{n=0}^{\infty}(n+2)(n+1)a_{n+2}x^n + \sum_{n=2}^{\infty} a_{n-2}x^n = 0,\ 2a_2 + 6a_3 x + \sum_{n=2}^{\infty}\left[(n+2)(n+1)a_{n+2} + a_{n-2}\right]x^n = 0.$

Hence, $a_2 = a_3 = 0$ and $a_{n+2} = \dfrac{-a_{n-2}}{(n+2)(n+1)}$ for $n \geq 2$. It follows that $a_6 = a_{10} = \cdots = a_{4k+2} = 0$ and $a_7 = a_{11} = \cdots = a_{4k+3} = 0$, and

$$a_4 = \frac{-a_0}{4 \cdot 3} \qquad\qquad\qquad a_5 = \frac{-a_1}{5 \cdot 4}$$

$$a_8 = \frac{-a_4}{8 \cdot 7} \qquad\qquad\qquad a_9 = \frac{-a_3}{9 \cdot 8}$$

$$\vdots \qquad\qquad\qquad\qquad\qquad \vdots$$

$$a_{4k} = \frac{-a_{4k-4}}{4k(4k-1)} \qquad\qquad a_{4k+1} = \frac{-a_{4k-3}}{(4k+1)(4k)}$$

$$a_{4k} = \frac{(-1)^k a_0}{4^k k! \left[3 \cdot 7 \cdots (4k-1)\right]} \qquad a_{4k+1} = \frac{(-1)^k a_1}{\left[5 \cdot 9 \cdots (4k+1)\right] 4^k k!}.$$

$$y = a_0 \left[1 + \sum_{k=1}^{\infty} \frac{(-1)^k x^{4k}}{2^{2k} k! \cdot 3 \cdot 7 \cdots (4k-1)}\right] + a_1 \left[x + \sum_{k=1}^{\infty} \frac{(-1)^k x^{4k+1}}{2^{2k} k! \cdot 5 \cdot 9 \cdots (4k+1)}\right] ; \text{ valid for all}$$
finite x.

15. $(1 + 2x^2)y'' + 3xy' - 3y = 0.$ $\displaystyle\sum_{n=2}^{\infty} n(n-1)a_n x^{n-2} + \sum_{n=0}^{\infty} \left[2n(n-1) + 3n - 3\right] a_n x^n = 0,$

$\displaystyle\sum_{n=0}^{\infty} \left[(n+2)(n+1)a_{n+2} + (2n+3)(n-1)a_n\right] x^n = 0.$ The recurrence relation is
$$a_{n+2} = \frac{-(2n+3)(n-1)a_n}{(n+2)(n+1)} \text{ for } n \geq 0.$$

$$a_2 = \frac{-3 \cdot (-1)a_0}{2 \cdot 1} \qquad\qquad\qquad a_3 = 0$$

$$a_4 = \frac{-7 \cdot 1 a_2}{4 \cdot 3} \qquad\qquad\qquad a_5 = 0$$

$$\vdots \qquad\qquad\qquad\qquad\qquad \vdots$$

$$a_{2k} = \frac{-(4k-1)(2k-3)a_{2k-2}}{2k(2k-1)} \qquad a_{2k+1} = 0 \text{ for } k \geq 1.$$

$$a_{2k} = \frac{(-1)^k \left[3 \cdot 7 \cdots (4k-1)\right]\left[(-1) \cdot 1 \cdots (2k-3)\right] a_0}{2^k k! \left[1 \cdot 3 \cdots (2k-1)\right]} = \frac{(-1)^{k+1} \left[3 \cdot 7 \cdots (4k-1)\right] a_0}{2^k k!(2k-1)}.$$

$$y = a_0 \left[1 + \sum_{k=1}^{\infty} \frac{(-1)^{k+1} 3 \cdot 7 \cdots (4k-1) x^{2k}}{2^k k!(2k-1)}\right] + a_1 x; \text{ valid for } |x| < 1/\sqrt{2}.$$

17. $y'' + xy' + 3y = x^2.$ We first solve the homogeneous equation.
$$\sum_{n=2}^{\infty} n(n-1)a_n x^{n-2} + \sum_{n=0}^{\infty} (n+3)a_n x^n = 0, \quad \sum_{n=0}^{\infty} \left[(n+2)(n+1)a_{n+2} + (n+3)a_n\right] x^n = 0.$$

The recurrence relation is $a_{n+2} = \dfrac{-(n+3)a_n}{(n+2)(n+1)}$ for $n \geq 0$.

$$a_2 = \frac{-3a_0}{2 \cdot 1} \qquad\qquad a_3 = \frac{-4a_1}{3 \cdot 2}$$

$$a_4 = \frac{-5a_2}{4 \cdot 3} \qquad\qquad a_5 = \frac{-6a_3}{5 \cdot 4}$$

$$\vdots \qquad\qquad\qquad \vdots$$

$$a_{2k} = \frac{-(2k+1)a_{2k-2}}{(2k)(2k-1)} \qquad\qquad a_{2k+1} = \frac{-(2k+2)a_{2k-1}}{(2k+1)(2k)}$$

$$a_{2k} = \frac{(-1)^k 3 \cdot 5 \cdots (2k+1)a_0}{2^k k! \cdot 1 \cdot 3 \cdots (2k-1)} \qquad\qquad a_{2k+1} = \frac{(-1)^k 4 \cdot 6 \cdots (2k+2)a_1}{\left[3 \cdot 5 \cdots (2k+1)\right]\left[2 \cdot 4 \cdots (2k)\right]}$$

$$a_{2k} = \frac{(-1)^k(2k+1)a_0}{2^k k!} \qquad\qquad a_{2k+1} = \frac{(-1)^k(k+1)a_1}{3 \cdot 5 \cdots (2k+1)}.$$

$$y_c = a_0 \left[1 + \sum_{k=1}^{\infty} \frac{(-1)^k(2k+1)x^{2k}}{2^k k!}\right] + a_1 \left[x + \sum_{k=1}^{\infty} \frac{(-1)^k(k+1)x^{2k+1}}{3 \cdot 5 \cdots (2k+1)}\right]. \text{ We now seek}$$

$y_p = A + Bx + Cx^2$, a solution to the nonhomogeneous equation. We need $2C + Bx + 2Cx^2 + 3A + 3Bx + 3Cx^2 = x^2$, or $5Cx^2 + 4Bx + (2C + 3A) = x^2$. It follows that $A = -\frac{2}{15}$, $B = 0$, $C = \frac{1}{5}$. Thus $y_p = -\frac{2}{15} + \frac{1}{5}x^2$. Finally, $y = y_c + y_p$; valid for all finite x.

19. $y'' + 3xy' + 7y = 0$. $\displaystyle\sum_{n=2}^{\infty} n(n-1)a_n x^{n-2} + \sum_{n=0}^{\infty}(3n+7)a_n x^n = 0$,

$$\sum_{n=0}^{\infty} \left[(n+2)(n+1)a_{n+2} + (3n+7)a_n\right]x^n = 0. \text{ The recurrence relation is}$$

$$a_{n+2} = \frac{-(3n+7)a_n}{(n+2)(n+1)} \text{ for } n \geq 2.$$

$$a_2 = \frac{-7a_0}{2 \cdot 1} \qquad\qquad a_3 = \frac{-10a_1}{3 \cdot 2}$$

$$a_4 = \frac{-13a_2}{4 \cdot 3} \qquad\qquad a_5 = \frac{-16a_3}{5 \cdot 4}$$

$$\vdots \qquad\qquad\qquad \vdots$$

$$a_{2k} = \frac{-(6k+1)a_{2k-2}}{2k(2k-1)} \qquad\qquad a_{2k+1} = \frac{-(6k+4)a_{2k-1}}{(2k+1)(2k)}$$

$$a_{2k} = \frac{(-1)^k 7 \cdot 13 \cdots (6k+1)a_0}{(2k)!} \qquad\qquad a_{2k+1} = \frac{(-1)^k 10 \cdot 16 \cdots (6k+4)a_1}{(2k+1)!}$$

$$y = a_0 \left[1 + \sum_{k=1}^{\infty} \frac{(-1)^k 7 \cdot 13 \cdots (6k+1)x^{2k}}{(2k)!}\right] + a_1 \left[x + \sum_{k=1}^{\infty} \frac{(-1)^k\, 10 \cdot 16 \cdots (6k+4)x^{2k+1}}{(2k+1)!}\right];$$

valid for all finite x.

21. $(x^2 + 4)y'' + xy' - 9y = 0.$ $\displaystyle\sum_{n=2}^{\infty} 4n(n-1)a_n x^{n-2} + \sum_{n=0}^{\infty} \big[n(n-1) + n - 9\big] a_n x^n = 0,$

$\displaystyle\sum_{n=0}^{\infty} \big[4(n+2)(n+1)a_{n+2} + (n-3)(n+3)a_n\big] x^n = 0.$ The recurrence relation is

$a_{n+2} = \dfrac{-(n+3)(n-3)a_n}{4(n+2)(n+1)}$ for $n \geq 0$.

$$a_2 = \frac{-3(-3)a_0}{4 \cdot 2 \cdot 1} \qquad\qquad\qquad a_3 = \frac{-4(-2)a_1}{4 \cdot 3 \cdot 2}$$

$$a_4 = \frac{-5(-1)a_2}{4 \cdot 4 \cdot 3} \qquad\qquad\qquad a_5 = 0$$

$$\vdots \qquad\qquad\qquad\qquad\qquad\qquad \vdots$$

$$a_{2k} = \frac{-(2k+1)(2k-5)a_{2k-2}}{4(2k)(2k-1)} \qquad\qquad a_{2k+1} = 0 \text{ for } k \geq 2$$

$$a_{2k} = \frac{3(-1)^k 3 \cdot 5 \cdots (2k+1)a_0}{2^{3k} k! \, (2k-1)(2k-3)}.$$

$$y = a_0 \left[1 + 3\sum_{k=1}^{\infty} \frac{(-1)^k 3 \cdot 5 \cdots (2k+1)x^{2k}}{2^{3k} k! \, (2k-1)(2k-3)}\right] + a_1 \big[x + \tfrac{1}{3}x^3\big]; \text{ valid for } |x| < 2.$$

23. $(1 + 9x^2)y'' - 18y = 0.$ $\displaystyle\sum_{n=2}^{\infty} n(n-1)a_n x^{n-2} + \sum_{n=0}^{\infty} \big[9n(n-1) - 18\big] a_n x^n = 0,$

$\displaystyle\sum_{n=0}^{\infty} \big[(n+2)(n+1)a_{n+2} + 9(n-2)(n+1)a_n\big] x^n = 0.$ The recurrence relation is

$a_{n+2} = \dfrac{-9(n-2)a_n}{n+2}$ for $n \geq 0$.

$$a_2 = \frac{-9(-2)a_0}{2} \qquad\qquad\qquad a_3 = \frac{-9(-1)a_1}{3}$$

$$a_4 = 0 \qquad\qquad\qquad\qquad\qquad a_5 = \frac{-9 \cdot 1 a_3}{5}$$

$$\vdots \qquad\qquad\qquad\qquad\qquad\qquad \vdots$$

$$a_{2k} = 0 \text{ for } k \geq 2 \qquad\qquad a_{2k+1} = \frac{-9(2k-3)a_{2k-1}}{2k+1}$$

$$a_{2k+1} = \frac{(-1)^{k+1} 3^{2k} a_1}{(2k-1)(2k+1)}$$

$$y = a_0[1 + 9x^2] + a_1 \left[x + \sum_{k=1}^{\infty} \frac{(-1)^{k+1} 3^{2k} x^{2k+1}}{4k^2 - 1}\right]; \text{ valid for } |x| < \tfrac{1}{3}.$$

25. $(1 + 2x^2)y'' + 11xy' + 9y = 0.$ $\displaystyle\sum_{n=2}^{\infty} n(n-1)a_n x^{n-2} + \sum_{n=0}^{\infty} \left[2n(n-1) + 11n + 9\right]a_n x^n = 0,$

$\displaystyle\sum_{n=0}^{\infty} \left[(n+2)(n+1)a_{n+2} + (2n+3)(n+3)a_n\right]x^n = 0.$ The recurrence relation is

$a_{n+2} = \dfrac{-(2n+3)(n+3)a_n}{(n+2)(n+1)}$ for $n \geq 0.$

$$a_2 = \frac{-3 \cdot 3a_0}{2 \cdot 1} \qquad\qquad a_3 = \frac{-5 \cdot 4a_1}{3 \cdot 2}$$

$$a_4 = \frac{-7 \cdot 5a_2}{4 \cdot 3} \qquad\qquad a_5 = \frac{-9 \cdot 6a_3}{5 \cdot 4}$$

$$\vdots \qquad\qquad\qquad\qquad \vdots$$

$$a_{2k} = \frac{-(4k-1)(2k+1)a_{2k-2}}{(2k)(2k-1)} \qquad\qquad a_{2k+1} = \frac{-(4k+1)(2k+2)a_{2k-1}}{(2k+1)(2k)}$$

$$a_{2k} = \frac{(-1)^k(2k+1)3 \cdot 7 \cdots (4k-1)a_0}{2^k k!},$$

$$a_{2k+1} = \frac{(-1)^k(k+1)5 \cdot 9 \cdots (4k+1)a_1}{3 \cdot 5 \cdots (2k+1)}.$$

$$y = a_0 \left[1 + \sum_{k=1}^{\infty} \frac{(-1)^k(2k+1)3 \cdot 7 \cdots (4k-1)x^{2k}}{2^k k!}\right]$$

$$+ a_1 \left[x + \sum_{k=1}^{\infty} \frac{(-1)^k(k+1)5 \cdot 9 \cdots (4k+1)x^{2k+1}}{3 \cdot 5 \cdots (2k+1)}\right]; \text{ valid for } |x| < 1/\sqrt{2}.$$

27. $y'' + (x-2)y = 0.$ We set $z = x - 2$ and obtain $\dfrac{d^2y}{dz^2} + zy = 0.$

$\displaystyle\sum_{n=2}^{\infty} n(n-1)a_n z^{n-2} + \sum_{n=0}^{\infty} a_n z^{n+1} = 0,$ $\displaystyle\sum_{n=0}^{\infty}(n+2)(n+1)a_{n+2}z^n + \sum_{n=1}^{\infty} a_{n-1}z^n = 0,$

$2a_2 + \displaystyle\sum_{n=1}^{\infty}\left[(n+2)(n+1)a_{n+2} + a_{n-1}\right]z^n = 0.$ Hence, $a_2 = 0$ and $a_{n+2} = \dfrac{-a_{n-1}}{(n+2)(n+1)}$ for $n \geq 1.$

$$a_3 = \frac{-a_0}{3 \cdot 2} \qquad\qquad a_4 = \frac{-a_1}{4 \cdot 3} \qquad\qquad a_5 = 0$$

$$a_6 = \frac{-a_3}{6 \cdot 5} \qquad\qquad a_7 = \frac{-a_4}{7 \cdot 6} \qquad\qquad a_8 = 0$$

$$\vdots \qquad\qquad\qquad\qquad \vdots \qquad\qquad\qquad\qquad \vdots$$

$$a_{3k} = \frac{-a_{3k-3}}{(3k)(3k-1)} \qquad\qquad a_{3k+1} = \frac{-a_{3k-2}}{(3k+1)(3k)} \qquad\qquad a_{3k+2} = 0$$

$$a_{3k} = \frac{(-1)^k a_0}{3^k \, k! \, 2 \cdot 5 \cdots (3k-1)} \quad a_{3k+1} = \frac{(-1)^k a_1}{3^k \, k! 4 \cdot 7 \cdots (3k+1)}.$$

$$y = a_0 \left[1 + \sum_{k=1}^{\infty} \frac{(-1)^k (x-2)^{3k}}{3^k \, k! \, 2 \cdot 5 \cdots (3k-1)} \right] + a_1 \left[(x-2) + \sum_{k=1}^{\infty} \frac{(-1)^k \, (x-2)^{3k+1}}{3^k \, k! \, 4 \cdot 7 \cdots (3k+1)} \right] ; \text{ valid for}$$

all finite x.

Chapter 18

Solutions Near Regular Singular Points

18.1 Regular Singular Points

1. $p(x) = \dfrac{1}{x^3}$; $q(x) = \dfrac{4}{x^2(x-1)}$.
   R.S.P. at $x = 1$; I.S.P. at $x = 0$.

3. No singular points in the finite plane.

5. $p(x) = \dfrac{1}{x^4}$. I.S.P. at $x = 0$.

7. $p(x) = \dfrac{3}{x^2}$; $q(x) = \dfrac{1}{x^2(x-2)}$.
   R.S.P. at $x = 2$; I.S.P. at $x = 0$.

9. $p(x) = \dfrac{1}{x^2}$; $q(x) = \dfrac{4}{x^2(x+2)}$.
   R.S.P. at $x = -2$; I.S.P. at $x = 0$.

11. $p(x) = 0$; $q(x) = \dfrac{4}{x^3}$. I.S.P. at $x = 0$.

13. $p(x) = \dfrac{6x}{1+4x^2}$; $q(x) = \dfrac{-9}{1+4x^2}$.
    R.S.P. at $x = \frac{1}{2}i, -\frac{1}{2}i$.

15. $p(x) = \dfrac{6x}{(1+4x^2)^2}$. I.S.P. at $x = \frac{1}{2}i, -\frac{1}{2}i$.

17. $p(x) = \dfrac{1}{(2x+1)^3}$. I.S.P. at $x = -\frac{1}{2}$.

19. $p(x) = \dfrac{-6x}{x^2+4}$; $q(x) = \dfrac{3}{x^2+4}$.
    R.S.P. at $x = 2i, -2i$.

21. No singular points in the finite plane.

23. $p(x) = \dfrac{-2x}{1+x^2}$; $q(x) = \dfrac{6}{1+x^2}$.
    R.S.P. at $x = i, -i$.

25. $p(x) = 0$; $q(x) = \dfrac{1+2x}{x^2(1-x)^3}$.
    R.S.P. at $x = 0$; I.S.P. at $x = 1$.

27. $p(x) = \dfrac{3}{x(x^2-9)}$; $q(x) = \dfrac{-1}{x^2(x^2-9)}$.
    R.S.P. at $x = 0, 3, -3$.

29. $p(x) = \dfrac{-1}{(2x+1)(x-3)}$; $q(x) = \dfrac{1}{x-3}$.
    R.S.P. at $x = -\frac{1}{2}, 3$.

31. $p(x) = 0$; $q(x) = \dfrac{1}{1+4x^2}$.
    R.S.P. at $x = \frac{1}{2}i, -\frac{1}{2}i$.

18.4 Difference of Roots Nonintegral

1. $2x(x+1)y'' + 3(x+1)y' - y = 0$. We substitute $y = \displaystyle\sum_{n=0}^{\infty} a_n x^{n+c}$ and obtain

$$\sum_{n=0}^{\infty}[2(n+c)(n+c-1)+3(n+c)-1]a_n x^{n+c} + \sum_{n=0}^{\infty}[2(n+c)(n+c-1)+3(n+c)]a_n x^{n+c-1} = 0,$$

$$\sum_{n=0}^{\infty}(2n+2c-1)(n+c+1)a_nx^{n+c} + \sum_{n=0}^{\infty}(n+c)(2n+2c+1)a_nx^{n+c-1} = 0,$$

$$c(2c+1)a_0x^{c-1} + \sum_{n=1}^{\infty}\left[(n+c)(2n+2c+1)a_n + (n+c)(2n+2c-3)a_{n-1}\right]x^{n+c-1} = 0. \text{ The}$$

indicial equation is $c(2c+1) = 0$ and the recurrence relation is $a_n = \dfrac{-(2n+c-3)a_{n-1}}{2n+2c+1}$ for

$n \geq 1$. Taking $c = -\frac{1}{2}$ yields $a_n = \dfrac{-(n-2)a_{n-1}}{n}$, so that $a_1 = a_0$ and $a_n = 0$ for $n \geq 2$. Thus

one solution is $y_1 = x^{-1/2} + x^{1/2}$. Taking $c = 0$, we have $a_n = \dfrac{-(2n-3)\,a_{n-1}}{2n+1}$,

$$a_1 = \frac{-(-1)a_0}{3}$$

$$a_2 = \frac{-(1)a_1}{5}$$

$$\vdots$$

$$a_n = \frac{(-1)^n(-1)\cdot 1\cdots(2n-3)a_0}{3\cdot 5\cdots(2n+1)} = \frac{(-1)^{n+1}a_0}{4n^2-1}.$$

We therefore obtain a second solution $y_2 = 1 + \displaystyle\sum_{n=1}^{\infty}\dfrac{(-1)^{n+1}x^n}{4n^2-1}.$

3. $4x^2y''+4xy'-(4x^2+1)y = 0.$ $\displaystyle\sum_{n=0}^{\infty}\left[4(n+c)(n+c-1)+4(n+c)-1\right]a_nx^{n+c} - \sum_{n=0}^{\infty}4a_nx^{n+c+2} = 0,$

$$\sum_{n=0}^{\infty}(2n+2c-1)(2n+2c+1)a_nx^{n+c} - \sum_{n=2}^{\infty}4a_{n-2}x^{n+c} = 0, \quad (2c-1)(2c+1)a_0x^c$$

$$+ (2c+1)(2c+3)a_1x^{c+1} + \sum_{n=2}^{\infty}\left[(2n+2c-1)(2n+2c+1)a_n - 4a_{n-2}\right]x^{n+c} = 0. \text{ The indicial}$$

equation is $(2c-1)(2c+3) = 0$, and the recurrence relation is $a_n = \dfrac{4a_{n-2}}{(2n+2c-1)(2n+2c+1)}.$

We take $a_1 = 0$, and therefore $a_{2k+1} = 0$ for $k \geq 0$. For $c = \frac{1}{2}$ we have $a_n = \dfrac{a_{n-2}}{n(n+1)}.$

$$a_2 = \frac{a_0}{2\cdot 3}$$

$$a_4 = \frac{a_2}{4\cdot 5}$$

$$\vdots$$

$$a_{2k} = \frac{a_{2k-1}}{2k(2k+1)} = \frac{a_0}{(2k+1)!} \text{ for } k \geq 0.$$

Therefore one solution is $y_1 = \displaystyle\sum_{k=0}^{\infty} \frac{x^{2k+1/2}}{(2k+1)!} = \frac{\sinh x}{\sqrt{x}}$. For $c = -\frac{1}{2}$, we have $a_n = \dfrac{a_{n-2}}{(n-1)n}$.

$$a_2 = \frac{a_0}{1 \cdot 2}$$

$$a_4 = \frac{a_2}{3 \cdot 4}$$

$$\vdots$$

$$a_{2k} = \frac{a_{2k-2}}{(2k-1)(2k)} = \frac{a_0}{(2k)!} \text{ for } k \geq 0.$$

Therefore a second solution is $y_2 = \displaystyle\sum_{k=0}^{\infty} \frac{x^{2k-1/2}}{(2k)!} = \frac{\cosh x}{\sqrt{x}}$.

5. $2x^2(1-x)y'' - x(1+7x)y' + y = 0.$

$$\sum_{n=0}^{\infty} \big[2(n+c)(n+c-1) - (n+c) + 1]\big]a_n x^{n+c} - \sum_{n=0}^{\infty} \big[2(n+c)(n+c-1) + 7(n+c)\big]a_n x^{n+c+1} = 0,$$

$$\sum_{n=0}^{\infty} (2n+2c-1)(n+c-1)a_n x^{n+c} - \sum_{n=0}^{\infty} (n+c)(2n+2c+5)a_n x^{n+c+1} = 0,$$

$$(2c-1)(c-1)a_0 + \sum_{n=1}^{\infty} \big[(2n+2c-1)(n+c-1)a_n - (n+c-1)(2n+2c+3)a_{n-1}\big]x^{n+c} = 0. \text{ The}$$

indicial equation is $(c-1)(2c-1) = 0$, and the recurrence relation is $a_n = \dfrac{(2n+2c+3)a_{n-1}}{2n+2c-1}$

for $n \geq 1$. For $c = 1$, $a_n = \dfrac{(2n+5)a_{n-1}}{2n+1}$ for $n \geq 1$.

$$a_1 = \frac{7a_0}{3}$$

$$a_2 = \frac{9a_1}{5}$$

$$\vdots$$

$$a_n = \frac{(2n+5)a_{n-1}}{2n+1} = \frac{7 \cdot 9 \cdots (2n+5)a_0}{3 \cdot 5 \cdots (2n+1)} = \tfrac{1}{15}(2n+3)(2n+5)a_0.$$

Hence one solution is $y_1 = x + \frac{1}{15} \sum_{n=1}^{\infty} (2n+3)(2n+5)x^{n+1}$.

For $c = \frac{1}{2}$, $a_n = \frac{(n+2)a_{n-1}}{n}$ for $n \geq 1$.

$$a_1 = \frac{3a_0}{1}$$

$$a_2 = \frac{4a_1}{2}$$

$$\vdots$$

$$a_n = \frac{(n+2)a_{n-1}}{n} = \frac{3 \cdot 4 \cdots (n+2)a_0}{1 \cdot 2 \cdots n} = \tfrac{1}{2}(n+1)(n+2)a_0.$$

Thus a second solution is $y_2 = x^{1/2} + \frac{1}{2} \sum_{n=1}^{\infty} (n+1)(n+2)x^{n+1/2}$.

7. $8x^2 y'' + 10xy' - (1+x)y = 0$. $\sum_{n=0}^{\infty} \big[8(n+c)(n+c-1) + 10(n+c) - 1\big]a_n x^{n+c}$

$- \sum_{n=0}^{\infty} a_n x^{n+c+1} = 0$, $\sum_{n=0}^{\infty} (4n+4c-1)(2n+2c+1)a_n x^{n+c} - \sum_{n=1}^{\infty} a_{n-1} x^{n+c} = 0$,

$(4c-1)(2c+1)a_0 x^c + \sum_{n=1}^{\infty} \big[(4n+4c-1)(2n+2c+1)a_n - a_{n-1}\big]x^{n+c} = 0$. The indicial equation

is $(4c-1)(2c+1) = 0$ and $a_n = \dfrac{a_{n-1}}{(4n+4c-1)(2n+2c+1)}$ for $n \geq 1$. For $c = \frac{1}{4}$, we have

$a_n = \dfrac{a_{n-1}}{2n(4n+3)}$ for $n \geq 1$.

$$a_1 = \frac{a_0}{2 \cdot 1 \cdot 7}$$

$$a_2 = \frac{a_1}{2 \cdot 2 \cdot 11}$$

$$\vdots$$

$$a_n = \frac{a_{n-1}}{2n(4n+3)} = \frac{a_0}{2^n \, n! \, \big[7 \cdot 11 \cdots (4n+3)\big]}.$$

Thus one solution is $y_1 = x^{1/4} + \sum_{n=1}^{\infty} \dfrac{x^{n+1/4}}{2^n \, n! \, \left[7 \cdot 11 \cdots (4n+3)\right]}$. For $c = -\frac{1}{2}$ we take

$a_n = \dfrac{a_{n-1}}{2n(4n-3)}$ for $n \geq 1$.

$$a_1 = \frac{a_0}{2 \cdot 1 \cdot 1}$$
$$a_2 = \frac{a_1}{2 \cdot 2 \cdot 5}$$
$$\vdots$$
$$a_n = \frac{a_{n-1}}{2n(4n-3)} = \frac{a_0}{2^n \, n! \, \left[1 \cdot 5 \cdots (4n-3)\right]}.$$

Thus a second solution is $y_2 = x^{-1/2} + \sum_{n=1}^{\infty} \dfrac{x^{n-1/2}}{2^n \, n! \, \left[1 \cdot 5 \cdots (4n-3)\right]}$.

9. $2x(x+3)y'' - 3(x+1)y' + 2y = 0$.

$$\sum_{n=0}^{\infty} \left[2(n+c)(n+c-1) - 3(n+c) + 2\right] a_n x^{n+c} + \sum_{n=0}^{\infty} \left[6(n+c)(n+c-1) - 3(n+c)\right] a_n x^{n+c-1} = 0,$$

$$\sum_{n=0}^{\infty} (2n + 2c - 1)(n+c-2) a_n x^{n+c} + \sum_{n=0}^{\infty} 3(n+c)(2n+2c-3) a_n x^{n+c-1} = 0,$$

$$\sum_{n=1}^{\infty} (2n + 2c - 3)(n+c-3) a_{n-1} x^{n+c-1} + \sum_{n=0}^{\infty} 3(n+c)(2n+2c-3) a_n x^{n+c-1} = 0,$$

$$3c(2c-3) a_0 x^{c-1} + \sum_{n=1}^{\infty} \left[3(n+c)(2n+2c-3) a_n + (2n+2c-3)(n+c-3) a_{n-1}\right] x^{n+c-1} = 0.$$

The indicial equation is $c(2c-3) = 0$ and $a_n = \dfrac{(n+c-3) a_{n-1}}{3(n+c)}$ for $n \geq 1$. For $c = \frac{3}{2}$, we have

$a_n = \dfrac{-(2n-3) a_{n-1}}{3(2n+3)}$.

$$a_1 = \frac{-(-1) a_0}{3 \cdot 5}$$
$$a_2 = \frac{-1 \, a_1}{3 \cdot 7}$$
$$\vdots$$
$$a_n = \frac{(-1)^n \left[(-1) \cdot 1 \cdots (2n-3)\right]}{3^n \left[5 \cdot 7 \cdots (2n+3)\right]} = \frac{(-1)^{n+1} a_0}{3^{n-1}(2n-1)(2n+1)(2n+3)}.$$

One solution is $y_1 = x^{3/2} + \sum_{n=1}^{\infty} \dfrac{(-1)^{n+1} x^{n+3/2}}{3^{n-1}(2n-1)(2n+1)(2n+3)}$. For $c = 0$ we get

$a_n = \dfrac{-(n-3) a_{n-1}}{3n}$. Thus $a_1 = \dfrac{-(-2) a_0}{3} = \frac{2}{3} a_0$, $\quad a_2 = \dfrac{-(-1) a_1}{3 \cdot 2} = \frac{1}{9} a_0$, $\quad a_n = 0$ for $n \geq 3$.

A second solution is $y_2 = 1 + \frac{2}{3} x + \frac{1}{9} x^2$.

11. $x(4-x)y'' + (2-x)y' + 4y = 0.$

$$\sum_{n=0}^{\infty}\left[4(n+c)(n+c-1)+2(n+c)\right]a_n x^{n+c-1} + \sum_{n=0}^{\infty}\left[-(n+c)(n+c-1)-(n+c)+4\right]a_n x^{n+c} = 0,$$

$$\sum_{n=0}^{\infty} 2(n+c)(2n+2c-1)a_n x^{n+c-1} - \sum_{n=0}^{\infty}(n+c+2)(n+c-2)a_n x^{n+c} = 0,$$

$$2c(2c-1)a_0 + \sum_{n=1}^{\infty}\left[2(n+c)(2n+2c-1)a_n - (n+c+1)(n+c-3)a_{n-1}\right]x^{n+c-1} = 0. \text{ The}$$

indicial equation is $c(2c-1)=0$ and $a_n = \dfrac{(n+c+1)(n+c-3)a_{n-1}}{2(n+c)(2n+2c-1)}$. For $c = \frac{1}{2}$ we have

$$a_n = \frac{(2n+3)(2n-5)a_{n-1}}{8n(2n+1)}.$$

$$a_1 = \frac{5\cdot(-3)a_0}{8\cdot 1\cdot 3}$$

$$a_2 = \frac{7\cdot(-1)a_1}{8\cdot 2\cdot 5}$$

$$\vdots$$

$$a_n = \frac{(2n+3)(2n-5)a_{n-1}}{8n(2n+1)} = \frac{\left[5\cdot 7\cdots(2n+3)\right]\left[(-3)(-1)\cdots(2n-5)\right]a_0}{2^{3n}\,n!\,3\cdot 5\cdots(2n+1)}$$

$$= \frac{(2n+3)\left[(-3)(-1)\cdots(2n-5)\right]a_0}{3\cdot 2^{3n}\,n!}.$$

One solution is $y_1 = x^{1/2} + \displaystyle\sum_{n=1}^{\infty}\frac{(2n+3)\left[(-3)(-1)\cdots(2n-5)\right]x^{n+1/2}}{3\cdot 2^{3n}\,n!}.$

For $c = 0$ we get $a_n = \dfrac{(n+1)(n-3)a_{n-1}}{2n(2n-1)}$ for $n \geq 1.$

$$a_1 = \frac{2(-2)a_0}{2} = -2a_0$$

$$a_2 = \frac{3(-1)a_1}{2\cdot 2} = \frac{1}{2}a_0$$

$$a_n = 0 \text{ for } n \geq 3.$$

A second solution is $y_2 = 1 - 2x + \frac{1}{2}x^2.$

13. $2xy'' + (1+2x)y' + 4y = 0.$

$$\sum_{n=0}^{\infty}\left[2(n+c)(n+c-1)+(n+c)\right]a_n x^{n+c-1} + \sum_{n=0}^{\infty}\left[2(n+c)+4\right]a_n x^{n+c} = 0,$$

$$\sum_{n=0}^{\infty}(n+c)(2n+2c-1)a_n x^{n+c-1} + \sum_{n=1}^{\infty}2(n+c+2)a_{n-1}x^{n+c-1} = 0,$$

$$c(2c-1)a_o x^{c-1} + \sum_{n=1}^{\infty}\left[(n+c)(2n+2c-1)a_n + 2(n+c+1)a_{n-1}\right]x^{n+c-1} = 0. \text{ The indicial}$$

equation is $c(2c - 1) = 0$ and $a_n = \dfrac{-2(n + c + 1)a_{n-1}}{(n + c)(2n + 2c - 1)}$ for $n \geq 1$. For $c = \frac{1}{2}$ we have $a_n = \dfrac{-(2n + 3)a_{n-1}}{n(2n + 1)}$.

$$a_1 = \frac{-5a_0}{1 \cdot 3}$$
$$a_2 = \frac{-7a_1}{2 \cdot 5}$$
$$\vdots$$
$$a_n = \frac{-(2n + 3)a_{n-1}}{n(2n + 1)} = \frac{(-1)^n \left[5 \cdot 7 \cdots (2n + 3)\right] a_0}{n! \left[3 \cdot 5 \cdots (2n + 1)\right]} = \frac{(-1)^n (2n + 3)a_0}{3 \, n!}.$$

One solution is $y_1 = x^{1/2} + \displaystyle\sum_{n=1}^{\infty} \frac{(-1)^n (2n + 3)x^{n+1/2}}{3 \, n!}$. For $c = 0$ we get $a_n = \dfrac{-2(n + 1)a_{n-1}}{n(2n - 1)}$.

$$a_1 = \frac{-2 \cdot 2a_0}{1 \cdot 1}$$
$$a_2 = \frac{-2 \cdot 3a_1}{2 \cdot 3}$$
$$\vdots$$
$$a_n = \frac{(-2)(n + 1)a_{n-1}}{n(2n - 1)} = \frac{(-1)^n 2^n (n + 1)a_0}{1 \cdot 3 \cdots (2n - 1)}.$$

A second solution is $y_2 = 1 + \displaystyle\sum_{n=1}^{\infty} \frac{(-1)^n 2^n (n + 1)x^n}{1 \cdot 3 \cdots (2n - 1)}$.

15. $2x^2 y'' - 3x(1 - x)y' + 2y = 0$.

$$\sum_{n=0}^{\infty} \left[2(n + c)(n + c - 1) - 3(n + c) + 2\right] a_n x^{n+c} + \sum_{n=0}^{\infty} 3(n + c)a_n x^{n+c+1} = 0,$$

$$\sum_{n=0}^{\infty} (2n + c - 1)(n + c - 2)a_n x^{n+c} + \sum_{n=1}^{\infty} 3(n + c - 1)a_{n-1} x^{n+c} = 0,$$

$(2c - 1)(c - 2)a_0 x^c + \displaystyle\sum_{n=1}^{\infty} \left[(2n + 2c - 1)(n + c - 2)a_n + 3(n + c - 1)a_{n-1}\right] x^{n+c} = 0.$ The indicial equation is $(2c - 1)(c - 2) = 0$ and $a_n = \dfrac{-3(n + c - 1)a_{n-1}}{(2n + 2c - 1)(n + c - 2)}$ for $n \geq 1$. For $c = 2$ we have $a_n = \dfrac{-3(n+1)a_{n-1}}{n(2n+3)}$.

$$a_1 = \frac{-3 \cdot 2\, a_0}{1 \cdot 5}$$

$$a_2 = \frac{-3 \cdot 3\, a_1}{2 \cdot 7}$$

$$\vdots$$

$$a_n = \frac{-3(n+1)\, a_{n-1}}{n(2n+3)} = \frac{(-1)^n\, 3^n (n+1)\, a_0}{5 \cdot 7 \cdots (2n+3)}.$$

One solution is $y_1 = x^2 + \displaystyle\sum_{n=1}^{\infty} \frac{(-1)^n\, 3^n(n+1)x^{n+2}}{5 \cdot 7 \cdots (2n+3)}$. For $c = \frac{1}{2}$ we get $a_n = \dfrac{-3(2n-1)\, a_{n-1}}{2n(2n-3)}$.

$$a_1 = \frac{-3 \cdot 1\, a_0}{2 \cdot 1(-1)}$$

$$a_2 = \frac{-3 \cdot 3\, a_1}{2 \cdot 2 \cdot 1}$$

$$\vdots$$

$$a_n = \frac{-3(2n-1)\, a_{n-1}}{2n(2n-3)} = \frac{(-1)^n 3^n \big[1 \cdot 3 \cdots (2n-1)\big]\, a_0}{2^n\, n!\, \big[(-1) \cdot 1 \cdots (2n-3)\big]} = \frac{(-1)^{n+1} 3^n (2n-1)\, a_0}{2^n\, n!}.$$

A second solution is $y_2 = x^{1/2} + \displaystyle\sum_{n=1}^{\infty} \frac{(-1)^{n+1} 3^n (2n-1) x^{n+1/2}}{2^n\, n!}$.

17. $2xy'' - (1 + 2x^2)y' - xy = 0$.

$$\sum_{n=0}^{\infty} \big[2(n+c)(n+c-1) - (n+c)\big]x^{n+c-1} - \sum_{n=0}^{\infty} \big[2(n+c)+1\big]x^{n+c+1} = 0,$$

$$\sum_{n=0}^{\infty} (n+c)(2n+2c-3)a_n x^{n+c-1} - \sum_{n=0}^{\infty} (2n+2c+1)a_n x^{n+c+1} = 0,$$

$$c(2c-3)a_0 x^{c-1} + (c+1)(2c-1)a_1 x^c + \sum_{n=2}^{\infty} (2n+2c-3)\big[(n+c)a_n - a_{n-2}\big]x^{n+c-1} = 0.$$ We

take $c(2c-3) = 0$, $a_1 = 0$, and $a_n = \dfrac{a_{n-2}}{n+c}$ for $n \geq 2$. The choice $c = 0$ gives us $a_n = \dfrac{a_{n-2}}{n}$.

$$a_2 = \frac{a_0}{2} \qquad\qquad\qquad a_3 = 0$$

$$a_4 = \frac{a_2}{4} \qquad\qquad\qquad a_5 = 0$$

$$\vdots \qquad\qquad\qquad\qquad \vdots$$

$$a_{2k} = \frac{a_{2k-2}}{2k} = \frac{a_0}{2^k\, k!} \qquad\qquad a_{2k+1} = 0.$$

Thus one solution is $y_1 = 1 + \sum_{k=1}^{\infty} \dfrac{x^{2k}}{2^k \, k!} = \exp\left(\frac{1}{2}x^2\right)$. A second solution comes from $c = \frac{3}{2}$ and

$$a_n = \frac{2a_{n-2}}{2n+3}.$$

$$a_2 = \frac{2a_0}{7}$$

$$a_4 = \frac{2a_2}{11}$$

$$\vdots$$

$$a_{2k} = \frac{2a_{2k-2}}{4k+3} = \frac{2^k \, a_0}{7 \cdot 11 \cdots (4k+3)}.$$

Hence $y_2 = x^{3/2} + \sum_{k=1}^{\infty} \dfrac{2^k \, x^{2k+3/2}}{7 \cdot 11 \cdots (4k+3)}.$

19. $2x^2 y'' + xy' - y = 0.$ $\sum_{n=0}^{\infty} \left[2(n+c)(n+c-1) + (n+c) - 1\right] a_n x^{n+c} = 0,$

$\sum_{n=0}^{\infty} (2n+2c+1)(n+c-1) a_n x^{n+c} = 0,$ $(2c+1)(c-1)a_0 x^c + \sum_{n=1}^{\infty} (2n+2c+1)(n+c-1) a_n x^{n+c} = 0.$
We can make the first term vanish by choosing $c = 1$ or $c = -\frac{1}{2}$ while leaving a_0 arbitrary.
We are also forced to take $a_n = 0$ for $n \geq 1$. For $c = 1$ we get the solution $y_1 = x$. For $c = -\frac{1}{2}$
we get the solution $y_2 = x^{-1/2}$.

21. $9x^2 y'' + 2y = 0.$ $\sum_{n=0}^{\infty} \left[9(n+c)(n+c-1) + 2\right] a_n x^{n+c} = \sum_{n=0}^{\infty} (3n+3c-2)(3n+3c-1) a_n x^{n+c} =$

$(3c-2)(3c-1)a_0 x^c + \sum_{n=1}^{\infty} (3n+3c-2)(3n+3c-1) a_n x^{n+c} = 0.$ We choose $c = \frac{2}{3}$ or $c = \frac{1}{3}$ and
a_0 arbitrary. We are forced to take $a_n = 0$ for $n \geq 1$. For $c = \frac{2}{3}$ we get the solution $y_1 = x^{2/3}.$
For $c = \frac{1}{3}$ we get the solution $y_2 = x^{1/3}.$

23. If y is a function of x and we let $t = \ln x$ for $x > 0$ then
$$\frac{dy}{dx} = \frac{dy}{dt}\frac{dt}{dx} = \frac{1}{x}\frac{dy}{dt} = e^{-t}\frac{dy}{dt} \text{ and } \frac{d^2y}{dx^2} = e^{-t}\frac{d^2y}{dt^2}\frac{dt}{dx} - e^{-t}\frac{dy}{dt} = e^{-2t}\left(\frac{d^2y}{dt^2} - \frac{dy}{dt}\right).$$

25. $2x^2 y'' + xy' - y = 0.$ We set $t = \ln x$ and use the results of Exercise 23 to get the differential
equation $2\left(\dfrac{d^2y}{dt^2} - \dfrac{dy}{dt}\right) + \dfrac{dy}{dt} - y = 0$ or $(2D^2 - D - 1)y = (D-1)(2D+1)y = 0.$ It follows
that $y = c_1 e^t + c_2 e^{(-1/2)t}$ or $y = c_1 x + c_2 x^{(-1/2)}.$

27. $9x^2 y'' + 2y = 0.$ Letting $t = \ln x$ we obtain $9\left(\dfrac{d^2y}{dt^2} - \dfrac{dy}{dt}\right) + 2y = (9D^2 - 9D + 2)y =$
$(3D-2)(3D-1)y = 0.$ It follows that $y = c_1 e^{(2/3)t} + c_2 e^{(1/3)t} = c_1 x^{2/3} + c_2 x^{1/3}.$

29. $x^2 y'' + 2xy' - 12y = 0$. Substituting $t = \ln x$ we obtain $\left(\dfrac{d^2 y}{dt^2} - \dfrac{dy}{dt}\right) + 2\dfrac{dy}{dt} - 12 =$

$(D^2 + D - 12)y = (D - 3)(D + 4)y = 0$. Thus $y = c_1 e^{3t} + c_2 e^{-4t} = c_1 x^3 + c_2 x^{-4}$.

31. $x^2 y'' - 3xy' + 4y = 0$. Letting $t = \ln x$ we get $\left(\dfrac{d^2 y}{dt^2} - \dfrac{dy}{dt}\right) - 3\dfrac{dy}{dt} + 4y =$

$(D^2 - 4D + 4)y = 0$. The general solution is $y = (c_1 + c_2 t)e^{2t} = x^2(c_1 + c_2 \ln x)$.

33. $x^2 y'' + 5xy' + 5y = 0$. Substituting $t = \ln x$ we get $\left(\dfrac{d^2 y}{dt^2} - \dfrac{dy}{dt}\right) + 5\dfrac{dy}{dt} + 5y =$

$(D^2 + 4D + 5)y = 0$. The roots of the auxiliary equation are $-2 + i$ and $-2 - i$. The general solution is $y = e^{-2t}(c_1 \cos t + c_2 \sin t) = x^{-2}\big[c_1 \cos(\ln x) + c_2 \sin(\ln x)\big]$.

18.6 Equal Roots

1. $x^2 y'' - x(1 + x)y' + y = 0$. $\displaystyle\sum_{n=0}^{\infty}\big[(n+c)(n+c-1) - (n+c) + 1\big]a_n x^{n+c} - \sum_{n=0}^{\infty}(n+c)a_n x^{n+c+1} =$

$\displaystyle\sum_{n=0}^{\infty}(n+c-1)^2 a_n x^{n+c} - \sum_{n=1}^{\infty}(n+c-1)a_{n-1} x^{n+c} = 0$. We therefore have

$(c-1)^2 a_0 x^c + \displaystyle\sum_{n=1}^{\infty}\big[(n+c-1)^2 a_n (n+c-1)a_{n-1}\big]x^{n+c} = 0$. The indicial equation is $(c-1)^2 = 0$

and $a_n = \dfrac{a_{n-1}}{n+c-1}$ for $n \geq 1$.

$$a_1 = \frac{a_0}{c}$$
$$a_2 = \frac{a_1}{c+1}$$
$$\vdots$$
$$a_n = \frac{a_{n-1}}{n+c-1} = \frac{a_0}{c(c+1)\cdots(c+n-1)}.$$

Thus $y_c = x^c + \displaystyle\sum_{n=1}^{\infty} \frac{x^{n+c}}{c(c+1)\cdots(c+n-1)}$, and

$\dfrac{\partial y_c}{\partial c} = y_c \ln x + \displaystyle\sum_{n=1}^{\infty} \frac{x^{n+c}}{c(c+1)\cdots(c+n-1)}\left[-\frac{1}{c} - \frac{1}{c+1} - \cdots - \frac{1}{c+n-1}\right]$. Substituting $c = 1$

yields $y_1 = x + \displaystyle\sum_{n=1}^{\infty} \frac{x^{n+1}}{n!} = xe^x$ and $y_2 = y_1 \ln x - \displaystyle\sum_{n=1}^{\infty} \frac{H_n\, x^{n+1}}{n!}$.

3. $x^2 y'' + x(x - 3)y' + 4y = 0$. $\displaystyle\sum_{n=0}^{\infty}\big[(n+c)(n+c-1) - 3(n+c) + 4\big]a_n x^{n+c} + \sum_{n=0}^{\infty}(n+c)a_n x^{n+c+1} =$

$\displaystyle\sum_{n=0}^{\infty}(n+c-2)^2 a_n x^{n+c} + \sum_{n=1}^{\infty}(n+c-1)a_{n-1} x^{n+c} = 0$. We therefore have

$$(c-2)^2 a_0 x^c + \sum_{n=1}^{\infty} \left[(n+c-2)^2 a_n + (n+c-1)a_{n-1}\right] x^{n+c} = 0. \text{ The indicial equation is } (c-2)^2 = 0$$

and $a_n = \dfrac{-(n+c-1)\,a_{n-1}}{(n+c-2)^2}$.

$$a_1 = \frac{-c\,a_0}{(c-1)^2}$$

$$a_2 = \frac{-(c+1)\,a_1}{c^2}$$

$$\vdots$$

$$a_n = \frac{-(n+c-1)\,a_{n-1}}{(n+c-2)^2} = \frac{(-1)^n c(c+1)\cdots(n+c-1)\,a_0}{(c-1)^2 c^2 \cdots (n+c-2)^2} = \frac{(-1)^n (n+c-1)\,a_0}{(c-1)^2 c(c+1)\cdots(n+c-2)}.$$

Therefore $y_c = x^c + \displaystyle\sum_{n=1}^{\infty} \frac{(-1)^n (n+c-1)\,x^{n+c}}{(c-1)^2 c(c+1)\cdots(n+c-2)}$ and $\dfrac{\partial y_c}{\partial c} = y_c \ln x + $

$$\sum_{n=1}^{\infty} \frac{(-1)^n (n+c-1)x^{n+c}}{(c-1)^2 c(c+1)\cdots(n+c-2)} \left[\frac{1}{n+c-1} - \frac{2}{c-1} - \frac{1}{c} - \frac{1}{c+1} \cdots - \frac{1}{n+c-2}\right].$$

Substituting $c = 2$ we have $y_1 = x^2 + \displaystyle\sum_{n=1}^{\infty} \frac{(-1)^n (n+1)x^{n+2}}{n!}$ and

$$y_2 = y_1 \ln x + \sum_{n=1}^{\infty} \frac{(-1)^n (n+1)x^{n+2}}{n!} \left[\frac{1}{n+1} - \frac{1}{1} - H_n\right] =$$

$$y_1 \ln x + \sum_{n=1}^{\infty} \frac{(-1)^{n+1}\left[n + (n+1)H_n\right] x^{n+2}}{n!}.$$

5. $x(1+x)y'' + (1+5x)y' + 3y = 0.$ $\displaystyle\sum_{n=0}^{\infty}\left[(n+c)(n+c-1)+(n+c)\right]a_n x^{n+c-1} + \sum_{n=0}^{\infty}\left[(n+c)(n+\right.$

$c-1) + 5(n+c) + 3\big]a_n x^{n+c} = \displaystyle\sum_{n=0}^{\infty}(n+c)^2 a_n x^{n+c-1} + \sum_{n=1}^{\infty}(n+c+1)(n+c+3)a_n x^{n+c} =$

$$c^2 a_0 x^{c-1} + \sum_{n=1}^{\infty}\left[(n+c)^2 a_n + (n+c)(n+c+2)a_{n-1}\right]x^{n+c-1} = 0. \text{ The indicial equation is } c^2 = 0$$

and $a_n = \dfrac{-(n+c+2)\,a_{n-1}}{n+c}$ for $n \geq 1$.

$$a_1 = \frac{-(c+3)\,a_0}{c+1}$$

$$a_2 = \frac{-(c+4)\,a_1}{c+2}$$

$$\vdots$$

$$a_n = \frac{-(n+c+2)\,a_{n-1}}{c+n} = \frac{(-1)^n(c+3)\cdots(c+n+2)\,a_0}{(c+1)(c+2)\cdots(c+n)} = \frac{(-1)^n(n+c+1)(n+c+2)\,a_0}{(c+1)(c+2)}.$$

Therefore $y_c = x^c + \sum\limits_{n=1}^{\infty} \dfrac{(-1)^n(n+c+1)(n+c+2)x^{n+c}}{(c+1)(c+2)}$ and

$\dfrac{\partial y_c}{\partial c} = y_c \ln x +$

$$\sum_{n=1}^{\infty} \frac{(-1)^n(n+c+1)(n+c+2)x^{n+c}}{(c+1)(c+2)} \left[\frac{1}{n+c+1} + \frac{1}{n+c+2} - \frac{1}{c+1} - \frac{1}{c+2}\right].$$

Substituting $c = 0$ we have $y_1 = 1 + \dfrac{1}{2}\sum\limits_{n=1}^{\infty}(-1)^n(n+1)(n+2)x^n$ and

$$y_2 = y_1 \ln x + \frac{1}{2}\sum_{n=1}^{\infty}(-1)^n(n+1)(n+2)x^n \left[\frac{1}{n+1} + \frac{1}{n+2} - 1 - \frac{1}{2}\right] =$$

$$y_1 \ln x - \frac{3}{2}(y_1 - 1) + \frac{1}{2}\sum_{n=1}^{\infty}(-1)^n(2n+3)\, x^n.$$

7. $x^2y'' + x(x-1)y' + (1-x)y = 0.$ $\sum\limits_{n=0}^{\infty}\big[(n+c)(n+c-1) - (n+c) + 1\big]a_n x^{n+c} +$

$\sum\limits_{n=0}^{\infty}(n+c-1)a_n x^{n+c+1} = \sum\limits_{n=0}^{\infty}(n+c-1)^2 a_n x^{n+c} + \sum\limits_{n=1}^{\infty}(n+c-2)a_{n-1}x^{n+c} = (c-1)^2 a_0 x^c +$

$\sum\limits_{n=1}^{\infty}\big[(n+c-1)^2 a_n + (n+c-2)a_{n-1}\big]x^{n+c} = 0.$ The indicial equation is $(c-1)^2 = 0$ and

$a_n = \dfrac{-(n+c-2)\,a_{n-1}}{(n+c-1)^2}$ for $n \geq 1$.

$$a_1 = \frac{-(c-1)\,a_0}{c^2}$$

$$a_2 = \frac{-c\,a_1}{(c+1)^2}$$

$$\vdots$$

$$a_n = \frac{-(n+c-2)\,a_{n-1}}{(n+c-1)^2}.$$

There is a problem here since we eventually wish to substitute both $n = 1$ and $c = 1$ into our recurrence relation. We therefore keep the a_1 term separate from the infinite series. Thus $a_1 = \dfrac{-(c-1)\,a_0}{c^2}$ and $a_n = \dfrac{(-1)^n(c-1)c\cdots(n+c-2)\,a_{n-1}}{c^2(c+1)^2\cdots(n+c-1)^2}$ for $n \geq 2$. We have

$y_c = x^c - \dfrac{(c-1)x^{c+1}}{c^2} + \sum\limits_{n=2}^{\infty}\dfrac{(-1)^n(c-1)x^{n+c}}{\big[c(c+1)\cdots(n+c-2)\big](n+c-1)^2}$ and

$\dfrac{\partial y_c}{\partial c} = y_c \ln x - \dfrac{c-1}{c^2}\left[\dfrac{1}{c-1} - \dfrac{2}{c}\right]x^{c+1} + \sum\limits_{n=2}^{\infty}\dfrac{(-1)^n(c-1)x^{n+c}}{\big[c(c+1)\cdots(n+c-2)\big](n+c-1)^2}$ •

$$\left[\frac{1}{c-1} - \frac{1}{c} - \frac{1}{c+1} - \cdots - \frac{1}{n+c-2} - \frac{2}{n+c-1}\right]. \text{ Substituting } c = 1 \text{ we have } y_1 = x \text{ and}$$

$$y_2 = y_1 \ln x - x^2 + \sum_{n=2}^{\infty} \frac{(-1)^n x^{n+1}}{(n-1)! \, n^2} = y_1 \ln x + \sum_{n=1}^{\infty} \frac{(-1)^n \, x^{n+1}}{n \cdot n!}.$$

9. $x(x-2)y'' + 2(x-1)y' - 2y = 0$. We set $z = x - 2$ and obtain $z(z+2)\dfrac{d^2y}{dz^2} + 2(z+1)\dfrac{dy}{dz} - 2y = 0$, where $z = 0$ is a regular singular point. Setting $y = \displaystyle\sum_{n=0}^{\infty} a_n z^{n+c}$ we have $\displaystyle\sum_{n=0}^{\infty} [(n+c)(n+c-1) +$

$$2(n+c) - 2]a_n z^{n+c} + \sum_{n=0}^{\infty} [2(n+c)(n+c-1) + 2(n+c)]a_n z^{n+c-1} = \sum_{n=0}^{\infty} (n+c+2)(n+c-1)a_n z^{n+c} +$$

$$\sum_{n=0}^{\infty} 2(n+c)^2 a_n z^{n+c-1} = 2c^2 a_0 z^{c-1} + \sum_{n=1}^{\infty} \left[(n+c+1)(n+c-2)\, a_{n-1} + 2(n+c)^2 a_n\right] z^{n+c-1} = 0.$$

The indicial equation is $c^2 = 0$ and $a_n = \dfrac{-(n+c+1)(n+c-2)\, a_{n-1}}{2(n+c)^2}$ for $n \geq 1$. Here we must be careful to keep the $n = 1$ term separate because when $c = 0$, $a_n = 0$ for $n \geq 2$.

$$a_1 = \frac{-(c+2)(c-1)\, a_0}{2(c+1)^2}$$

$$a_2 = \frac{-(c+3)c\, a_1}{2(c+2)^2}$$

$$\vdots$$

$$a_n = \frac{-(n+c+1)(n+c-2)\, a_{n-1}}{2(c+n)^2} = \frac{(-1)^n (c-1)c(n+c+1)\, a_0}{2^n (c+1)(n+c-1)(n+c)}.$$

Thus $y_c = z^c - \dfrac{(c+2)(c-1)\, z^{c+1}}{2(c+1)^2} + \displaystyle\sum_{n=2}^{\infty} \frac{(-1)^n (c-1)c(n+c+1)\, z^{n+c}}{2^n (c+1)(n+c-1)(n+c)}$ and

$$\frac{\partial y_c}{\partial c} = y_c \ln z - \frac{(c+2)(c-1)}{2(c+1)^2} \left[\frac{1}{c+2} + \frac{1}{c-1} - \frac{2}{c+1}\right] z^{c+1} +$$

$$\sum_{n=2}^{\infty} \frac{(-1)^n (c-1)c(n+c+1)\, z^{c+n}}{2^n (c+1)(n+c-1)(n+c)} \left[\frac{1}{c-1} + \frac{1}{c} + \frac{1}{n+c+1} - \frac{1}{c+1} - \frac{1}{n+c-1} - \frac{1}{n+c}\right].$$

Substituting $c = 0$ we have $y_1 = 1 + z$ and $y_2 = y_1 \ln z + \left[\frac{1}{2} - 1 - 2\right] z - \displaystyle\sum_{n=2}^{\infty} \frac{(-1)^n (n+1)\, z^n}{2^n n(n-1)}$.

Thus $y_1 = 1 + (x-2)$ and $y_2 = y_1 \ln(x-2) - \dfrac{5}{2}(x-2) - \displaystyle\sum_{n=2}^{\infty} \frac{(-1)^n (n+1)(x-2)^n}{2^n n(n-1)}$.

11. $xy'' + y' + xy = 0.$ $\sum_{n=0}^{\infty}[(n+c)(n+c-1)+(n+c)]a_n x^{n+c-1} + \sum_{n=0}^{\infty} a_n x^{n+c+1} = 0.$

$$\sum_{n=0}^{\infty}(n+c)^2 a_n x^{n+c-1} + \sum_{n=2}^{\infty} a_{n-2} x^{n+c-1} =$$

$$c^2 a_0 x^{c-1} + (c+1)^2 a_1 x^c + \sum_{n=2}^{\infty}\left[(n+c)^2 a_n + a_{n-2}\right] x^{n+c-1} = 0.$$

We take $c^2 = 0$, $a_1 = 0$, and $a_n = \dfrac{-a_{n-2}}{(n+c)^2}$ for $n \geq 2$.

$$a_2 = \frac{-a_0}{(c+2)^2}$$

$$a_4 = \frac{-a_2}{(c+4)^2}$$

$$\vdots$$

$$a_{2k} = \frac{-a_{2k-2}}{(c+2k)^2} = \frac{(-1)^k a_0}{(c+2)^2 \cdots (c+2k)^2}.$$

$y_c = x^c + \sum_{k=1}^{\infty} \dfrac{(-1)^k x^{2k+c}}{(c+2)^2 \cdots (c+2k)^2}$; $\dfrac{\partial y_c}{\partial c} = y_c \ln x + \sum_{k=1}^{\infty} \dfrac{(-1)^k x^{2k+c}}{(c+2)^2 \cdots (c+2k)^2}$ •

$\left[\dfrac{-2}{c+2} - \cdots - \dfrac{-2}{c+2k}\right]$. Substituting $c = 0$ yields $y_1 = 1 + \sum_{k=1}^{\infty} \dfrac{(-1)^k x^{2k}}{2^{2k}\,(k!)^2}$ and

$y_2 = y_1 \ln x - \sum_{k=1}^{\infty} \dfrac{H_k(-1)^k x^{2k}}{2^{2k}\,(k!)^2}.$

13. $H_{2k} - \dfrac{1}{2}H_k = \left(1 + \dfrac{1}{2} + \cdots + \dfrac{1}{2k-1} + \dfrac{1}{2k}\right) - \dfrac{1}{2}\left(1 + \dfrac{1}{2} + \cdots + \dfrac{1}{k}\right) = 1 + \dfrac{1}{3} + \cdots + \dfrac{1}{2k-1}$. The

y_2 of Exercise 12 becomes $y_2 = y_1 \ln x + \sum_{k=1}^{\infty} \dfrac{1 \cdot 3 \cdots (2k-1)\left[H_{2k} - \frac{3}{2}H_k\right] x^{2k}}{2^{2k}\,(k!)^2}.$

15. $4x^2 y'' + 8x(x+1)y' + y = 0.$ $\sum_{n=0}^{\infty}\left[4(n+c)(n+c-1)+8(n+c)+1\right]a_n x^{n+c} + \sum_{n=0}^{\infty} 8(n+c)a_n x^{n+c+1} =$

$$\sum_{n=0}^{\infty}(2n+2c+1)^2 a_n x^{n+c} + \sum_{n=1}^{\infty} 8(n+c-1)a_{n-1} x^{n+c} =$$

$$(2c+1)^2 a_0 x^c + \sum_{n=1}^{\infty}\left[(2n+2c+1)^2 a_n + 8(n+c-1)a_{n-1}\right]x^{n+c} = 0.$$

The indicial equation is $(2c+1)^2 = 0$ and $a_n = \dfrac{-8(n+c-1)\,a_{n-1}}{(2n+2c+1)^2}$ for $n \geq 1$.

$$a_1 = \frac{-8c\,a_0}{(2c+3)^2}$$

$$a_2 = \frac{-8(c+1)\,a_1}{(2c+5)^2}$$

$$\vdots$$

$$a_n = \frac{-8(n+c-1)\,a_{n-1}}{(2n+2c+1)^2} = \frac{(-1)^n 8^n c(c+1)\cdots(n+c-1)\,a_0}{(2c+3)^2(2c+5)^2\cdots(2c+2n+1)^2}.$$

Hence $y_c = x^c + \displaystyle\sum_{n=1}^{\infty} \frac{(-1)^n 8^n c(c+1)\cdots(n+c-1)\,x^{n+c}}{(2c+3)^2(2c+5)^2\cdots(2c+2n+1)^2}$ and

$$\frac{\partial y_c}{\partial c} = y_c \ln x + \sum_{n=1}^{\infty} \frac{(-1)^n 8^n c(c+1)\cdots(n+c-1)\,x^{n+c}}{(2c+3)^2(2c+5)^2\cdots(2c+2n+1)^2} \bullet$$

$$\left[\frac{1}{c} + \cdots + \frac{1}{n+c-1} - \frac{4}{2c+3} - \cdots - \frac{4}{2c+2n+1}\right].$$

For $c = -\frac{1}{2}$, $y_1 = x^{-1/2} + \displaystyle\sum_{n=1}^{\infty} \frac{(-1)^n 8^n \left[(-\frac{1}{2})(\frac{1}{2})\cdots(n-\frac{3}{2})\right] x^{n-1/2}}{2^2 \cdot 4^2 \cdots (2n)^2} =$

$x^{-1/2} + \displaystyle\sum_{n=1}^{\infty} \frac{(-1)^n (-1) \cdot 1 \cdots (2n-3)\,x^{n-1/2}}{(n!)^2}$; and $y_2 = y_1 \ln x +$

$\displaystyle\sum_{n=1}^{\infty} \frac{(-1)^n (-1) \cdot 1 \cdots (2n-3)\,x^{n-1/2}}{(n!)^2}\left[-2 + 2 + \frac{2}{3} + \cdots + \frac{2}{2n-3} - \frac{4}{2} - \frac{4}{4} - \cdots - \frac{4}{2n}\right] =$

$y_1 \ln x + \displaystyle\sum_{n=1}^{\infty} \frac{(-1)^n(-1)\cdots(2n-3)\big(2H_{2n-2} - H_{n-1} - 2H_n - 2\big)x^{n-1/2}}{(n!)^2}.$

17. $xy'' + (1-x)y' - y = 0 \quad \displaystyle\sum_{n=0}^{\infty}\big[(n+c)(n+c-1)+(n+c)\big]a_n x^{n+c-1} - \sum_{n=0}^{\infty}(n+c+1)a_n x^{n+c} = \sum_{n=0}^{\infty}(n+$

$c)^2 a_n x^{n+c-1} - \displaystyle\sum_{n=1}^{\infty}(n+c)a_{n-1}x^{n+c-1} = c^2 a_0 x^{c-1} + \sum_{n=1}^{\infty}\big[(n+c)^2 a_n - (n+c)a_{n-1}\big]x^{n+c-1} = 0.$

The indicial equation is $c^2 = 0$ and $a_n = \dfrac{a_{n-1}}{n+c}$ for $n \geq 1$.

$$a_1 = \frac{a_0}{c+1}$$

$$a_2 = \frac{a_1}{c+2}$$

$$\vdots$$

$$a_n = \frac{a_{n-1}}{c+n} = \frac{a_0}{(c+1)(c+2)\cdots(c+n)}.$$

Hence $y_c = x^c + \displaystyle\sum_{n=1}^{\infty} \frac{x^{n+c}}{(c+1)(c+2)\cdots(c+n)}$ and

$\dfrac{\partial y_c}{\partial c} = y_c \ln x + \displaystyle\sum_{n=1}^{\infty} \frac{x^{n+c}}{(c+1)(c+2)\cdots(c+n)} \left[-\frac{1}{c+1} - \frac{1}{c+2} - \cdots - \frac{1}{c+n} \right]$. We have the

solutions $y_1 = 1 + \displaystyle\sum_{n=1}^{\infty} \frac{x^n}{n!} = e^x$ and $y_2 = y_1 \ln x - \displaystyle\sum_{n=1}^{\infty} \frac{H_n x^n}{n!}$.

18.7 Equal Roots, an Alternative

3. $L(y) = x^2 y'' + x(x-3)y' + 4y$. We previously determined the nonlogarithmic solution of
$L(y) = 0$ to be $y_1 = \displaystyle\sum_{n=0}^{\infty} \frac{(-1)^n(n+1)x^{n+2}}{n!}$. We now seek a second solution of the form

$$y_2 = y_1 \ln x + \sum_{n=1}^{\infty} b_n x^{n+2},$$

$$y_2' = y_1' \ln x + \frac{y_1}{x} + \sum_{n=1}^{\infty} (n+2)b_n x^{n+1},$$

$$y_2'' = y_1'' \ln x + \frac{2y_1'}{x} - \frac{y_1}{x^2} + \sum_{n=1}^{\infty} (n+2)(n+1)b_n x^n,$$

so that

$$L(y_2) = L(y_1)\ln x + 2xy_1' + (x-4)y_1' + \sum_{n=1}^{\infty}(n+2)(n+1)b_n x^{n+2} + \sum_{n=1}^{\infty}(n+2)b_n x^{n+3} -$$

$$3\sum_{n=1}^{\infty}(n+2)b_n x^{n+2} + 4\sum_{n=1}^{\infty}b_n x^{n+2}$$

$$= 2xy_1' + (x-4)y_1 + \sum_{n=1}^{\infty}n^2 b_n x^{n+2} + \sum_{n=2}^{\infty}(n+1)b_{n-1}x^{n+2}$$

$$= 2\sum_{n=0}^{\infty}\frac{(-1)^n(n+2)(n+1)x^{n+2}}{n!} + \sum_{n=0}^{\infty}\frac{(-1)^n(n+1)x^{n+3}}{n!} - 4\sum_{n=0}^{\infty}\frac{(-1)^n(n+1)x^{n+2}}{n!} +$$

$$b_1 x^3 + \sum_{n=2}^{\infty}\left[n^2 b_n + (n+1)b_{n-1}\right]x^{n+2}$$

$$= \sum_{n=0}^{\infty}\frac{(-1)^n 2n(n+1)x^{n+2}}{n!} + \sum_{n=1}^{\infty}\frac{(-1)^{n-1}nx^{n+2}}{(n-1)!} + b_1 x^3 + \sum_{n=2}^{\infty}\left[n^2 b_n + (n+1)b_{n-1}\right]x^{n+2}$$

$$= (b_1 - 3)x^3 + \sum_{n=2}^{\infty}\left[\frac{(-1)^n(n+2)}{(n-1)!} + n^2 b_n + (n+1)b_{n-1}\right]x^{n+2}.$$

In order to make $L(y_2) = 0$ we must take $b_1 = 3$ and $n^2 b_n + (n+1)b_{n-1} + \dfrac{(-1)^n(n+2)}{(n-1)!} = 0$
for $n \geq 2$.

5. We let $L(y) = x(1 + x)y'' + (1 + 5x)y' + 3y$. We previously determined the nonlogarithmic solution of $L(y) = 0$ to be $y_1 = 1 + \dfrac{1}{2}\sum_{n=1}^{\infty}(-1)^n(n+1)(n+2)x^n$. We now seek a second solution of the form

$$y_2 = y_1 \ln x + \sum_{n=1}^{\infty} b_n x^n,$$

$$y_2' = y_1' \ln x + \frac{y_1}{x} + \sum_{n=1}^{\infty} nb_n x^{n-1},$$

$$y_2'' = y_1'' \ln x + \frac{2y_1'}{x} - \frac{y_1}{x^2} + \sum_{n=1}^{\infty} n(n-1)b_n x^{n-2},$$

so that

$$L(y_2) = L(y_1)\ln x + 2(x+1)y_1' + 4y_1 + \sum_{n=1}^{\infty} n(n-1)b_n x^{n-1} + \sum_{n=1}^{\infty} n(n-1)b_n x^n +$$

$$\sum_{n=1}^{\infty} nb_n x^{n-1} + \sum_{n=1}^{\infty} 5nb_n x^n + \sum_{n=1}^{\infty} 3b_n x^n$$

$$= 2(x+1)y_1' + 4y_1 + \sum_{n=1}^{\infty} n^2 b_n x^{n-1} + \sum_{n=1}^{\infty}(n+1)(n+3)b_n x^n$$

$$= 2(x+1)y_1' + 4y_1 + b_1 + \sum_{n=2}^{\infty} \left[n^2 b_n + n(n+2)b_{n-1}\right]x^{n-1}$$

$$= \sum_{n=1}^{\infty}(-1)^n n(n+1)(n+2)x^n + \sum_{n=1}^{\infty}(-1)^n n(n+1)(n+2)x^{n-1} + 4 +$$

$$2\sum_{n=1}^{\infty}(-1)^n(n+1)(n+2)x^n + b_1 + \sum_{n=2}^{\infty}\left[n^2 b^n + n(n+2)b_{n-1}\right]x^{n-1}$$

$$= (b_1 - 2) + \sum_{n=1}^{\infty}(-1)^n(n+1)(n+2)^2 x^n +$$

$$\sum_{n=2}^{\infty}\left[(-1)^n n(n+1)(n+2) + n^2 b_n + n(n+2)b_{n-1}\right]x^{n-1}$$

$$= (b_1 - 2) + \sum_{n=2}^{\infty}(-1)^{n-1}n(n+1)^2 x^{n-1} +$$

$$\sum_{n=2}^{\infty}\left[(-1)^n n(n+1)(n+2) + n^2 b_n + n(n+2)b_{n-1}\right]x^{n-1}$$

$$= (b_1 - 2) + \sum_{n=2}^{\infty}\left[(-1)^n n(n+1) + n^2 b_n + n(n+2)b_{n-1}\right]x^{n-1}.$$

In order to make $L(y_2) = 0$ we must take $b_1 = 2$ and $nb_n + (n+2)b_{n-1} + (-1)^n(n+1) = 0$ for $n \geq 2$.

7. We let $L(y) = x^2 y'' + x(x - 1)y' + (1 - x)y$. We previously determined the nonlogarithmic solution of $L(y) = 0$ to be $y_1 = x$. We now seek a second solution of the form

$$y_2 = y_1 \ln x + \sum_{n=1}^{\infty} b_n x^{n+1},$$

$$y_2' = y_1' \ln x + \frac{y_1}{x} + \sum_{n=1}^{\infty} (n+1)b_n x^n,$$

$$y_2'' = y_1'' \ln x + \frac{2y_1'}{x} - \frac{y_1}{x^2} + \sum_{n=1}^{\infty} n(n+1)b_n x^{n-1},$$

so that

$$L(y_2) = L(y_1) \ln x + 2xy_1' + (x-2)y_1 + \sum_{n=1}^{\infty} n(n+1)b_n x^{n+1} + \sum_{n=1}^{\infty} (n+1)b_n x^{n+2} -$$

$$\sum_{n=1}^{\infty} (n+1)b_n x^{n+1} + \sum_{n=1}^{\infty} b_n x^{n+1} - \sum_{n=1}^{\infty} b_n x^{n+2}$$

$$= x^2 + \sum_{n=1}^{\infty} n^2 b_n x^{n+1} + \sum_{n=1}^{\infty} n b_n x^{n+2}$$

$$= (b_1 + 1)x^2 + \sum_{n=2}^{\infty} [n^2 b_n + (n-1)b_{n-1}] x^{n+1}.$$

In order to make $L(y_2) = 0$ we must take $b_1 = -1$ and $n^2 b_n + (n-1)b_{n-1} = 0$ for $n \geq 2$. In this exercise the two-term recurrence relation can be solved in the usual way to obtain the solution $y_2 = y_1 \ln x + \sum_{n=1}^{\infty} \frac{(-1)^n x^{n+1}}{n \cdot n!}$.

18.8   Nonlogarithmic Case

1. Set $L(y) = x^2 y'' + 2x(x-2)y' + 2(2 - 3x)y$. Then

$$L(y) = \sum_{n=0}^{\infty} (n+c)(n+c-1)a_n x^{n+c} + 2\sum_{n=0}^{\infty} (n+c)a_n x^{n+c+1} -$$

$$4\sum_{n=0}^{\infty} (n+c)a_n x^{n+c} + 4\sum_{n=0}^{\infty} a_n x^{n+c} - 6\sum_{n=0}^{\infty} a_n x^{n+c+1}$$

$$= \sum_{n=0}^{\infty} [(n+c)^2 - 5(n+c) + 4]a_n x^{n+c} + 2\sum_{n=0}^{\infty} (n+c-3)a_n x^{n+c+1}$$

$$= \sum_{n=0}^{\infty} (n+c-1)(n+c-4)a_n x^{n+c} + 2\sum_{n=1}^{\infty} (n+c-4)a_{n-1} x^{n+c}$$

$$= (c-1)(c-4)a_0 x^c + \sum_{n=1}^{\infty} [(n+c-1)(n+c-4)a_n + 2(n+c-4)a_{n-1}] x^{n+c}.$$

We choose the smaller of the two roots of the indicial equation, namely $c = 1$, and consider

$$
\begin{array}{llll}
n(n-3)a_n & +2(n-3)a_{n-1} = 0, & \text{for } n \geq 1. \\
1(-2)a_1 & +2(-2)a_0 & = 0, & a_1 = -2a_0, \\
2(-1)a_2 & +2(-1)a_1 & = 0, & a_2 = -a_1, \\
3(0)a_3 & +2(0)a_2 & = 0, & a_3 \text{ arbitrary,}
\end{array}
$$

$$
a_n = \frac{-2\,a_{n-1}}{n}, \text{ for } n \geq 4,
$$

$$
a_4 = \frac{-2a_3}{4},
$$

$$
\vdots
$$

$$
a_n = \frac{(-2)^{n-3}a_3}{4 \cdot 5 \cdots n} = \frac{6(-2)^{n-3}\,a_3}{n!}.
$$

Thus $y = a_0\left[x - 2x^2 + 2x^3\right] + a_3\left[x^4 + \displaystyle\sum_{n=4}^{\infty} \frac{6(-2)^{n-3}x^{n+1}}{n!}\right]$.

3. Set $L(y) = x^2 y'' + x(2 + 3x)y' - 2y$. Then

$$
\begin{aligned}
L(y) &= \sum_{n=0}^{\infty}\left[(n+c)(n+c-1) + 2(n+c) - 2\right]a_n x^{n+c} + \sum_{n=0}^{\infty} 3(n+c)a_n x^{n+c+1} \\
&= \sum_{n=0}^{\infty}(n+c+2)(n+c-1)a_n x^{n+c} + \sum_{n=1}^{\infty} 3(n+c-1)a_{n-1} x^{n+c} \\
&= (c+2)(c-1)a_0 x^c + \sum_{n=1}^{\infty}(n+c-1)\left[(n+c+2)a_n + 3a_{n-1}\right]x^{n+c}.
\end{aligned}
$$

We choose the smaller of the two roots of the indicial equation, namely $c = -2$, and consider

$$
\begin{array}{llll}
n(n-3)a_n & +3(n-3)a_{n-1} = 0, & \text{for } n \geq 1. \\
1(-2)a_1 & +3(-2)a_0 & = 0, & a_1 = -3a_0, \\
2(-1)a_2 & +3(-1)a_1 & = 0, & a_2 = \frac{9}{2}a_0, \\
3(0)a_3 & +3(0)a_2 & = 0, & a_3 \text{ arbitrary,}
\end{array}
$$

$$
a_n = \frac{-3\,a_{n-1}}{n} \text{ for } n \geq 4.
$$

$$
a_4 = \frac{-3a_3}{4}
$$

$$
\vdots
$$

$$
a_n = \frac{(-1)^{n-3}3^{n-3}\,a_3}{4 \cdot 5 \cdots n} = \frac{2(-1)^{n-3}3^{n-2}\,a_3}{n!}.
$$

Thus $y = a_0\left[x^{-2} - 3x^{-1} + \dfrac{9}{2}\right] + a_3\left[x + \displaystyle\sum_{n=4}^{\infty} \frac{2(-1)^{n-3}3^{n-2}x^{n-2}}{n!}\right]$.

5. Set $L(y) = x(1 + x)y'' + (x + 5)y' - 4y$. Then

$$L(y) = \sum_{n=0}^{\infty} \left[(n + c)(n + c - 1) + 5(n + c)\right]a_n x^{n+c-1} + \sum_{n=0}^{\infty} \left[(n + c)(n + c - 1) + (n + c) - 4\right]a_n x^{n+c}$$

$$= \sum_{n=0}^{\infty} (n + c)(n + c + 4)a_n x^{n+c-1} + \sum_{n=0}^{\infty} (n + c + 2)(n + c - 2)a_n x^{n+c}$$

$$= c(c + 4)a_0 x^{c-1} + \sum_{n=1}^{\infty} \left[(n + c)(n + c + 4)a_n + (n + c + 1)(n + c - 3)a_{n-1}\right]x^{n+c-1}.$$

We choose the smaller of the two roots of the indicial equation, namely $c = -4$, and consider

$$\begin{array}{llll}
n(n - 4)a_n & +(n - 3)(n - 7)a_{n-1} = 0, & \text{for } n \geq 1. \\
1(-3)a_1 & +(-2)(-6)a_0 & = 0, & a_1 = 4a_0, \\
2(-2)a_2 & +(-1)(-5)a_1 & = 0, & a_2 = 5a_0, \\
3(-1)a_3 & +(0)(-4)a_2 & = 0, & a_3 = 0, \\
4(0)a_4 & +(1)(-3)a_3 & = 0, & a_3 = 0, \quad a_4 \text{ arbitrary}, \\
5(1)a_5 & +(2)(-2)a_4 & = 0, & a_5 = \frac{4}{5}a_4, \\
6(2)a_6 & +(3)(-1)a_5 & = 0, & a_6 = \frac{1}{5}a_4, \\
7(3)a_7 & +(4)(0)a_6 & = 0, & a_7 = 0.
\end{array}$$

All subsequent a's are proportional to a_7 and are therefore zero. Finally, we have

$$y = a_0 \left[x^{-4} + 4x^{-3} + 5x^{-2}\right] + a_4 \left[1 + \frac{4}{5}x + \frac{1}{5}x^2\right].$$

7. Set $L(y) = x^2 y'' + x^2 y' - 2y$. Then

$$L(y) = \sum_{n=0}^{\infty} \left[(n + c)(n + c - 1) - 2\right]a_n x^{n+c} + \sum_{n=0}^{\infty} (n + c)a_n x^{n+c+1}$$

$$= \sum_{n=0}^{\infty} (n + c + 1)(n + c - 2)a_n x^{n+c} + \sum_{n=1}^{\infty} (n + c - 1)a_{n-1} x^{n+c}$$

$$= (c + 1)(c - 2)a_0 x^c + \sum_{n=1}^{\infty} \left[(n + c + 1)(n + c - 2)a_n + (n + c - 1)a_{n-1}\right]x^{n+c}.$$

We choose the smaller of the two roots of the indicial equation, namely $c = -1$, and consider

$$\begin{array}{llll}
n(n - 3)a_n & + (n - 2)a_{n-1} = 0, & \text{for } n \geq 1. \\
1(-2)a_1 & + (-1)a_0 & = 0, & a_1 = -\frac{1}{2}a_0, \\
2(-1)a_2 & + (0)a_1 & = 0, & a_2 = 0, \\
3(0)a_3 & + (1)a_2 & = 0, & a_2 = 0, \quad a_3 \text{ arbitrary},
\end{array}$$

$$a_n = \frac{-(n-2)\,a_{n-1}}{n(n-3)} \text{ for } n \geq 4.$$

$$a_4 = \frac{-2\,a_3}{4\cdot 1}$$

$$a_5 = \frac{-3\,a_4}{5\cdot 2}$$

$$\vdots$$

$$a_n = \frac{(-1)^{n-3}[2\cdot 3\cdots(n-2)]\,a_3}{[4\cdot 5\cdots n][1\cdot 2\cdots(n-3)]} = \frac{6(-1)^{n-3}\,a_3}{n(n-1)\cdot(n-3)!} \text{ for } n \geq 4.$$

Thus $y = a_0\left[x^{-1} - \frac{1}{2}\right] + a_3\left[x^2 + \sum_{n=4}^{\infty}\frac{6(-1)^{n-3}x^{n-1}}{n(n-1)\cdot(n-3)!}\right].$

9. We set $z = x - 1$ and obtain $L(y) = z(z+1)\dfrac{d^2y}{dz^2} + 3\dfrac{dy}{dz} - 2y = 0.$

$$L(y) = \sum_{n=0}^{\infty}\big[(n+c)(n+c-1) - 2\big]a_n z^{n+c} + \sum_{n=0}^{\infty}\big[(n+c)(n+c-1) + 3(n+c)\big]a_n z^{n+c-1}$$

$$= \sum_{n=0}^{\infty}(n+c+1)(n+c-2)a_n z^{n+c} + \sum_{n=0}^{\infty}(n+c)(n+c+2)a_n z^{n+c-1}$$

$$= c(c+2)a_0 z^{c-1} + \sum_{n=1}^{\infty}\big[(n+c)(n+c+2)a_n + (n+c)(n+c-3)a_{n-1}\big]z^{n+c-1}.$$

We choose the smaller of the two roots of the indicial equation, namely $c = -2$, and consider

$$n(n-2)a_n + (n-2)(n-5)a_{n-1} = 0, \quad \text{for } n \geq 1.$$
$$1(-1)a_1 \quad + (-1)(-4)a_0 \quad\quad = 0, \quad a_1 = 4a_0,$$
$$2(0)a_2 \quad + (0)(-3)a_1 \quad\quad = 0, \quad a_2 \text{ arbitrary,}$$

We have $a_n = \dfrac{-(n-5)\,a_{n-1}}{n}$ for $n \geq 3$, so that $a_3 = \frac{2}{3}a_2$, $a_4 = \frac{1}{6}a_2$, and $a_n = 0$ for $n \geq 5$.

Thus $y = a_0\left[(x-1)^{-2} + 4(x-1)^{-1}\right] + a_2\left[1 + \frac{2}{3}(x-1) + \frac{1}{6}(x-1)^2\right].$

11. Set $L(y) = xy'' - 2(x+2)y' + 4y.$ Then

$$L(y) = \sum_{n=0}^{\infty}\big[(n+c)(n+c-1) - 4(n+c)\big]a_n x^{n+c-1} - 2\sum_{n=0}^{\infty}(n+c-2)a_n x^{n+c}$$

$$= \sum_{n=0}^{\infty}(n+c)(n+c-5)a_n x^{n+c-1} - 2\sum_{n=1}^{\infty}(n+c-3)a_{n-1} x^{n+c-1}$$

$$= c(c-5)a_0 x^{c-1} + \sum_{n=1}^{\infty}\big[(n+c)(n+c-5)a_n - 2(n+c-3)a_{n-1}\big]x^{n+c-1}.$$

We choose the smaller of the two roots of the indicial equation, namely $c = 0$, and consider

$$
\begin{aligned}
n(n-5)a_n \quad -2(n-3)a_{n-1} &= 0, \text{ for } n \geq 1. \\
1(-4)a_1 \quad -2(-2)a_0 &= 0, \ a_1 = a_0, \\
2(-3)a_2 \quad -2(-1)a_1 &= 0, \ a_2 = \tfrac{1}{3}a_0, \\
3(-2)a_3 \quad -2(0)a_2 &= 0, \ a_3 = 0, \\
4(-1)a_4 \quad -2(1)a_3 &= 0, \ a_4 = 0, \\
5(0)a_5 \quad -2(2)a_4 &= 0, \ a_4 = 0, \quad a_5 \text{ arbitrary,}
\end{aligned}
$$

$$
a_n = \frac{2(n-3)\, a_{n-1}}{n(n-5)} \text{ for } n \geq 6.
$$

$$
a_6 = \frac{2 \cdot 3\, a_5}{6 \cdot 1}
$$

$$
a_7 = \frac{2 \cdot 4\, a_6}{7 \cdot 2}
$$

$$
\vdots
$$

$$
a_n = \frac{2^{n-5}\left[3 \cdot 4 \cdots (n-3)\right] a_5}{\left[6 \cdot 7 \cdots n\right]\left[1 \cdot 2 \cdots (n-5)\right]} = \frac{60 \cdot 2^{n-5}\, a_5}{(n-5)!\, n(n-1)(n-2)} \text{ for } n \geq 6.
$$

Thus $y = a_0\left[1 + x + \dfrac{1}{3}x^2\right] + a_5\left[x^5 + \displaystyle\sum_{n=6}^{\infty} \dfrac{60 \cdot 2^{n-5}x^n}{(n-5)!\, n(n-1)(n-2)}\right].$

13. Set $L(y) = x(x+3)y'' - 9y' - 6y$. Then

$$
\begin{aligned}
L(y) &= \sum_{n=0}^{\infty}\left[(n+c)(n+c-1) - 6\right]a_n x^{n+c} + \sum_{n=0}^{\infty}\left[3(n+c)(n+c-1) - 9(n+c)\right]a_n x^{n+c-1} \\
&= \sum_{n=0}^{\infty}(n+c-3)(n+c+2)a_n x^{n+c} + \sum_{n=0}^{\infty}3(n+c)(n+c-4)a_n x^{n+c-1} \\
&= 3c(c-4)a_0 x^{c-1} + \sum_{n=1}^{\infty}\left[3(n+c)(n+c-4)a_n + (n+c-4)(n+c+1)a_{n-1}\right]x^{n+c-1}.
\end{aligned}
$$

We choose the smaller of the two roots of the indicial equation, namely $c = 0$, and consider

$$
\begin{aligned}
3n(n-4)a_n \quad +(n-4)(n+1)a_{n-1} &= 0, \quad \text{for } n \geq 1. \\
3 \cdot 1(-3)a_1 \quad +(-3) \cdot 2a_0 &= 0, \quad a_1 = -\tfrac{2}{3}a_0, \\
3 \cdot 2(-2)a_2 \quad +(-2) \cdot 3a_1 &= 0, \quad a_2 = \tfrac{1}{3}a_0, \\
3 \cdot 3(-1)a_3 \quad +(-1) \cdot 4a_2 &= 0, \quad a_3 = -\tfrac{4}{27}a_0, \\
3 \cdot 4(0)a_4 \quad +(0) \cdot 5a_3 &= 0, \quad a_4 \text{ arbitrary.}
\end{aligned}
$$

$$a_n = \frac{-(n+1)\,a_{n-1}}{3n} \text{ for } n \geq 5.$$

$$a_5 = \frac{-6\,a_4}{3 \cdot 5}$$

$$a_6 = \frac{-7\,a_5}{3 \cdot 6}$$

$$\vdots$$

$$a_n = \frac{(-1)^{n-4}\left[6 \cdot 7 \cdots (n+1)\right]\,a_4}{3^{n-4}(5 \cdot 6 \cdots n)} = \frac{(-1)^{n-4}(n+1)\,a_4}{5 \cdot 3^{n-4}} \text{ for } n \geq 5.$$

Finally, $y = a_0 \left[1 - \frac{2}{3}x + \frac{1}{3}x^2 - \frac{4}{27}x^3\right] + a_4 \left[x^4 + \sum_{n=5}^{\infty} \frac{(-1)^n(n+1)x^n}{5 \cdot 3^{n-4}}\right].$

15. Set $L(y) = xy'' + (x^3 - 1)y' + x^2 y$. Then

$$\begin{aligned}
L(y) &= \sum_{n=0}^{\infty} \left[(n+c)(n+c-1) - (n+c)\right]a_n x^{n+c-1} + \sum_{n=0}^{\infty}(n+c+1)a_n x^{n+c+2} \\
&= \sum_{n=0}^{\infty}(n+c)(n+c-2)a_n x^{n+c-1} + \sum_{n=3}^{\infty}(n+c-2)a_{n-3}x^{n+c-1} \\
&= c(c-2)a_0 x^{c-1} + (c+1)(c-1)a_1 x^c + (c+2)ca_2 x^{c+1} \\
&\quad + \sum_{n=3}^{\infty}(n+c-2)\left[(n+c)a_n + a_{n-3}\right]x^{n+c-1}.
\end{aligned}$$

We choose the smaller of the two roots of the indicial equation, namely $c = 0$ and are forced to choose $a_1 = 0$. However, a_2 can be left arbitrary and $a_n = \frac{-a_{n-3}}{n}$ for $n \geq 3$.

$a_3 = \dfrac{-a_0}{3}$	$a_4 = 0$	$a_5 = \dfrac{-a_2}{5}$
$a_6 = \dfrac{-a_3}{6}$	$a_7 = 0$	$a_8 = \dfrac{-a_5}{8}$
$\vdots$	$\vdots$	$\vdots$
$a_{3k} = \dfrac{-a_{3k-3}}{3k}$	$a_{3k+1} = 0$	$a_{3k+2} = \dfrac{-a_{3k-1}}{3k+2}$
$a_{3k} = \dfrac{(-1)^k\,a_0}{3^k k!}$	$a_{3k+1} = 0$	$a_{3k+2} = \dfrac{(-1)^k\,a_2}{\left[5 \cdot 8 \cdots (3k+2)\right]}$

Finally, $y = a_0 \left[1 + \sum_{k=1}^{\infty} \frac{(-1)^k x^{3k}}{3^k\,k!}\right] + a_2 \left[x^2 + \sum_{k=1}^{\infty} \frac{(-1)^k x^{3k+2}}{\left[5 \cdot 8 \cdots (3k+2)\right]}\right].$

18.9 Logarithmic Case

1. Set $L(y) = xy'' + y$. Then

$$L(y) = \sum_{n=0}^{\infty} (n+c)(n+c-1)a_n x^{n+c-1} + \sum_{n=0}^{\infty} a_n x^{n+c}$$

$$= \sum_{n=0}^{\infty} (n+c)(n+c-1)a_n x^{n+c-1} + \sum_{n=1}^{\infty} a_{n-1} x^{n+c-1}$$

$$= c(c-1)a_0 x^{c-1} + \sum_{n=1}^{\infty} [(n+c)(n+c-1)a_n + a_{n-1}] x^{n+c-1}.$$

The indicial equation is $c(c-1) = 0$ and $a_n = \dfrac{-a_{n-1}}{(n+c)(n+c-1)}$ for $n \geq 1$.

$$a_1 = \frac{-a_0}{(c+1)c}$$

$$a_2 = \frac{-a_1}{(c+2)(c+1)}$$

$$\vdots$$

$$a_n = \frac{-a_{n-1}}{(c+n)(c+n-1)} = \frac{(-1)^n\, a_0}{(c+1)\cdots(c+n)c(c+1)\cdots(c+n-1)}.$$

$$y_c = a_0 \left[x^c + \sum_{n=1}^{\infty} \frac{(-1)^n x^{n+c}}{(c+1)\cdots(c+n)c(c+1)\cdots(c+n-1)} \right].$$

We wish to take $a_0 = c$, so we write

$$y_c = cx^c - \frac{x^{c+1}}{c+1} + \sum_{n=2}^{\infty} \frac{(-1)^n x^{n+c}}{(c+1)\cdots(c+n)(c+1)\cdots(c+n-1)},$$

$$\frac{\partial y_c}{\partial c} = y_c \ln x + x^c + \frac{x^{c+1}}{(c+1)^2} + \sum_{n=2}^{\infty} \frac{(-1)^n x^{n+c}}{(c+1)\cdots(c+n)(c+1)\cdots(c+n-1)} \bullet$$

$$\left[-\frac{1}{c+1} - \cdots - \frac{1}{c+n} - \frac{1}{c+1} - \cdots - \frac{1}{c+n-1} \right].$$

Substituting $c = 0$ we obtain

$$y_1 = -x + \sum_{n=2}^{\infty} \frac{(-1)^n x^n}{n!\,(n-1)!},$$

$$y_2 = y_1 \ln x + 1 + x - \sum_{n=2}^{\infty} \frac{(-1)^n (H_n + H_{n-1}) x^n}{n!\,(n-1)!}.$$

3. Set $L(y) = 2xy'' + 6y' + y$. Then

$$L(y) = \sum_{n=0}^{\infty}\left[2(n+c)(n+c-1)+6(n+c)\right]a_n x^{n+c-1} + \sum_{n=0}^{\infty} a_n x^{n+c}$$

$$= \sum_{n=0}^{\infty} 2(n+c)(n+c+2)a_n x^{n+c-1} + \sum_{n=1}^{\infty} a_{n-1} x^{n+c-1}$$

$$= 2c(c+2)a_0 x^{c-1} + \sum_{n=1}^{\infty}\left[\left[2(n+c)(n+c+2)a_n + a_{n-1}\right]x^{n+c-1}\right].$$

The indicial equation is $c(c+2) = 0$ and $a_n = \dfrac{-a_{n-1}}{2(n+c)(n+c+2)}$ for $n \geq 1$.

$$a_1 = \frac{-a_0}{2(c+1)(c+3)}$$

$$a_2 = \frac{-a_1}{2(c+2)(c+4)}$$

$$\vdots$$

$$a_n = \frac{-a_{n-1}}{2(c+n)(c+n+2)} = \frac{(-1)^n\, a_0}{2^n(c+1)\cdots(c+n)(c+3)\cdots(c+n+2)}.$$

$$y_c = a_0\left[x^c + \sum_{n=1}^{\infty}\frac{(-1)^n x^{n+c}}{2^n(c+1)\cdots(c+n)(c+3)\cdots(c+n+2)}\right].$$

We need to take $a_0 = c+2$, so we write

$$y_c = (c+2)x^c - \frac{(c+2)x^{c+1}}{2(c+1)(c+3)} + \frac{x^{c+2}}{2^2(c+1)(c+3)(c+4)}$$
$$+ \sum_{n=3}^{\infty}\frac{(-1)^n x^{n+c}}{2^n(c+1)\left[(c+3)\cdots(c+n)\right]\left[(c+3)\cdots(c+n+2)\right]},$$

$$\frac{\partial y_c}{\partial c} = y_c \ln x + x^c - \frac{(c+2)x^{c+1}}{2(c+1)(c+3)}\left[\frac{1}{c+2}-\frac{1}{c+1}-\frac{1}{c+3}\right]$$
$$+ \frac{x^{c+2}}{2^2(c+1)(c+3)(c+4)}\left[-\frac{1}{c+1}-\frac{1}{c+3}-\frac{1}{c+4}\right]$$
$$+ \sum_{n=3}^{\infty}\frac{(-1)^n x^{n+c}}{2^n(c+1)\left[(c+3)\cdots(c+n)\right]\left[(c+3)\cdots(c+n+2)\right]} \bullet$$
$$\left[-\frac{1}{c+1}-\left\{\frac{1}{c+3}+\cdots+\frac{1}{c+n}\right\}-\left\{\frac{1}{c+3}+\cdots+\frac{1}{c+n+2}\right\}\right].$$

Substituting $c = -2$ we obtain

$$y_1 = -\frac{1}{8} + \sum_{n=3}^{\infty} \frac{(-1)^n x^{n-2}}{2^n(-1)(n-2)! \, n!} = \sum_{n=2}^{\infty} \frac{(-1)^{n+1} x^{n-2}}{2^n \, n! \, (n-2)!},$$

$$y_2 = y_1 \ln x + x^{-2} + \frac{1}{2}x^{-1} + \frac{1}{16} + \sum_{n=3}^{\infty} \frac{(-1)^n \left(-1 + H_{n-2} + H_n\right) x^{n-2}}{2^n \, n! \, (n-2)!}$$

$$= y_1 \ln x + y_1 + x^{-2} + \frac{1}{2}x^{-1} + \sum_{n=2}^{\infty} \frac{(-1)^n \left(H_n + H_{n-2}\right) x^{n-2}}{2^n \, n! \, (n-2)!}.$$

5. Set $L(y) = x^2 y'' - x(6 + x)y' + 10y$. Then

$$L(y) = \sum_{n=0}^{\infty} \left[(n+c)(n+c-1) - 6(n+c) + 10\right] a_n x^{n+c} - \sum_{n=0}^{\infty} (n+c) a_n x^{n+c+1}$$

$$= \sum_{n=0}^{\infty} (n+c-2)(n+c-5) a_n x^{n+c} - \sum_{n=1}^{\infty} (n+c-1) a_{n-1} x^{n+c}$$

$$= (c-2)(c-5) a_0 x^c + \sum_{n=1}^{\infty} \left[(n+c-2)(n+c-5) a_n - (n+c-1) a_{n-1}\right] x^{n+c}.$$

The indicial equation is $(c-2)(c-5) = 0$ and $a_n = \dfrac{(n+c-1) \, a_{n-1}}{(n+c-2)(n+c-5)}$ for $n \geq 1$.

$$a_1 = \frac{c \, a_0}{(c-1)(c-4)}$$

$$a_2 = \frac{(c+1) \, a_1}{c(c-3)} = \frac{(c+1) \, a_0}{(c-1)(c-4)(c-3)}$$

$$a_3 = \frac{(c+2) \, a_2}{(c+1)(c-2)} = \frac{(c+2) \, a_0}{(c-1)(c-4)(c-3)(c-2)}$$

$$\vdots$$

$$a_n = \frac{(n+c-1) \, a_{n-1}}{(n+c-2)(n+c-5)} = \frac{(c+n-1) \, a_0}{(c-1)(c-4)(c-3)(c-2) \cdots (c+n-5)}.$$

$$y_c = a_0 x^c + \frac{c a_0 x^{c+1}}{(c-1)(c-4)} + \frac{(c+1) a_0 x^{c+2}}{(c-1)(c-4)(c-3)} + \frac{(c+2) a_0 x^{c+3}}{(c-1)(c-4)(c-3)(c-2)}$$

$$+ \sum_{n=4}^{\infty} \frac{(c+n-1) a_0 x^{n+c}}{(c-1)(c-4)(c-3)(c-2) \cdots (c+n-5)}.$$

We need to take $a_0 = c - 2$, so we write

$$y_c = (c-2)x^c + \frac{c(c-2)x^{c+1}}{(c-1)(c-4)} + \frac{(c+1)(c-2)x^{c+2}}{(c-1)(c-4)(c-3)} + \frac{(c+2)x^{c+3}}{(c-1)(c-4)(c-3)}$$
$$+ \sum_{n=4}^{\infty} \frac{(c+n-1)x^{n+c}}{(c-1)(c-4)(c-3)[(c-1)\cdots(c+n-5)]},$$

$$\frac{\partial y_c}{\partial c} = y_c \ln x + x^c + \frac{c(c-2)x^{c+1}}{(c-1)(c-4)}\left[\frac{1}{c} + \frac{1}{c-2} - \frac{1}{c-1} - \frac{1}{c-4}\right]$$
$$+ \frac{(c+1)(c-2)x^{c+2}}{(c-1)(c-4)(c-3)}\left[\frac{1}{c+1} + \frac{1}{c-2} - \frac{1}{c-1} - \frac{1}{c-4} - \frac{1}{c-3}\right]$$
$$+ \frac{(c+2)x^{c+3}}{(c-1)(c-4)(c-3)}\left[\frac{1}{c+2} - \frac{1}{c-1} - \frac{1}{c-4} - \frac{1}{c-3}\right]$$
$$+ \sum_{n=4}^{\infty} \frac{(c+n-1)x^{n+c}}{(c-1)(c-4)(c-3)[(c-1)\cdots(c+n-5)]} \bullet$$
$$\left[\frac{1}{c+n-1} - \frac{1}{c-1} - \frac{1}{c-4} - \frac{1}{c-3} - \left\{\frac{1}{c-1} - \cdots - \frac{1}{c+n-5}\right\}\right].$$

Substituting $c = 2$ we obtain

$$y_1 = 2x^5 + \sum_{n=4}^{\infty} \frac{(n+1)x^{n+2}}{2(n-3)!} = \sum_{n=3}^{\infty} \frac{(n+1)x^{n+2}}{2(n-3)!},$$

$$y_2 = y_1 \ln x + x^2 - x^3 + \frac{3}{2}x^4 + \frac{3}{2}x^5 + \sum_{n=4}^{\infty} \frac{(n+1)x^{n+2}}{2(n-3)!}\left[\frac{1}{n+1} - \frac{1}{1} + \frac{1}{2} + 1 - H_{n-3}\right],$$

$$= y_1 \ln x + \frac{1}{2}y_1 + x^2 - x^3 + \frac{3}{2}x^4 + \sum_{n=3}^{\infty} \frac{[1 - (n+1)H_{n-3}]x^{n+2}}{2(n-3)!}.$$

7. Set $L(y) = x(1-x)y'' + 2(1-x)y' + 2y$. Then

$$L(y) = \sum_{n=0}^{\infty} \left[(n+c)(n+c-1) + 2(n+c)\right]a_n x^{n+c-1}$$
$$- \sum_{n=0}^{\infty} (n+c)(n+c-1) + 2(n+c) - 2]a_n x^{n+c}$$
$$= \sum_{n=0}^{\infty} (n+c)(n+c+1)a_n x^{n+c-1} - \sum_{n=0}^{\infty} (n+c+2)(n+c-1)a_n x^{n+c}$$
$$= c(c+1)a_0 x^{c-1} + \sum_{n=1}^{\infty} \left[(n+c)(n+c+1)a_n - (n+c+1)(n+c-2)a_{n-1}\right]x^{n+c-1}.$$

The indicial equation is $c(c+1) = 0$ and $a_n = \dfrac{(n+c-2)\, a_{n-1}}{n+c}$ for $n \geq 1$.

$$a_1 = \frac{(c-1)\, a_0}{c+1}$$

$$a_2 = \frac{c\, a_1}{c+2} = \frac{(c-1)c\, a_0}{(c+1)(c+2)}$$

$$\vdots$$

$$a_n = \frac{(n+c-2)\, a_{n-1}}{c+n} = \frac{(c-1)c\cdots(c+n-2)\, a_0}{(c+1)\cdots(c+n)} = \frac{(c-1)c\, a_0}{(c+n-1)(c+n)}.$$

We need to take $a_0 = c+1$, so we have

$$y_c = (c+1)x^c + (c-1)x^{c+1} + \frac{(c-1)cx^{c+2}}{c+2} + \sum_{n=3}^{\infty} \frac{(c-1)c(c+1)x^{n+c}}{(c+n-1)(c+n)},$$

$$\frac{\partial y_c}{\partial c} = y_c \ln x + x^c + x^{c+1} + \frac{(c-1)cx^{c+2}}{(c+2)}\left[\frac{1}{c-1} + \frac{1}{c} - \frac{1}{c+2}\right]$$

$$+ \sum_{n=3}^{\infty} \frac{(c-1)c(c+1)x^{n+c}}{(c+n-1)(c+n)}\left[\frac{1}{c-1} + \frac{1}{c} + \frac{1}{c+1} - \frac{1}{c+n-1} - \frac{1}{c+n}\right].$$

Substituting $c = -1$ we obtain

$$y_1 = -2 + 2x,$$

$$y_2 = y_1 \ln x + x^{-1} + 1 - 5x + \sum_{n=3}^{\infty} \frac{2x^{n-1}}{(n-1)(n-2)}.$$

9. $x(1-x)y'' + 2(1-x)y' + 2y$. Let $z = x - 1$ to get $L(y) = z(z+1)\dfrac{d^2y}{dz^2} + 2z\dfrac{dy}{dz} - 2y = 0$. Then

$$L(y) = \sum_{n=0}^{\infty} \left[(n+c)(n+c-1) + 2(n+c) - 2\right]a_n z^{n+c} + \sum_{n=0}^{\infty}(n+c)(n+c-1)a_n z^{n+c-1}$$

$$= \sum_{n=0}^{\infty}(n+c+2)(n+c-1)a_n z^{n+c} + \sum_{n=0}^{\infty}(n+c)(n+c-1)a_n z^{n+c-1}$$

$$= c(c-1)a_0 z^{c-1} + \sum_{n=1}^{\infty}\left[(n+c)(n+c-1)a_n + (n+c+1)(n+c-2)a_{n-1}\right]z^{n+c-1}.$$

The indicial equation is $c(c-1) = 0$ and $a_n = \dfrac{-(n+c+1)(n+c-2)\,a_{n-1}}{(n+c)(n+c-1)}$ for $n \geq 1$.

$$a_1 = \frac{-(c+2)(c-1)\,a_0}{(c+1)c}$$

$$a_2 = \frac{-(c+3)c\,a_1}{(c+2)(c+1)} = \frac{(c-1)(c+3)\,a_0}{(c+1)^2}$$

$$\vdots$$

$$a_n = \frac{-(n+c+1)(n+c-2)\,a_{n-1}}{(n+c)(n+c-1)} = \frac{(-1)^n\big[(c+2)\cdots(c+n+1)\big]\big[(c-1)\cdots(c+n-2)\big]\,a_0}{\big[(c+1)\cdots(c+n)\big]\big[c\cdots(c+n-1)\big]}$$

$$= \frac{(-1)^n(c+n+1)(c-1)\,a_0}{(c+1)(c+n-1)}.$$

We need to take $a_0 = c$, so we have

$$y_c = cz^c - \frac{(c+2)(c-1)z^{c+1}}{c+1} + \sum_{n=2}^{\infty} \frac{(-1)^n(c+n+1)(c-1)cz^{n+c}}{(c+1)(c+n-1)},$$

$$\frac{\partial y_c}{\partial c} = y_c \ln z + z^c - \frac{(c+2)(c-1)z^{c+1}}{c+1}\left[\frac{1}{c+2} + \frac{1}{c-1} - \frac{1}{c+1}\right]$$

$$+ \sum_{n=2}^{\infty} \frac{(-1)^n(c+n+1)(c-1)cz^{n+c}}{(c+1)(c+n-1)}\left[\frac{1}{c+n+1} + \frac{1}{c-1} + \frac{1}{c} - \frac{1}{c+1} - \frac{1}{c+n-1}\right].$$

Substituting $c = 0$ we obtain

$$y_1 = 2z = 2(x-1),$$

$$y_2 = y_1 \ln(x-1) + 1 - 3(x-1) + \sum_{n=2}^{\infty} \frac{(-1)^{n+1}(n+1)(x-1)^n}{n-1}.$$

11. Set $L(y) = x^2 y'' - 5xy' + (8+5x)y$. Then

$$L(y) = \sum_{n=0}^{\infty} \big[(n+c)(n+c-1) - 5(n+c) + 8\big]a_n x^{n+c} + \sum_{n=0}^{\infty} 5a_n x^{n+c+1}$$

$$= \sum_{n=0}^{\infty} (n+c-2)(n+c-4)a_n x^{n+c} + \sum_{n=1}^{\infty} 5a_{n-1} x^{n+c}$$

$$= (c-2)(c-4)a_0 x^c + \sum_{n=1}^{\infty} \big[(n+c-2)(n+c-4)a_n + 5a_{n-1}\big]x^{n+c}.$$

The indicial equation is $(c - 2)(c - 4) = 0$ and $a_n = \dfrac{-5\,a_{n-1}}{(n + c - 2)(n + c - 4)}$ for $n \geq 1$.

$$a_1 = \frac{-5\,a_0}{(c - 1)(c - 3)}$$

$$a_2 = \frac{-5\,a_1}{c(c - 2)} = \frac{5^2\,a_0}{c(c - 1)(c - 3)(c - 2)}$$

$$\vdots$$

$$a_n = \frac{-5\,a_{n-1}}{(n + c - 2)(n + c - 4)} = \frac{(-1)^n 5^n\,a_0}{\big[(c - 1)\cdots(c + n - 2)\big]\big[(c - 3)\cdots(c + n - 4)\big]}.$$

We need to take $a_0 = c - 2$, so we have

$$y_c = (c - 2)x^c - \frac{5(c - 2)x^{c+1}}{(c - 1)(c - 3)} + \frac{5^2 x^{c+2}}{c(c - 1)(c - 3)}$$
$$+ \sum_{n=3}^{\infty} \frac{(-1)^n 5^n x^{n+c}}{\big[(c - 1)c\cdots(c + n - 2)\big](c - 3)\big[(c - 1)\cdots(c + n - 4)\big]},$$

$$\frac{\partial y_c}{\partial c} = y_c \ln x + x^c - \frac{5(c - 2)x^{c+1}}{(c - 1)(c - 3)}\left[\frac{1}{c - 2} - \frac{1}{c - 1} - \frac{1}{c - 3}\right]$$
$$+ \frac{5^2 x^{c+2}}{c(c - 1)(c - 3)}\left[-\frac{1}{c} - \frac{1}{c - 1} - \frac{1}{c - 3}\right]$$
$$+ \sum_{n=3}^{\infty} \frac{(-1)^n 5^n x^{n+c}}{\big[(c - 1)c\cdots(c + n - 2)\big](c - 3)\big[(c - 1)\cdots(c + n - 4)\big]} \bullet$$
$$\left[-\left\{\frac{1}{c - 1} + \cdots + \frac{1}{c + n - 2}\right\} - \frac{1}{c - 3} - \left\{\frac{1}{c - 1} + \cdots + \frac{1}{c + n - 4}\right\}\right].$$

Substituting $c = 2$ we obtain

$$y_1 = -\frac{25}{2}x^4 + \sum_{n=3}^{\infty} \frac{(-1)^{n+1} 5^n x^{n+2}}{n!\,(n - 2)!} = \sum_{n=2}^{\infty} \frac{(-1)^{n+1} 5^n x^{n+2}}{n!\,(n - 2)!},$$

$$y_2 = y_1 \ln x + x^2 + 5x^3 + \sum_{n=2}^{\infty} \frac{(-1)^n 5^n \big(H_n + H_{n-2} - 1\big)x^{n+2}}{n!\,(n - 2)!}$$

$$= y_1 \ln x + y_1 + x^2 + 5x^3 + \sum_{n=2}^{\infty} \frac{(-1)^n 5^n \big(H_n + H_{n-2}\big)x^{n+2}}{n!\,(n - 2)!}.$$

13. Set $L(y) = 9x^2 y'' - 15xy' + 7(1 + x)y$. Then

$$L(y) = \sum_{n=0}^{\infty} \big[9(n + c)(n + c - 1) - 15(n + c) + 7\big]a_n x^{n+c} + \sum_{n=0}^{\infty} 7a_n x^{n+c+1}$$

$$= \sum_{n=0}^{\infty}(3n + 3c - 7)(3n + 3c - 1)a_n x^{n+c} + \sum_{n=1}^{\infty} 7a_{n-1}x^{n+c}$$

$$= (3c - 7)(3c - 1)a_0 x^c + \sum_{n=1}^{\infty}\big[(3n + 3c - 7)(3n + 3c - 1)a_n + 7a_{n-1}\big]x^{n+c}.$$

The indicial equation is $(3c - 7)(3c - 1) = 0$ and $a_n = \dfrac{-7\,a_{n-1}}{(3n + 3c - 7)(3n + 3c - 1)}$ for $n \ge 1$.

$$a_1 = \frac{-7\,a_0}{(3c - 4)(3c + 2)}$$

$$a_2 = \frac{-7\,a_1}{(3c - 1)(3c + 5)} = \frac{49\,a_0}{(3c - 4)(3c - 1)(3c + 2)(3c + 5)}$$

$$\vdots$$

$$a_n = \frac{-7\,a_{n-1}}{(3n + 3c - 7)(3n + 3c - 1)} = \frac{(-1)^n 7^n\,a_0}{\big[(3c - 4)\cdots(3c + 3n - 7)\big]\big[(3c + 2)\cdots(3c + 3n - 1)\big]}.$$

We need to take $a_0 = 3c - 1$, so we have

$$y_c = (3c - 1)x^c - \frac{7(3c - 1)x^{c+1}}{(3c - 4)(3c + 2)}$$

$$+ \sum_{n=2}^{\infty} \frac{(-1)^n 7^n x^{n+c}}{(3c - 4)\big[(3c + 2)\cdots(3c + 3n - 7)\big]\big[(3c + 2)\cdots(3c + 3n - 1)\big]},$$

$$\frac{\partial y_c}{\partial c} = y_c \ln x + 3x^c - \frac{7(3c - 1)x^{c+1}}{(3c - 4)(3c + 2)}\left[\frac{3}{3c - 1} - \frac{3}{3c - 4} - \frac{3}{3c + 2}\right]$$

$$+ \sum_{n=2}^{\infty} \frac{(-1)^{n+1} 7^n x^{n+c}}{(3c - 4)\big[(3c + 2)\cdots(3c + 3n - 7)\big]\big[(3c + 2)\cdots(3c + 3n - 1)\big]} \bullet$$

$$\left[\frac{3}{3c - 4} + \left\{\frac{3}{3c + 2} + \cdots + \frac{3}{3c + 3n - 7}\right\} + \left\{\frac{3}{3c + 2} + \cdots + \frac{3}{3c + 3n - 1}\right\}\right].$$

Substituting $c = \frac{1}{3}$ we obtain

$$y_1 = \sum_{n=2}^{\infty} \frac{(-1)^n 7^n x^{n+1/3}}{(-3)\big[3\cdots(3n - 6)\big]\big[3\cdots(3n)\big]} = \sum_{n=2}^{\infty} \frac{(-1)^{n+1} 7^n x^{n+1/3}}{3^{2n-1}\,n!\,(n - 2)!},$$

$$y_2 = y_1 \ln x + 3x^{1/3} + \frac{7}{3}x^{4/3} + \sum_{n=2}^{\infty} \frac{(-7)^n\big(H_n + H_{n-2} - 1\big)x^{n+1/3}}{3^{2n-1}\,n!\,(n - 2)!}$$

18.10   Solution for Large x

1. $x^3(x-1)y'' + (x-1)y' + 4xy = 0$. Using the substitutions suggested in the text we get

$$\frac{1}{w^3}\left(\frac{1}{w}-1\right)\left(w^4\frac{d^2y}{dw^2} + 2w^3\frac{dy}{dw}\right) + \left(\frac{1}{w}-1\right)\left(-w^2\frac{dy}{dw}\right) + \frac{4}{w}y = 0,$$

$$w(1-w)\frac{d^2y}{dw^2} + (2-2w-w^2+w^3)\frac{dy}{dw} + 4y = 0.$$

Since $w = 0$ is a R.S.P. of this equation the point at infinity is a R.S.P. of the original equation.

3. $y'' + xy = 0$. Again the change of variable $w = \dfrac{1}{x}$ gives

$$w^4\frac{d^2y}{dw^2} + 2w^3\frac{dy}{dw} + \frac{1}{w}y = 0,$$

$$w^5\frac{d^2y}{dw^2} + 2w^4\frac{dy}{dw} + y = 0.$$

$w = 0$ is an I.S.P. of this equation. It follows that $x = 0$ is an I.S.P. of the original equation.

5. $x^4y'' + y = 0$. Setting $w = \dfrac{1}{x}$ yields

$$\frac{1}{w^4}\left(w^4\frac{d^2y}{dw^2} + 2w^3\frac{dy}{dw}\right) + y = 0,$$

$$w\frac{d^2y}{dw^2} + 2\frac{dy}{dw} + wy = 0.$$

$w = 0$ is a R.S.P. of this equation. It follows that $x = 0$ is a R.S.P. of the original equation.

7. $x^4y'' + x(1+2x^2)y' + 5y = 0$. We set $w = 1/x$ and get $L(y) = \dfrac{d^2y}{dw^2} - w\dfrac{dy}{dw} + 5y = 0$. $w = 0$ is an ordinary point so we put $y = \displaystyle\sum_{n=0}^{\infty} a_n w^n$.

$$L(y) = \sum_{n=2}^{\infty} n(n-1)a_n w^{n-2} - \sum_{n=1}^{\infty} na_n w^n + \sum_{n=0}^{\infty} 5a_n w^n$$

$$= \sum_{n=0}^{\infty}(n+2)(n+1)a_{n+2}w^n + 5a_0 - \sum_{n=1}^{\infty}(n-5)a_n w^n$$

$$= 5a_0 + 2a_2 + \sum_{n=1}^{\infty}\left[(n+2)(n+1)a_{n+2} - (n-5)a_n\right]w^n$$

$$= \sum_{n=0}^{\infty}\left[(n+2)(n+1)a_{n+2} - (n-5)a_n\right]w^n.$$

The recurrence relation is $a_{n+2} = \dfrac{(n-5)\, a_n}{(n+1)(n+2)}$ for $n \geq 0$.

$$a_2 = \frac{-5\, a_0}{1 \cdot 2} \qquad\qquad a_3 = \frac{-4\, a_1}{2 \cdot 3} = -\frac{2}{3}\, a_1$$

$$a_4 = \frac{-3\, a_2}{3 \cdot 4} \qquad\qquad a_5 = \frac{-2\, a_3}{4 \cdot 5} = \frac{1}{15}\, a_1$$

$$\vdots \qquad\qquad\qquad\qquad \vdots$$

$$a_{2k} = \frac{(2k-7)\, a_{2k-2}}{(2k-1)(2k)} \qquad\qquad a_{2k+1} = 0 \text{ for } k \geq 3$$

$$a_{2k} = \frac{(-5)(-3)\cdots(2k-7)\, a_0}{1 \cdot 3 \cdots (2k-1)\, 2^k\, k!} = \frac{-15\, a_0}{(2k-5)(2k-3)(2k-1)\, 2^k\, k!}. \text{ It follows that}$$

$$y_1 = \sum_{k=0}^{\infty} \frac{-15 w^{2k}}{2^k\, k!\, (2k-5)(2k-3)(2k-1)} = \sum_{k=0}^{\infty} \frac{-15 x^{-2k}}{2^k\, k!\, (2k-5)(2k-3)(2k-1)},$$

$$y_2 = w - \frac{2}{3} w^3 + \frac{1}{15} w^5 = x^{-1} - \frac{2}{3} x^{-3} + \frac{1}{15} x^{-5}.$$

9. $x(1-x)y'' - 3y' + 2y = 0$. We set $w = 1/x$ and get

$$L(y) = (w^3 - w^2)\frac{d^2 y}{dw^2} + (5w^2 - 2w)\frac{dy}{dw} + 2y = 0.$$

$w = 0$ is a regular singular point so we put $y = \displaystyle\sum_{n=0}^{\infty} a_n w^{n+c}$.

$$L(y) = \sum_{n=0}^{\infty} \left[(n+c)(n+c-1) + 5(n+c)\right] a_n w^{n+c+1}$$

$$- \sum_{n=0}^{\infty} \left[(n+c)(n+c-1) + 2(n+c) - 2\right] a_n w^{n+c}$$

$$= \sum_{n=0}^{\infty} (n+c)(n+c+4) a_n w^{n+c+1} - \sum_{n=0}^{\infty} (n+c+2)(n+c-1) a_n w^{n+c}$$

$$= -(c+2)(c-1)a_0 w^c$$

$$+ \sum_{n=1}^{\infty} \left[(n+c-1)(n+c+3)a_{n-1} - (n+c+2)(n+c-1)a_n\right] w^{n+c}.$$

The indicial equation is $(c+2)(c-1) = 0$. Choosing the smaller root, namely $c = -2$, we must satisfy the recurrence relation $n(n-3)\,a_n = (n+1)(n-3)\,a_{n-1}$ for $n \geq 1$.

$$1(-2)a_1 = 2(-2)a_0, \qquad\qquad a_1 = 2a_0,$$
$$2(-1)a_2 = 3(-1)a_1, \qquad\qquad a_2 = 3a_0,$$
$$3(0)a_3 = 4(0)a_2, \qquad\qquad a_3 \text{ arbitrary}$$

$$a_n = \frac{(n+1)\,a_{n-1}}{n} \text{ for } n \geq 4$$

$$a_4 = \frac{5\,a_3}{4}$$

$$a_5 = \frac{6\,a_4}{5}$$

$$\vdots$$

$$a_n = \frac{5 \cdot 6 \cdots (n+1)\,a_{n-1}}{4 \cdot 5 \cdots n} = \frac{n+1}{4}\,a_3.$$

$$y = a_0\left(w^{-2} + 2w^{-1} + 3\right) + a_3\left[w + \sum_{n=4}^{\infty} \frac{n+1}{4} w^{n-2}\right]$$

$$= a_0\left(x^2 + 2x + 3\right) + a_3\left[x^{-1} + \sum_{n=4}^{\infty} \frac{n+1}{4} x^{2-n}\right]$$

$$= a_0\left(x^2 + 2x + 3\right) + \frac{1}{4}a_3 \sum_{n=0}^{\infty} (n+4)x^{-n-1}.$$

11. $2x^2(x-1)y'' + x(5x-3)y' + (x+1)y = 0$. We set $w = 1/x$ and get

$$L(y) = 2w^2(1-w)\frac{d^2y}{dw^2} - w(w+1)\frac{dy}{dw} + (w+1)y = 0.$$

$w = 0$ is a regular singular point so we put $y = \sum_{n=0}^{\infty} a_n w^{n+c}$.

$$L(y) = -\sum_{n=0}^{\infty} \left[2(n+c)(n+c-1) + (n+c) - 1\right]a_n w^{n+c+1}$$

$$+ \sum_{n=0}^{\infty} \left[2(n+c)(n+c-1) - (n+c) + 1\right]a_n w^{n+c}$$

$$= -\sum_{n=0}^{\infty} (2n+2c+1)(n+c-1)a_n w^{n+c+1} + \sum_{n=0}^{\infty} (2n+2c-1)(n+c-1)a_n w^{n+c}$$

$$= (2c-1)(c-1)a_0 w^c$$

$$+ \sum_{n=1}^{\infty} \left[(2n+2c-1)(n+c-1)a_n - (2n+2c-1)(n+c-2)a_{n-1}\right]w^{n+c}.$$

The indicial equation is $(2c - 1)(c - 1) = 0$ and $a_n = \dfrac{(n + c - 2)\, a_{n-1}}{(n + c - 1)}$ for $n \geq 1$. For $c = 1$

we have $a_n = \dfrac{(n - 1)\, a_{n-1}}{n}$ for $n \geq 1$. Thus $a_1 = 0$ and $a_n = 0$ for all $n \geq 1$. That is,

$y_1 = w = x^{-1}$. For $c = 1/2$ we have $a_n = \dfrac{(2n - 3)\, a_{n-1}}{(2n - 1)}$ for $n \geq 1$.

$$a_1 = \frac{(-1)\, a_0}{1}$$

$$a_2 = \frac{1 \cdot a_1}{3}$$

$$\vdots$$

$$a_n = \frac{(-1) \cdot 1 \cdots (2n - 3)\, a_0}{1 \cdot 3 \cdots (2n - 1)} = \frac{-a_0}{2n - 1}.$$

Thus $y_2 = w^{1/2} - \displaystyle\sum_{n=1}^{\infty} \frac{w^{n+1/2}}{2n - 1} = -\sum_{n=0}^{\infty} \frac{x^{-n-1/2}}{2n - 1}.$

13. $2x^2(1 - x)y'' - 5x(1 + x)y' + (5 - x)y = 0$. We set $w = 1/x$ and get

$$L(y) = 2w^2(w - 1)\frac{d^2y}{dw^2} + w(9w + 1)\frac{dy}{dw} + (5w - 1)y = 0.$$

$w = 0$ is a regular singular point so we put $y = \displaystyle\sum_{n=0}^{\infty} a_n w^{n+c}.$

$$L(y) = \sum_{n=0}^{\infty} \big[2(n + c)(n + c - 1) + 9(n + c) + 5\big]a_n w^{n+c+1}$$

$$- \sum_{n=0}^{\infty} \big[2(n + c)(n + c - 1) - (n + c) + 1\big]a_n w^{n+c}$$

$$= \sum_{n=0}^{\infty} (2n + 2c + 5)(n + c + 1)a_n w^{n+c+1} - \sum_{n=0}^{\infty} (2n + 2c - 1)(n + c - 1)a_n w^{n+c}$$

$$= (2c - 1)(c - 1)a_0 w^c$$

$$+ \sum_{n=1}^{\infty} \big[(2n + 2c - 1)(n + c - 1)a_n - (2n + 2c + 3)(n + c)a_{n-1}\big]w^{n+c}.$$

The indicial equation is $(2c-1)(c-1) = 0$ and $a_n = \dfrac{(2n+2c+3)(n+c)\,a_{n-1}}{(2n+2c-1)(n+c-1)}$ for $n \geq 1$. For

$c = 1$ we have $a_n = \dfrac{(2n+5)(n+1)\,a_{n-1}}{n(2n+1)}$ for $n \geq 1$.

$$a_1 = \frac{7 \cdot 2\, a_0}{1 \cdot 3}$$

$$a_2 = \frac{9 \cdot 3\, a_1}{2 \cdot 5}$$

$$\vdots$$

$$a_n = \frac{\left[7 \cdot 9 \cdots (2n+5)\ (n+1)!\right] a_0}{n!\ 3 \cdot 5 \cdots (2n+1)} = \frac{(n+1)(2n+3)(2n+5)\,a_0}{15}.$$

Thus $y_1 = w + \dfrac{1}{15} \displaystyle\sum_{n=1}^{\infty}(n+1)(2n+3)(2n+5)w^{n+1} = \dfrac{1}{15}\displaystyle\sum_{n=0}^{\infty}(n+1)(2n+3)(2n+5)x^{-n-1}$. For

$c = 1/2$ we have $a_n = \dfrac{(n+2)(2n+1)\,a_{n-1}}{n(2n-1)}$ for $n \geq 1$.

$$a_1 = \frac{3 \cdot 3\, a_0}{1 \cdot 1}$$

$$a_2 = \frac{4 \cdot 5\, a_1}{2 \cdot 3}$$

$$\vdots$$

$$a_n = \frac{\left[3 \cdot 4 \cdots (n+2)\right]\left[3 \cdot 5 \cdots (2n+1)\right] a_0}{n!\ 1 \cdot 3 \cdots (2n-1)} = \frac{(n+1)(n+2)(2n+1)\,a_0}{2}.$$

Thus $y_2 = w^{1/2} + \dfrac{1}{2}\displaystyle\sum_{n=1}^{\infty}(n+1)(n+2)(2n+1)w^{n+1/2} = \dfrac{1}{2}\displaystyle\sum_{n=0}^{\infty}(n+1)(n+2)(2n+1)x^{-n-1/2}$.

15. $x(1+x)y'' + (1+5x)y' + 3y = 0$. We set $w = 1/x$ and get

$$L(y) = w^2(w+1)\frac{d^2y}{dw^2} + w(w-3)\frac{dy}{dw} + 3y = 0.$$

$w = 0$ is a regular singular point so we put $y = \displaystyle\sum_{n=0}^{\infty} a_n w^{n+c}$.

$$L(y) = \sum_{n=0}^{\infty}\left[(n+c)(n+c-1) + (n+c)\right]a_n w^{n+c+1}$$

$$+ \sum_{n=0}^{\infty}\left[(n+c)(n+c-1) - 3(n+c) + 3\right]a_n w^{n+c}$$

$$= \sum_{n=0}^{\infty}(n+c)^2 a_n w^{n+c+1} + \sum_{n=0}^{\infty}(n+c-1)(n+c-3)a_n w^{n+c}$$

$$= (c-1)(c-3)a_0 w^c + \sum_{n=1}^{\infty} \left[(n+c-1)(n+c-3)a_n + (n+c-1)^2 a_{n-1} \right] w^{n+c}.$$

The indicial equation is $(c-1)(c-3) = 0$ and $a_n = \dfrac{-(n+c-1)\,a_{n-1}}{(n+c-3)}$ for $n \geq 1$.

$$a_1 = \frac{-c\,a_0}{c-2}$$

$$a_2 = \frac{-(c+1)\,a_1}{c-1}$$

$$\vdots$$

$$a_n = \frac{(-1)^n c(c+1)\cdots(c+n-1)\,a_0}{(c-2)(c-1)\cdots(c+n-3)} = \frac{(-1)^n (n+c-2)(n+c-1)\,a_0}{(c-2)(c-1)}.$$

We need to take $a_0 = c - 1$, so we have

$$y_c = (c-1)w^c - \frac{c(c-1)w^{c+1}}{c-2} + \sum_{n=2}^{\infty} \frac{(-1)^n(n+c-2)(n+c-1)w^{n+c}}{c-2},$$

$$\frac{\partial y_c}{\partial c} = y_c \ln w + w^c - \frac{c(c-1)w^{c+1}}{c-2}\left[\frac{1}{c} + \frac{1}{c-1} - \frac{1}{c-2} \right]$$

$$+ \sum_{n=2}^{\infty} \frac{(-1)^n(n+c-2)(n+c-1)w^{n+c}}{c-2}\left[\frac{1}{n+c-2} + \frac{1}{n+c-1} - \frac{1}{c-2} \right].$$

Substituting $c = 1$ we obtain

$$y_1 = \sum_{n=2}^{\infty} (-1)^{n+1} n(n-1)w^{n+1} = \sum_{n=2}^{\infty} (-1)^{n+1} n(n-1)x^{-n-1},$$

$$y_2 = y_1 \ln w + w + w^2 + \sum_{n=2}^{\infty} (-1)^{n+1} n(n-1)w^{n+1}\left[1 + \frac{1}{n} + \frac{1}{n-1} \right]$$

$$= y_1 \ln(1/x) + x^{-1} + x^{-2} + \sum_{n=2}^{\infty} (-1)^{n+1}(n^2+n-1)x^{-n-1}.$$

17. $x(1-x)y'' + (1-4x)y' - 2y = 0$. We set $w = 1/x$ and get

$$L(y) = w^2(w-1)\frac{d^2y}{dw^2} + w(w+2)\frac{dy}{dw} - 2y = 0.$$

$w = 0$ is a regular singular point so we put $y = \displaystyle\sum_{n=0}^{\infty} a_n w^{n+c}$.

$$L(y) = \sum_{n=0}^{\infty} \left[(n+c)(n+c-1) + (n+c) \right] a_n w^{n+c+1}$$

$$-\sum_{n=0}^{\infty}\left[(n+c)(n+c-1)-2(n+c)+2\right]a_n w^{n+c}$$

$$=\sum_{n=0}^{\infty}(n+c)^2 a_n w^{n+c+1}-\sum_{n=0}^{\infty}(n+c-1)(n+c-2)a_n w^{n+c}$$

$$=(c-1)(c-2)a_0 w^c+\sum_{n=1}^{\infty}\left[(n+c-1)(n+c-2)a_n-(n+c-1)^2 a_{n-1}\right]w^{n+c}.$$

The indicial equation is $(c-1)(c-2)=0$ and $a_n=\dfrac{(n+c-1)\,a_{n-1}}{n+c-2}$ for $n\geq 1$.

$$a_1=\frac{c\,a_0}{c-1}$$

$$a_2=\frac{(c+1)\,a_1}{c}$$

$$\vdots$$

$$a_n=\frac{c(c+1)\cdots(c+n-1)\,a_0}{(c-1)c\cdots(c+n-2)}=\frac{(n+c-1)\,a_0}{c-1}.$$

We need to take $a_0=c-1$, so we have

$$y_c=(c-1)w^c+\sum_{n=1}^{\infty}(n+c-1)w^{n+c},$$

$$\frac{\partial y_c}{\partial c}=y_c\ln w+w^c+\sum_{n=1}^{\infty}w^{n+c}.$$

Substituting $c=1$ we obtain

$$y_1=\sum_{n=1}^{\infty}n w^{n+1}=\sum_{n=1}^{\infty}n x^{-n-1},$$

$$y_2=y_1\ln w+w+\sum_{n=1}^{\infty}w^{n+1}=y_1\ln(1/x)+\sum_{n=0}^{\infty}x^{-n-1}.$$

19. $x^4 y''+2x^3 y'+4y=0$. The substitution $w=\dfrac{1}{x}$ and a bit of simplification gives us $\dfrac{d^2 y}{dw^2}+4y=0$. It follows that

$$y=c_1\cos 2w+c_2\sin 2w$$

$$=c_1\cos\left(2x^{-1}\right)+c_2\sin\left(2x^{-1}\right).$$

18.11 Many-Term Recurrence Relations

1. Set $L(y) = x^2 y'' + 3xy' + (1 + x + x^3)y$. Then

$$L(y) = \sum_{n=0}^{\infty} \left[(n+c)(n+c-1) + 3(n+c) + 1 \right] a_n x^{n+c} + \sum_{n=0}^{\infty} a_n x^{n+c+1} + \sum_{n=0}^{\infty} a_n x^{n+c+3}$$

$$= \sum_{n=0}^{\infty} (n+c+1)^2 a_n x^{n+c} + \sum_{n=1}^{\infty} a_{n-1} x^{n+c} + \sum_{n=3}^{\infty} a_{n-3} x^{n+c}$$

$$= (c+1)^2 a_0 x^c + \left[(c+2)^2 a_1 + a_0 \right] x^{c+1} + \left[(c+3)^2 a_2 + a_1 \right] x^{c+2}$$

$$+ \sum_{n=3}^{\infty} \left[(n+c+1)^2 a_n + a_{n-1} + a_{n-3} \right] x^{n+c}.$$

The repeated root $c = -1$ of the indicial equation will be used to obtain one solution with $a_0 = 1$. Equating each coefficient to zero yields

$$a_1 + a_0 = 0,$$
$$4a_2 + a_1 = 0,$$
$$n^2 a_n + a_{n-1} + a_{n-3} = 0, \text{ for } n \geq 3.$$

Thus $y_1 = \sum_{n=0}^{\infty} a_n x^{n-1}$ in which $a_0 = 1$, $a_1 = -1$, $a_2 = \frac{1}{4}$, and $a_n = \dfrac{-(a_{n-1} + a_{n-3})}{n^2}$ for $n \geq 3$.

The second solution will be of the form

$$y_2 = y_1 \ln x + \sum_{n=1}^{\infty} b_n x^{n-1},$$

$$y_2' = y_1' \ln x + x^{-1} y_1 + \sum_{n=1}^{\infty} (n-1) b_n x^{n-2},$$

$$y_2'' = y_1'' \ln x + 2x^{-1} y_1' - x^{-2} y_1 + \sum_{n=1}^{\infty} (n-1)(n-2) b_n x^{n-3}.$$

Substituting into the differential equation and simplifying gives us

$$\sum_{n=1}^{\infty} 2n a_n x^{n-1} + \sum_{n=1}^{\infty} n^2 b_n x^{n-1} + \sum_{n=2}^{\infty} b_{n-1} x^{n-1} + \sum_{n=4}^{\infty} b_{n-3} x^{n-1} = 0,$$

$$(2a_1 + b_1) + (4a_2 + 4b_2 + b_1)x + (6a_3 + 9b_3 + b_2)x^2 + \sum_{n=4}^{\infty} \left[2n a_n + n^2 b_n + b_{n-1} + b_{n-3} \right] x^{n-1} = 0.$$

Equating each coefficient to zero leads to

$$b_1 = -2a_1 = 2,$$

$$b_2 = \frac{-4a_2 - b_1}{4} = -\frac{3}{4},$$

$$b_3 = \frac{-b_2 - 6a_3}{9} = \frac{19}{108},$$

$$b_n = \frac{-(b_{n-1} + b_{n-3})}{n^2} - \frac{2a_n}{n} \text{ for } n \geq 4.$$

3. Set $L(y) = xy'' + y' + x(1 + x)y$. We put $y = \sum_{n=0}^{\infty} a_n x^{n+c}$.

$$L(y) = \sum_{n=0}^{\infty} \left[(n+c)(n+c-1) + (n+c)\right]a_n x^{n+c-1} + \sum_{n=0}^{\infty} a_n x^{n+c+1} + \sum_{n=0}^{\infty} a_n x^{n+c+2}$$

$$= \sum_{n=0}^{\infty} (n+c)^2 a_n x^{n+c-1} + \sum_{n=2}^{\infty} a_{n-2} x^{n+c-1} + \sum_{n=3}^{\infty} a_{n-3} x^{n+c-1}$$

$$= c^2 a_0 x^{c-1} + (c+1)^2 a_1 x^c + \left[(c+2)^2 a_2 + a_0\right] x^{c+1}$$

$$+ \sum_{n=3}^{\infty} \left[(n+c)^2 a_n + a_{n-2} + a_{n-3}\right] x^{n+c-1}.$$

We choose $c = 0$ and obtain one solution of the form $y_1 = \sum_{n=0}^{\infty} a_n x^n$, where we arbitrarily take $a_0 = 1$, and then must take $a_1 = 0$, $a_2 = -\frac{1}{4}$, and $a_n = -\frac{a_{n-2} + a_{n-3}}{n^2}$ for $n \geq 3$.
A second solution will be be of the form

$$y_2 = y_1 \ln x + \sum_{n=1}^{\infty} b_n x^n,$$

$$y_2' = y_1' \ln x + x^{-1} y_1 + \sum_{n=1}^{\infty} n b_n x^{n-1},$$

$$y_2'' = y_1'' \ln x + 2x^{-1} y_1' - x^{-2} y_1 + \sum_{n=2}^{\infty} n(n-1) b_n x^{n-2}.$$

Substituting into the differential equation and simplifying gives us

$$\sum_{n=1}^{\infty} n^2 b_n x^{n-1} + \sum_{n=3}^{\infty} b_{n-2} x^{n-1} + \sum_{n=4}^{\infty} b_{n-3} x^{n-1} = -2y_1' = -2 \sum_{n=1}^{\infty} n a_n x^{n-1},$$

$$(b_1 + 2a_1) + (4b_2 + 4a_2)x + (9b_3 + b_1 + 6a_3)x^2 + \sum_{n=4}^{\infty} \left[n^2 b_n + b_{n-2} + b_{n-3} + 2na_n\right] x^{n-1} = 0.$$

Equating each coefficient to zero leads to

$$b_1 = -2a_1 = 0,$$

$$b_2 = -a_2 = \frac{1}{4},$$

$$b_3 = -\frac{1}{9}b_1 - \frac{6}{9}a_3 = \frac{2}{27},$$

$$b_n = -\frac{b_{n-2} + b_{n-3}}{n^2} - \frac{2a_n}{n} \text{ for } n \geq 4.$$

5. $y = \sum_{n=0}^{\infty} a_n x^{n-1}$, a_0 arbitrary, $a_1 = 2a_0$, a_2 arbitrary, and $a_n = -\dfrac{(n+1)a_{n-1} - 6a_{n-2}}{n(n-2)}$ for $n \geq 3$.

$$a_3 = -\frac{4a_2 - 6a_1}{3} = -\frac{4}{3}a_2 + 4a_0,$$

$$a_4 = -\frac{5a_3 - 6a_2}{8} = \frac{19}{12}a_2 - \frac{5}{2}a_0,$$

$$a_5 = -\frac{6a_4 - 6a_3}{15} = -\frac{7}{6}a_2 + \frac{13}{5}a_0,$$

$$a_6 = -\frac{7a_5 - 6a_4}{24} = \frac{53}{72}a_2 - \frac{83}{60}a_0.$$

Thus $y = a_0\left(x^{-1} + 2 + 4x^2 - \frac{5}{2}x^3 + \frac{13}{5}x^4 - \frac{83}{60}x^5 + \cdots\right) + a_2\left(x - \frac{4}{3}x^2 + \frac{19}{12}x^3 - \frac{7}{6}x^4 + \frac{53}{72}x^5 + \cdots\right).$

7. We note that $a_0 = 1$, $a_1 = -1$, $a_2 = \dfrac{1}{2}$, $a_3 = -\dfrac{1}{6}$, $a_4 = \dfrac{1}{24}$, and $a_n = -\dfrac{(n+1)a_{n-1} + a_{n-5}}{n(n+1)}$ for $n \geq 5$. Indeed

$$a_5 = -\frac{6a_4 + a_0}{30} = -\frac{1}{24},$$

$$a_6 = -\frac{7a_5 + a_1}{42} = \frac{31}{1008},$$

$$y_1 = x - x^2 + \frac{1}{2}x^3 - \frac{1}{6}x^4 + \frac{1}{24}x^5 - \frac{1}{24}x^6 + \frac{31}{1008}x^7 - \cdots.$$

Note that

$$a_{2k} = -\left[\frac{(2k+1)a_{2k-1} + a_{2k-5}}{2k(2k+1)}\right] \text{ and } a_{2k+1} = -\left[\frac{(2k+2)a_{2k} + a_{2k-4}}{(2k+1)(2k+2)}\right].$$

Observe that each a_{2k} for $k = 0, 1, 2, 3$ is positive while each a_{2k+1} for $k = 0, 1, 2, 3$ is negative. From the recurrence relation for a_{2k} we see that each successive a_{2k} will be the negative of the sum of two negative numbers; that is, a_{2k} will always be positive. Similarly, a_{2k+1} will always be the negative of the sum of two positive numbers, so that a_{2k+1} will always be negative.

9. $x(x-2)^2 y'' - 2(x-2)y' + 2y = 0$. We set $z = x - 2$ and get $L(y) = z^2(z+2)\dfrac{d^2 y}{dz^2} - 2z\dfrac{dy}{dz} + 2y = 0$,

where $z = 0$ is a regular singular point. Setting $y = \sum\limits_{n=0}^{\infty} a_n z^{n+c}$ we obtain

$$
\begin{aligned}
L(y) &= \sum_{n=0}^{\infty}(n+c)(n+c-1)a_n z^{n+c+1} + \sum_{n=0}^{\infty}\big[2(n+c)(n+c-1) - 2(n+c) + 2\big]a_n z^{n+c} \\
&= \sum_{n=0}^{\infty}(n+c)(n+c-1)a_n z^{n+c+1} + \sum_{n=0}^{\infty}2(n+c-1)^2 a_n z^{n+c} \\
&= 2(c-1)^2 a_0 z^c + \sum_{n=1}^{\infty}\big[2(n+c-1)^2 a_n + (n+c-1)(n+c-2)a_{n-1}\big] z^{n+c} = 0.
\end{aligned}
$$

The indicial equation is $(c-1)^2 = 0$ and $a_n = \dfrac{-(n+c-2)\,a_{n-1}}{2(n+c-1)}$ for $n \geq 1$.

$$
a_1 = \frac{-(c-1)\,a_0}{2c}
$$

$$
a_2 = \frac{-c\,a_1}{2(c+1)}
$$

$$
\vdots
$$

$$
a_n = \frac{(-1)^n(c-1)c\cdots(c+n-2)\,a_0}{2^n c(c+1)\cdots(c+n-1)} = \frac{(-1)^n(c-1)\,a_0}{2^n(c+n-1)},
$$

$$
y_c = z^c + \sum_{n=1}^{\infty} \frac{(-1)^n(c-1)z^{n+c}}{2^n(n+c-1)}
$$

$$
\frac{\partial y_c}{\partial c} = y_c \ln z + \sum_{n=1}^{\infty} \frac{(-1)^n(c-1)z^{n+c}}{2^n(c+n-1)}\left[\frac{1}{c-1} - \frac{1}{c+n-1}\right].
$$

Substituting $c = 1$ we obtain two solutions

$$
y_1 = z = x - 2;
$$

$$
y_2 = y_1 \ln z + \sum_{n=1}^{\infty} \frac{(-1)^n z^{n+1}}{2^n n} = y_1 \ln(x-2) + \sum_{n=1}^{\infty} \frac{(-1)^n(x-2)^{n+1}}{2^n n}.
$$

11. Let $L(y) = 2xy'' + (1-x)y' - (1+x)y$. Then $L(e^x) = \big[2x + (1-x) - (1+x)\big]e^x = 0$; that is, e^x is a solution of the differential equation $L(y) = 0$. Moreover, the general solution is $y = c_1 y_1 + c_2 y_2$ where y_1 and y_2 were obtained in Exercise 10. It follows that the solution e^x must be a linear combination of y_1 and y_2; $e^x = c_1 y_1 + c_2 y_2$. Differentiation of this equation

yields $e^x = c_1 y_1' + c_2 y_2'$. We look at the first few terms of the series for y_1, y_2, y_1', y_2' and have

$$e^x = c_1 \left[x^{1/2} + \frac{1}{2}x^{3/2} + \cdots \right] + c_2 \left[1 + x + \frac{1}{2}x^2 + \cdots \right]$$

$$e^x = c_1 \left[\frac{1}{2}x^{-1/2} + \frac{3}{4}x^{1/2} + \cdots \right] + c_2 \left[1 + x + \cdots \right].$$

We examine the behavior of these series at $x = 0$. From the first equation we see that $c_2 = 1$. From the second equation we must take $c_1 = 0$. Hence $y_2 = e^x$. With this solution of the differential equation we use the method of reduction of order to produce the general solution. We set $y = ve^x$ and obtain $2xv'' + (3x + 1)v' = 0$. Separating the variables and integrating yields the first-order equation $v' = \dfrac{e^{(-3/2)x}}{\sqrt{x}}$, so that finally

$$y_3 = vx = e^x \int_0^x \frac{e^{(-3/2)\gamma}}{\sqrt{\gamma}} \, d\gamma,$$

as another solution of the differential equation. In this integral we substitute $\gamma = \beta^2$ and obtain $y_3 = 2e^x \int_0^{\sqrt{x}} e^{-\frac{3}{2}\beta^2} \, d\beta$. It follows that $y_3 = k_1 y_1 + k_2 y_2$. As before we can examine the first few terms of the power series expansion for both sides of this last equation for $x = 0$ and see that we must have $k_2 = 0$ and $k_1 = 2$. That is, $y_1 = e^x \int_0^{\sqrt{x}} e^{-(3/2)\beta^2} \, d\beta$.

Miscellaneous Exercises

1. Set $L(y) = xy'' - (2 + x)y' - y$. Then

$$L(y) = \sum_{n=0}^{\infty} \left[(n+c)(n+c-1) - 2(n+c) \right] a_n x^{n+c-1} - \sum_{n=0}^{\infty} \left[(n+c) + 1 \right] a_n x^{n+c}$$

$$= \sum_{n=0}^{\infty} (n+c)(n+c-3) a_n x^{n+c-1} - \sum_{n=1}^{\infty} (n+c) a_{n-1} x^{n+c-1}$$

$$= c(c-3)a_0 x^{c-1} + \sum_{n=1}^{\infty} \left[(n+c)(n+c-3)a_n - (n+c)a_{n-1} \right] x^{n+c-1}.$$

The indicial equation is $c(c - 3) = 0$ and $a_n = \dfrac{a_{n-1}}{n + c - 3}$ for $n \geq 1$. Solving the recurrence relation we have $a_n = \dfrac{a_0}{(c - 2)(c - 1)c \cdots (c + n - 3)}$. We now take $a_0 = c$ and get

$$y_c = cx^c + \frac{cx^{c+1}}{c - 2} + \frac{cx^{c+2}}{(c - 2)(c - 1)} + \frac{x^{c+3}}{(c - 2)(c - 1)}$$

$$+ \sum_{n=4}^{\infty} \frac{x^{n+c}}{(c - 2)(c - 1)\big[(c + 1) \cdots (c + n - 3)\big]},$$

$$\frac{\partial y_c}{\partial c} = y_c \ln x + x^c + \frac{cx^{c+1}}{c - 2}\left[\frac{1}{c} - \frac{1}{c - 2}\right] + \frac{cx^{c+2}}{(c - 2)(c - 1)}\left[\frac{1}{c} - \frac{1}{c - 2} - \frac{1}{c - 1}\right]$$

$$+ \frac{x^{c+3}}{(c - 2)(c - 1)}\left[-\frac{1}{c - 2} - \frac{1}{c - 1}\right] + \sum_{n=4}^{\infty} \frac{x^{n+c}}{(c - 2)(c - 1)\big[(c + 1) \cdots (c + n - 3)\big]} \cdot$$

$$\left[-\frac{1}{c - 2} - \frac{1}{c - 1} - \frac{1}{c + 1} - \cdots - \frac{1}{c + n - 3}\right].$$

Setting $c = 0$ gives the solutions

$$y_1 = \frac{1}{2}x^3 + \frac{1}{2}\sum_{n=4}^{\infty} \frac{x^n}{(n - 3)!},$$

$$y_2 = y_1 \ln x + 1 - \frac{1}{2}x + \frac{1}{2}x^2 + \frac{3}{4}x^3 + \frac{1}{2}\sum_{n=4}^{\infty} \frac{x^n}{(n - 3)!}\left[\frac{3}{2} - H_{n-3}\right].$$

3. Set $L(y) = x^2 y'' + 2x^2 y' - 2y$. Then

$$L(y) = \sum_{n=0}^{\infty} \big[(n + c)(n + c - 1) - 2\big]a_n x^{n+c} + \sum_{n=0}^{\infty} 2(n + c)a_n x^{n+c+1}$$

$$= \sum_{n=0}^{\infty} (n + c - 2)(n + c + 1)a_n x^{n+c} + \sum_{n=1}^{\infty} 2(n + c - 1)a_{n-1} x^{n+c}$$

$$= (c - 2)(c + 1)a_0 x^c + \sum_{n=1}^{\infty} \big[(n + c - 2)(n + c + 1)a_n + 2(n + c - 1)a_{n-1}\big]x^{n+c}.$$

The indicial equation is $(c - 2)(c + 1) = 0$ and $(n + c - 2)(n + c + 1)a_n + 2(n + c - 1)a_{n-1} = 0$ for $n \geq 1$. Using the smaller root of the indicial equation, $c = -1$, we have

$$(-2) \cdot 1\, a_1 + 2 \cdot (-1)\, a_0 = 0, \quad a_1 = -a_0,$$
$$(-1) \cdot 2\, a_2 + 2 \cdot (0)\, a_1 = 0, \quad a_2 = 0,$$
$$(0) \cdot 3\, a_3 + 2 \cdot (1)\, a_2 = 0, \quad a_3 \text{ arbitrary},$$

$$a_n = \frac{-2(n - 2)\, a_{n-1}}{(n - 3)n} \text{ for } n \geq 4.$$

Solving this recurrence relation we get $a_n = \dfrac{6(-1)^{n-3}2^{n-3}(n-2)\,a_3}{n!}$. We therefore have the solutions

$$y_1 = x^{-1} - 1,$$

$$y_2 = x^2 + 6 \sum_{n=4}^{\infty} \frac{(-1)^{n-3}2^{n-3}(n-2)x^{n-1}}{n!}.$$

5. Set $L(y) = x^2(1+x^2)y'' + 2x(3+x^2)y' + 6y$. Then

$$L(y) = \sum_{n=0}^{\infty} \big[(n+c)(n+c-1) + 6(n+c) + 6\big]a_n x^{n+c}$$

$$+ \sum_{n=0}^{\infty} \big[(n+c)(n+c-1) + 2(n+c)\big]a_n x^{n+c+2}$$

$$= \sum_{n=0}^{\infty}(n+c+2)(n+c+3)a_n x^{n+c} + \sum_{n=0}^{\infty}(n+c)(n+c+1)a_n x^{n+c+2}$$

$$= (c+2)(c+3)a_0 x^c + (c+3)(c+4)a_1 x^{c+1}$$

$$+ \sum_{n=2}^{\infty} \big[(n+c+2)(n+c+3)a_n + (n+c-2)(n+c-1)a_{n-2}\big]x^{n+c}.$$

The indicial equation is $(c+2)(c+3) = 0$ and $(n+c+2)(n+c+3)a_n + (n+c-2)(n+c-1)a_{n-2} = 0$ for $n \geq 2$. Using the smaller root of the indicial equation, $c = -3$, we have can leave both a_0 and a_1 arbitrary and $a_n = \dfrac{-(n-5)(n-4)\,a_{n-2}}{n(n-1)}$ for $n \geq 4$. Thus $a_n = 0$ for $n \geq 4$ and we have the solutions $y_1 = x^{-3} - 3x^{-1}$ and $y_2 = x^{-2} - \frac{1}{3}$.

7. Set $L(y) = 2xy'' + (1+2x)y' - 3y$. Then

$$L(y) = \sum_{n=0}^{\infty} \big[2(n+c)(n+c-1) + (n+c)\big]a_n x^{n+c-1} + \sum_{n=0}^{\infty} \big[2(n+c) - 3\big]a_n x^{n+c}$$

$$= \sum_{n=0}^{\infty}(n+c)(2n+2c-1)a_n x^{n+c-1} + \sum_{n=1}^{\infty}(2n+2c-5)a_{n-1}x^{n+c-1}$$

$$= c(2c-1)a_0 x^{c-1} + \sum_{n=1}^{\infty} \big[(n+c)(2n+2c-1)a_n + (2n+2c-5)a_{n-1}\big]x^{n+c-1}.$$

The indicial equation is $c(2c-1) = 0$ and $a_n = \dfrac{-(2n+2c-5)\,a_{n-1}}{(n+c)(2n+2c-1)}$ for $n \geq 1$. Choosing $c = 0$ and solving the recurrence relation yields

$$a_n = \frac{3(-1)^n\,a_0}{n!\,(2n-1)(2n-3)},$$

$$y_1 = 1 + 3 \sum_{n=1}^{\infty} \frac{(-1)^n x^n}{n!\,(2n-1)(2n-3)}.$$

Choosing $c = 1/2$ gives us $a_n = \dfrac{-2(n-2)\,a_{n-1}}{n(2n+1)}$ for $n \geq 1$. Thus $a_1 = \frac{2}{3}\,a_0$ and $a_n = 0$ for $n \geq 2$. Finally, $y_2 = x^{1/2} + \frac{2}{3}x^{3/2}$.

9. Set $L(y) = x(1 - x^2)y'' - (7 + x^2)y' + 4xy$. Then

$$
\begin{aligned}
L(y) &= \sum_{n=0}^{\infty} \big[(n+c)(n+c-1) - 7(n+c)\big]a_n x^{n+c-1} \\
&\quad - \sum_{n=0}^{\infty} \big[(n+c)(n+c-1) + (n+c) - 4\big]a_n x^{n+c+1} \\
&= \sum_{n=0}^{\infty} (n+c)(n+c-8)a_n x^{n+c-1} - \sum_{n=0}^{\infty} (n+c-2)(n+c+2)a_n x^{n+c+1} \\
&= c(c-8)a_0 x^{c-1} + (c+1)(c-7)a_1 x^{c} \\
&\quad + \sum_{n=2}^{\infty} \big[(n+c)(n+c-8)a_n - (n+c-4)(n+c)a_{n-2}\big]x^{n+c-1}.
\end{aligned}
$$

The indicial equation is $c(c - 8) = 0$. Using the smaller root of this equation, $c = 0$, we also need $a_1 = 0$ and $(n - 8)a_n = (n - 4)a_{n-1}$ for $n \geq 2$.

$$
\begin{array}{ll}
(-3)\,a_2 = (-1)\,a_0, & a_2 = \tfrac{1}{3}\,a_0, \\
(-2)\,a_4 = 0, & a_4 = 0, \\
(-1)\,a_6 = (1)\,a_4, & a_6 = 0, \\
(0)\,a_8 = (2)\,a_6, & a_6 = 0.
\end{array}
$$

We can therefore take a_8 arbitrary and $a_{2k} = \dfrac{(k-2)\,a_{2k-2}}{k-4} = \dfrac{(k-3)(k-2)\,a_8}{2}$ for $k \geq 5$.

Finally, $y = a_0\left[1 + \frac{1}{3}x^2\right] + a_8\left[x^8 + \displaystyle\sum_{k=5}^{\infty} \frac{(k-3)(k-2)x^{2k}}{2}\right]$.

11. Set $L(y) = 4x^2 y'' - 2x(2 + x)y' + (3 + x)y$. Then

$$
\begin{aligned}
L(y) &= \sum_{n=0}^{\infty} \big[4(n+c)(n+c-1) - 4(n+c) + 3\big]a_n x^{n+c} - \sum_{n=0}^{\infty} \big[2(n+c) - 1\big]a_n x^{n+c+1} \\
&= \sum_{n=0}^{\infty} (2n+2c-1)(2n+2c-3)a_n x^{n+c} - \sum_{n=1}^{\infty} (2n+2c-3)a_{n-1}x^{n+c} \\
&= (2c-1)(2c-3)a_0 x^{c} + \sum_{n=1}^{\infty} (2n+2c-3)\big[(2n+2c-1)a_n - a_{n-1}\big]x^{n+c}.
\end{aligned}
$$

The indicial equation is $(2c-1)(2c-3) = 0$. Choosing the smaller root of the indicial equation, $c = \frac{1}{2}$, yields $2(n-1)(2na_n - a_{n-1}) = 0$ for $n \geq 1$. This leaves both a_0 and a_1 arbitrary and

$a_n = \dfrac{a_{n-1}}{2n}$ for $n \geq 2$. One solution is $y_1 = x^{1/2}$. Solving the recurrence relation yields

$a_n = \dfrac{a_1}{2^{n-1}\, n!}$. Thus a second solution is $y_2 = x^{3/2} + \displaystyle\sum_{n=2}^{\infty} \dfrac{x^{n+1/2}}{2^{n-1}\, n!} = x^{3/2} + \sum_{n=1}^{\infty} \dfrac{x^{n+3/2}}{2^n\,(n+1)!}$.

Note that the answer in the text is correct, but perhaps misleading. If we put $a_0 = 1$ and $a_1 = -1/2$ in the answer in the text we get the solution $y = x^{1/2}$.

13. Set $L(y) = 2xy'' + y' + y$. Then

$$L(y) = \sum_{n=0}^{\infty} \big[2(n+c)(n+c-1) + (n+c)\big]a_n x^{n+c-1} + \sum_{n=0}^{\infty} a_n x^{n+c}$$

$$= \sum_{n=0}^{\infty}(n+c)(2n+2c-1)a_n x^{n+c-1} + \sum_{n=1}^{\infty} a_{n-1} x^{n+c-1}$$

$$= c(2c-1)a_0 x^{c-1} + \sum_{n=1}^{\infty}\big[(n+c)(2n+2c-1)a_n + a_{n-1}\big]x^{n+c-1}.$$

The indicial equation is $c(2c-1) = 0$ and $a_n = \dfrac{-a_{n-1}}{(n+c)(2n+2c-1)}$ for $n \geq 1$. Choosing $c = 0$ and solving the recurrence relation yields

$$a_n = \frac{(-1)^n\, a_0}{n!\big[1\cdot 3\cdots(2n-1)\big]},$$

$$y_1 = 1 + \sum_{n=1}^{\infty} \frac{(-1)^n x^n}{n!\,\big[1\cdot 3\cdots(2n-1)\big]}.$$

Choosing $c = 1/2$ and solving the recurrence relation gives us

$$a_n = \frac{(-1)^n\, a_0}{n!\,\big[3\cdot 5\cdots(2n+1)\big]},$$

$$y_2 = x^{1/2} + \sum_{n=1}^{\infty} \frac{(-1)^n x^{n+1/2}}{n!\,\big[3\cdot 5\cdots(2n+1)\big]}.$$

15. Set $L(y) = 4x^2 y'' - x^2 y' + y$. Then

$$L(y) = \sum_{n=0}^{\infty}\big[4(n+c)(n+c-1) + 1\big]a_n x^{n+c} - \sum_{n=0}^{\infty}(n+c)a_n x^{n+c+1}$$

$$= \sum_{n=0}^{\infty}(2n+2c-1)^2 a_n x^{n+c} - \sum_{n=1}^{\infty}(n+c-1)a_{n-1} x^{n+c}$$

$$= (2c-1)^2 a_0 x^c + \sum_{n=1}^{\infty}\big[(2n+2c-1)^2 a_n - (n+c-1)a_{n-1}\big]x^{n+c}.$$

The indicial equation is $(2c - 1)^2 = 0$ and $a_n = \dfrac{(n + c - 1)\, a_{n-1}}{(2n + 2c - 1)^2}$ for $n \geq 1$. Solving the recurrence relation we have $a_n = \dfrac{c(c + 1) \cdots (c + n - 1)\, a_0}{(2c + 1)^2 (2c + 3)^2 \cdots (2c + 2n - 1)^2}$. Therefore

$$y_c = x^c + \sum_{n=1}^{\infty} \frac{c \cdot (c + 1) \cdots (c + n - 1) x^{n+c}}{(2c + 1)^2 \cdot (2c + 3)^2 \cdots (2c + 2n - 1)^2},$$

$$\frac{\partial y_c}{\partial c} = y_c \ln x + \sum_{n=1}^{\infty} \frac{c \cdot (c + 1) \cdots (c + n - 1) x^{n+c}}{(2c + 1)^2 \cdot (2c + 3)^2 \cdots (2c + 2n - 1)^2} \bullet$$

$$\left[\frac{1}{c} + \cdots + \frac{1}{c + n - 1} - \frac{2}{2c + 1} - \cdots - \frac{2}{2c + 2n - 1} \right].$$

Setting $c = 1/2$ gives the solutions

$$y_1 = x^{1/2} + \sum_{n=1}^{\infty} \frac{1 \cdot 3 \cdots (2n - 1) x^{n+1/2}}{2^{3n} (n!)^2},$$

$$y_2 = y_1 \ln x + \sum_{n=1}^{\infty} \frac{1 \cdot 3 \cdots (2n - 1) x^{n+1/2}}{2^{3n} (n!)^2} \left[\frac{2}{1} + \frac{2}{3} + \cdots + \frac{2}{2n - 1} - H_n \right].$$

17. Set $L(y) = 2x^2 y'' - x(1 + 2x)y' + (1 + 3x)y$. Then

$$L(y) = \sum_{n=0}^{\infty} \left[2(n + c)(n + c - 1) - (n + c) + 1 \right] a_n x^{n+c} - \sum_{n=0}^{\infty} \left[2(n + c) - 3 \right] a_n x^{n+c+1}$$

$$= \sum_{n=0}^{\infty} (2n + 2c - 1)(n + c - 1) a_n x^{n+c} - \sum_{n=1}^{\infty} (2n + 2c - 5) a_{n-1} x^{n+c}$$

$$= (2c - 1)(c - 1) a_0 x^c + \sum_{n=1}^{\infty} \left[(2n + 2c - 1)(n + c - 1) a_n - (2n + 2c - 5) a_{n-1} \right] x^{n+c}.$$

The indicial equation is $(2c - 1)(c - 1) = 0$ and $a_n = \dfrac{(2n + 2c - 5) a_{n-1}}{(2n + 2c - 1)(n + c - 1)}$ for $n \geq 1$.

Choosing $c = 1/2$ and $a_n = \dfrac{2(n - 2)\, a_{n-1}}{n(2n - 1)}$ we have $a_1 = -2a_0$ and $a_n = 0$ for $n \geq 2$.

Therefore one solution is $y_1 = x^{1/2} - 2x^{3/2}$. Choosing $c = 1$ and solving the recurrence relation gives us

$$a_n = \frac{-a_0}{n!\, (2n - 1)(2n + 1)},$$

$$y_2 = x - \sum_{n=1}^{\infty} \frac{x^{n+1}}{n!\, (2n - 1)(2n + 1)} = \sum_{n=0}^{\infty} \frac{-x^{n+1}}{n!\, (2n - 1)(2n + 1)}.$$

19. Set $L(y) = 4x^2y'' + 3x^2y' + (1 + 3x)y$. Then

$$L(y) = \sum_{n=0}^{\infty} \left[4(n+c)(n+c-1)+1\right]a_n x^{n+c} + \sum_{n=0}^{\infty} 3(n+c+1)a_n x^{n+c+1}$$

$$= \sum_{n=0}^{\infty} (2n+2c-1)^2 a_n x^{n+c} + \sum_{n=1}^{\infty} 3(n+c)a_{n-1} x^{n+c}$$

$$= (2c-1)^2 a_0 x^c + \sum_{n=1}^{\infty} \left[(2n+2c-1)^2 a_n + 3(n+c)a_{n-1}\right] x^{n+c}.$$

The indicial equation is $(2c-1)^2 = 0$ and $a_n = \dfrac{-3(n+c)\,a_{n-1}}{(2n+2c-1)^2}$ for $n \geq 1$. Solving this recurrence relation we get

$$a_n = \frac{(-1)^n 3^n (c+1) \cdots (c+n)\, a_0}{(2c+1)^2 \cdots (2c+2n-1)^2},$$

$$y_c = x^c + \sum_{n=1}^{\infty} \frac{(-1)^n 3^n (c+1) \cdots (c+n)\, x^{n+c}}{(2c+1)^2 \cdots (2c+2n-1)^2},$$

$$\frac{\partial y_c}{\partial c} = y_c \ln x + \sum_{n=1}^{\infty} \frac{(-1)^n 3^n (c+1) \cdots (c+n)\, x^{n+c}}{(2c+1)^2 \cdots (2c+2n-1)^2} \bullet$$

$$\left[\frac{1}{c+1} + \cdots + \frac{1}{c+n} - \frac{4}{2c+1} - \cdots - \frac{4}{2c+2n-1}\right].$$

Substituting $c = 1/2$ gives the solutions

$$y_1 = x^{1/2} + \sum_{n=1}^{\infty} \frac{(-1)^n 3^n \left[\frac{3}{2} \cdot \frac{5}{2} \cdots \frac{2n+1}{2}\right] x^{n+1/2}}{2^2 \cdot 4^2 \cdots (2n)^2}$$

$$= x^{1/2} + \sum_{n=1}^{\infty} \frac{(-1)^n 3^n \left[3 \cdot 5 \cdots (2n+1)\right] x^{n+1/2}}{2^{3n}\, (n!)^2},$$

$$y_2 = y_1 \ln x + \sum_{n=1}^{\infty} \frac{(-1)^n 3^n \left[3 \cdot 5 \cdots (2n+1)\right] x^{n+1/2}}{2^{3n-1}(n!)^2} \left[\frac{1}{3} + \cdots + \frac{1}{2n+1} - \frac{2}{2} - \cdots - \frac{2}{2n}\right]$$

$$= y_1 \ln x + \sum_{n=1}^{\infty} \frac{(-1)^n 3^n \left[3 \cdot 5 \cdots (2n+1)\right] x^{n+1/2}}{2^{3n-1}(n!)^2} \left[\frac{1}{3} + \cdots + \frac{1}{2n+1} - H_n\right].$$

21. Set $L(y) = 4x^2y'' + 2x^2y' - (x+3)y$. Then

$$L(y) = \sum_{n=0}^{\infty} \left[4(n+c)(n+c-1) - 3\right]a_nx^{n+c} = \sum_{n=0}^{\infty} \left[2(n+c) - 1\right]a_nx^{n+c+1}$$

$$= \sum_{n=0}^{\infty} (2n+2c+1)(2n+2c-3)a_nx^{n+c} + \sum_{n=1}^{\infty} (2n+2c-3)a_{n-1}x^{n+c}$$

$$= (2c+1)(2c-3)a_0x^c + \sum_{n=1}^{\infty} \left[(2n+2c+1)(2n+2c-3)a_n + (2n+2c-3)a_{n-1}\right]x^{n+c}.$$

The indicial equation is $(2c+1)(2c-3) = 0$. Choosing the smaller root of the indicial equation, $c = -1/2$, and $2n(n-2)a_n + (n-2)a_{n-1} = 0$ for $n \geq 1$, we have

$$2 \cdot 1 \cdot (-1)\, a_1 + (-1) \cdot a_0 = 0, \quad a_1 = \frac{-a_0}{2},$$

$$2 \cdot 2 \cdot 0\, a_2 + 0 \cdot a_1 = 0, \quad a_2 \text{ arbitrary,}$$

and for $n \geq 3$, $a_n = \dfrac{(-1)^{n-2}\, a_2}{2^{n-2}\, 3 \cdot 4 \cdots n}$. Thus

$$y = a_0\left[x^{(-1/2)} - \tfrac{1}{2}x^{1/2}\right] + a_2\left[x^{3/2} + \sum_{n=3}^{\infty} \frac{(-1)^{n-2}\, x^{n-(1/2)}}{2^{n-2}\, 3 \cdots n}\right].$$

Note that if we take $a_0 = 1$ and $a_2 = \tfrac{1}{8}$ we have the solution

$$y = x^{-1/2}\left[1 - \frac{x}{2} + \frac{x^2}{2^2 2!} + \sum_{n=3}^{\infty} \frac{(-1)^n x^n}{2^n\, n!}\right] = x^{(-1/2)}e^{(-x/2)}.$$

That is, $x^{(-1/2)}e^{(-x/2)}$ is also a solution of the differential equation.

23. Set $L(y) = x^2y'' + x(3+x)y' + (1+2x)y$. Then

$$L(y) = \sum_{n=0}^{\infty} \left[(n+c)(n+c-1) + 3(n+c) + 1\right]a_nx^{n+c} + \sum_{n=0}^{\infty} (n+c+2)a_nx^{n+c+1}$$

$$= \sum_{n=0}^{\infty} (n+c+1)^2a_nx^{n+c} + \sum_{n=1}^{\infty} (n+c+1)a_{n-1}x^{n+c}$$

$$= (c+1)^2a_0x^c + \sum_{n=1}^{\infty} \left[(n+c+1)^2a_n + (n+c+1)a_{n-1}\right]x^{n+c}.$$

The indicial equation is $(c+1)^2 = 0$ and $a_n = \dfrac{-a_{n-1}}{n+c+1}$ for $n \geq 1$. Solving this recurrence relation we get

$$a_n = \frac{(-1)^n \, a_0}{(c+2)\cdots(c+n+1)},$$

$$y_c = x^c + \sum_{n=1}^{\infty} \frac{(-1)^n \, x^{n+c}}{(c+2)\cdots(c+n+1)},$$

$$\frac{\partial y_c}{\partial c} = y_c \ln x + \sum_{n=1}^{\infty} \frac{(-1)^n \, x^{n+c}}{(c+2)\cdots(c+n+1)} \left[-\frac{1}{c+2} - \cdots - \frac{1}{c+n+1} \right].$$

Substituting $c = -1$ gives the solutions

$$y_1 = x^{-1} + \sum_{n=1}^{\infty} \frac{(-1)^n \, x^{n-1}}{n!} = \sum_{n=0}^{\infty} \frac{(-1)^n \, x^{n-1}}{n!},$$

$$y_2 = y_1 \ln x + \sum_{n=1}^{\infty} \frac{(-1)^{n+1} H_n \, x^{n-1}}{n!}.$$

25. Set $L(y) = x(1-2x)y'' - 2(2+x)y' + 18y$. Then

$$L(y) = \sum_{n=0}^{\infty} \left[(n+c)(n+c-1) - 4(n+c) \right] a_n x^{n+c-1}$$

$$- \sum_{n=0}^{\infty} \left[2(n+c)(n+c-1) + 2(n+c) - 18 \right] a_n x^{n+c}$$

$$= \sum_{n=0}^{\infty} (n+c)(n+c-5) a_n x^{n+c-1} - \sum_{n=0}^{\infty} 2(n+c-3)(n+c+3) a_n x^{n+c}$$

$$= c(c-5)a_0 x^{c-1} + \sum_{n=1}^{\infty} \left[(n+c)(n+c-5)a_n - 2(n+c-4)(n+c+2)a_{n-1} \right] x^{n+c-1}.$$

Choosing $c = 0$ and $n(n-5)a_n = 2(n-4)(n+2)a_{n-1}$ for ≥ 1,

$1 \cdot (-4) \, a_1 = 2 \cdot (-3) \cdot 3 \, a_0,$	$a_1 = \frac{9}{2} \, a_0,$
$2 \cdot (-3) \, a_2 = 2 \cdot (-2) \cdot 4 \, a_1,$	$a_2 = 12 \, a_0,$
$3 \cdot (-2) \, a_3 = 2 \cdot (-1) \cdot 5 \, a_2,$	$a_3 = 20 \, a_0,$
$4 \cdot (-1) \, a_4 = 2 \cdot (0) \cdot 6 \, a_3,$	$a_4 = 0,$
$5 \cdot (0) \, a_5 = 2 \cdot (1) \cdot 7 \, a_4,$	a_5 arbitrary.

For $n \geq 6$ we solve the recurrence relation and get $a_n = \dfrac{2^{n-5}(n-4)(n+1)(n+2)\, a_5}{6 \cdot 7}$. Finally,

$$y = a_0 \left[1 + \tfrac{9}{2}x + 12x^2 + 20x^3\right] + a_5 \left[x^5 + \sum_{n=6}^{\infty} \frac{2^{n-5}(n-4)(n+1)(n+2)x^n}{42}\right]$$

$$= \frac{1}{2}a_0 \left[2 + 9x + 24x^2 + 40x^3\right] + \frac{a_5}{2^5 \cdot 42} \sum_{n=5}^{\infty} 2^n(n-4)(n+1)(n+2)x^n.$$

27. Set $L(y) = x^2 y'' - 3xy' + 4(1+x)y$. Then

$$L(y) = \sum_{n=0}^{\infty} \left[(n+c)(n+c-1) - 3(n+c) + 4\right]a_n x^{n+c} + \sum_{n=0}^{\infty} 4a_n x^{n+c+1}$$

$$= \sum_{n=0}^{\infty} (n+c-2)^2 a_n x^{n+c} + \sum_{n=1}^{\infty} 4a_{n-1} x^{n+c}$$

$$= (c-2)^2 a_0 x^c + \sum_{n=1}^{\infty} \left[(n+c-2)^2 a_n + 4a_{n-1}\right]x^{n+c}.$$

The indicial equation is $(c-2)^2 = 0$ and $a_n = \dfrac{-4\, a_{n-1}}{(n+c-2)^2}$ for $n \geq 1$. Solving this recurrence relation we get

$$a_n = \frac{(-1)^n 4^n\, a_0}{(c-1)^2 \cdots (c+n-2)^2},$$

$$y_c = x^c + \sum_{n=1}^{\infty} \frac{(-1)^n 4^n\, x^{n+c}}{(c-1)^2 \cdots (c+n-2)^2},$$

$$\frac{\partial y_c}{\partial c} = y_c \ln x + \sum_{n=1}^{\infty} \frac{(-1)^n 4^n\, x^{n+c}}{(c-1)^2 \cdots (c+n-2)^2} \left[-\frac{2}{c-1} - \cdots - \frac{2}{c+n-2}\right].$$

Substituting $c = 2$ gives the solutions

$$y_1 = x^2 + \sum_{n=1}^{\infty} \frac{(-1)^n 4^n\, x^{n+2}}{(n!)^2} = \sum_{n=0}^{\infty} \frac{(-1)^n 4^n\, x^{n+2}}{(n!)^2},$$

$$y_2 = y_1 \ln x - 2 \sum_{n=1}^{\infty} \frac{(-1)^n 4^n H_n\, x^{n+2}}{(n!)^2}.$$

29. Set $L(y) = 4x^2 y'' + 2x(x-4)y' + (5-3x)y$. Then

$$L(y) = \sum_{n=0}^{\infty} \left[4(n+c)(n+c-1) - 8(n+c) + 5\right] a_n x^{n+c} + \sum_{n=0}^{\infty} \left[2(n+c) - 3\right] a_n x^{n+c+1}$$

$$= \sum_{n=0}^{\infty} (2n+2c-1)(2n+2c-5) a_n x^{n+c} + \sum_{n=1}^{\infty} (2n+2c-5) a_{n-1} x^{n+c}$$

$$= (2c-1)(2c-5) a_0 x^c + \sum_{n=1}^{\infty} \left[(2n+2c-5)(2n+2c-1) a_n + (2n+2c-5) a_{n-1}\right] x^{n+c}.$$

Choosing $c = 1/2$ and $2n(n-2) a_n + (n-2) a_{n-1} = 0$ for $n \geq 1$.

$$2 \cdot 1 \cdot (-1) \cdot a_1 + (-1) \cdot a_0 = 0, \qquad\qquad a_1 = -\tfrac{1}{2} a_0,$$
$$2 \cdot 2 \cdot (0) \cdot a_2 + (0) \cdot a_1 = 0, \qquad\qquad a_2 \text{ arbitrary.}$$

For $n \geq 3$ we solve the recurrence relation and get $a_n = \dfrac{(-1)^n a_2}{2^{n-3} n!}$. Finally,

$$y = a_0 \left[x^{1/2} - \tfrac{1}{2} x^{3/2}\right] + a_2 \sum_{n=2}^{\infty} \frac{(-1)^n x^{n+1/2}}{2^{n-3} n!}$$

$$= a_0 \left[x^{1/2} - \frac{1}{2} x^{3/2}\right] + a_2 \sum_{n=0}^{\infty} \frac{(-1)^n x^{n+5/2}}{2^{n-1} (n+2)!}.$$

31. Set $L(y) = x(1-x)y'' - (4+x)y' + 4y$. Then

$$L(y) = \sum_{n=0}^{\infty} \left[(n+c)(n+c-1) - 4(n+c)\right] a_n x^{n+c-1}$$

$$- \sum_{n=0}^{\infty} \left[(n+c)(n+c-1) + (n+c) - 4\right] a_n x^{n+c}$$

$$= \sum_{n=0}^{\infty} (n+c)(n+c-5) a_n x^{n+c-1} - \sum_{n=0}^{\infty} (n+c-2)(n+c+2) a_n x^{n+c}$$

$$= c(c-5) a_0 x^{c-1} + \sum_{n=1}^{\infty} \left[(n+c)(n+c-5) a_n - (n+c-3)(n+c+1) a_{n-1}\right] x^{n+c-1}.$$

Choosing $c = 0$ and $n(n-5) a_n - (n-3)(n+1) a_{n-1} = 0$ for $n \geq 1$ we have

$$1 \cdot (-4) \, a_1 - (-2) \cdot 2 \, a_0 = 0, \qquad\qquad a_1 = a_0,$$
$$2 \cdot (-3) \, a_2 - (-1) \cdot 3 \, a_1 = 0, \qquad\qquad a_2 = \tfrac{1}{2} a_0,$$
$$3 \cdot (-2) \, a_3 - (0) \cdot 4 \, a_2 = 0, \qquad\qquad a_3 = 0,$$
$$4 \cdot (-1) \, a_4 - (1) \cdot 5 \, a_3 = 0, \qquad\qquad a_4 = 0,$$
$$5 \cdot (0) \, a_5 - (2) \cdot 6 \, a_4 = 0, \qquad\qquad a_5 \text{ arbitrary.}$$

We solve the recurrence relation $a_n = \dfrac{(n-3)(n+1)\, a_{n-1}}{n(n-5)}$ and get

$a_n = \dfrac{(n-4)(n-3)(n+1)\, a_5}{12}$ for $n \geq 6$. Finally,

$$y = a_0 \left[1 + x + \tfrac{1}{2}x^2\right] + \frac{1}{12}a_5 \sum_{n=5}^{\infty} (n-4)(n-3)(n+1)x^n.$$

33. $x(1-x)y'' - (4+x)y' + 4y = 0$. We let $z = x-1$ and get $L(y) = z(z+1)\dfrac{d^2y}{dz^2} + (z+5)\dfrac{dy}{dz} - 4y = 0$. Then

$$L(y) = \sum_{n=0}^{\infty} \left[(n+c)(n+c-1) + (n+c) - 4\right]a_n z^{n+c}$$

$$+ \sum_{n=0}^{\infty} \left[(n+c)(n+c-1) + 5(n+c)\right]a_n z^{n+c-1}$$

$$= \sum_{n=0}^{\infty} (n+c-2)(n+c+2)a_n z^{n+c} + \sum_{n=0}^{\infty} (n+c)(n+c+4)a_n z^{n+c-1}$$

$$= c(c+4)a_0 z^{c-1} + \sum_{n=1}^{\infty} \left[(n+c)(n+c+4)a_n + (n+c-3)(n+c+1)a_{n-1}\right] z^{n+c-1}.$$

Choosing $c = -4$ and $n(n-4)a_n + (n-7)(n-3)a_{n-1} = 0$ for $n \geq 1$ we have

$1 \cdot (-3)\, a_1 + (-6) \cdot (-2)\, a_0 = 0,$	$a_1 = 4\, a_0,$
$2 \cdot (-2)\, a_2 + (-5) \cdot (-1)\, a_1 = 0,$	$a_2 = 5\, a_0,$
$3 \cdot (-1)\, a_3 + (-4) \cdot (0)\, a_2 = 0,$	$a_3 = 0,$
$4 \cdot (0)\, a_4 + (-3) \cdot (1)\, a_3 = 0,$	a_4 arbitrary,
$5 \cdot (1)\, a_5 + (-2) \cdot (2)\, a_4 = 0,$	$a_5 = \tfrac{4}{5}\, a_4,$
$6 \cdot (2)\, a_6 + (-1) \cdot (3)\, a_5 = 0,$	$a_6 = \tfrac{1}{5}\, a_4,$
$7 \cdot (3)\, a_7 + (0) \cdot (4)\, a_6 = 0,$	$a_n = 0$ for $n \geq 7.$

Finally,

$$y = a_0\left[z^{-4} + 4z^{-3} + 5z^{-2}\right] + a_4\left[1 + \tfrac{4}{5}z + \tfrac{1}{5}z^2\right]$$

$$= a_0\left[(x-1)^{-4} + 4(x-1)^{-3} + 5(x-1)^{-2}\right] + a_4\left[1 + \tfrac{4}{5}(x-1) + \tfrac{1}{5}(x-1)^2\right].$$

35. Set $L(y) = x(1-x)y'' + (1-4x)y' - 2y$. Then

$$L(y) = \sum_{n=0}^{\infty} \left[(n+c)(n+c-1) + (n+c) \right] a_n x^{n+c-1}$$

$$- \sum_{n=0}^{\infty} \left[(n+c)(n+c-1) + 4(n+c) + 2 \right] a_n x^{n+c}$$

$$= \sum_{n=0}^{\infty} (n+c)^2 a_n x^{n+c-1} - \sum_{n=0}^{\infty} (n+c+1)(n+c+2) a_n x^{n+c}$$

$$= c^2 a_0 x^{c-1} + \sum_{n=1}^{\infty} \left[(c+n)^2 a_n - (c+n)(c+n+1) a_{n-1} \right] x^{n+c-1}.$$

The indicial equation is $c^2 = 0$ and $a_n = \dfrac{(c+n+1)\, a_{n-1}}{c+n}$ for $n \geq 1$. Solving this recurrence relation we get

$$a_n = \frac{(c+n+1)\, a_0}{c+1},$$

$$y_c = x^c + \sum_{n=1}^{\infty} \frac{(c+n+1)\, x^{n+c}}{c+1},$$

$$\frac{\partial y_c}{\partial c} = y_c \ln x + \sum_{n=1}^{\infty} \frac{(c+n+1)\, x^{n+c}}{c+1} \left[\frac{1}{c+n+1} - \frac{1}{c+1} \right].$$

Substituting $c = 0$ gives the solutions

$$y_1 = 1 + \sum_{n=1}^{\infty} (n+1)\, x^n = \sum_{n=0}^{\infty} (n+1)\, x^n,$$

$$y_2 = y_1 \ln x + \sum_{n=1}^{\infty} (n+1)\, x^n \left[\frac{1}{n+1} - 1 \right] = y_1 \ln x - \sum_{n=1}^{\infty} n x^n.$$

37. Set $L(y) = xy'' + (1-x)y' + 3y$. Then

$$L(y) = \sum_{n=0}^{\infty} \left[(n+c)(n+c-1) + (n+c) \right] a_n x^{n+c-1} - \sum_{n=0}^{\infty} (n+c-3) a_n x^{n+c}$$

$$= \sum_{n=0}^{\infty} (n+c)^2 a_n x^{n+c-1} - \sum_{n=1}^{\infty} (n+c-4) a_{n-1} x^{n+c-1}$$

$$= c^2 a_0 x^{c-1} + \sum_{n=1}^{\infty} \left[(c+n)^2 a_n - (c+n-4) a_{n-1} \right] x^{n+c-1}.$$

The indicial equation is $c^2 = 0$ and $a_n = \dfrac{(c+n-4)\, a_{n-1}}{(c+n)^2}$ for $n \geq 1$. Solving this recurrence

relation we get $a_1 = \dfrac{(c-3)\, a_0}{(c+1)^2}$, $a_2 = \dfrac{(c-3)(c-2)\, a_0}{(c+1)^2(c+2)^2}$, $a_3 = \dfrac{(c-3)(c-2)(c-1)\, a_0}{(c+1)^2(c+2)^2(c+3)^2}$, and for

$n \geq 4$, $a_n = \dfrac{(c-3)(c-2)(c-1)c\, a_0}{[(c+1)\cdots(c+n)](c+n-3)(c+n-2)(c+n-1)(c+n)}$.

$$y_c = x^c + \frac{(c-3)x^{c+1}}{(c+1)^2} + \frac{(c-3)(c-2)x^{c+2}}{(c+1)^2(c+2)^2} + \frac{(c-3)(c-2)(c-1)x^{c+3}}{(c+1)^2(c+2)^2(c+3)^2}$$

$$+ \sum_{n=4}^{\infty} \frac{(c-3)(c-2)(c-1)c\, x^{c+n}}{[(c+1)\cdots(c+n)](c+n-3)(c+n-2)(c+n-1)(c+n)}$$

$$\frac{\partial y_c}{\partial c} = y_c \ln x + \frac{(c-3)x^{c+1}}{(c+1)^2}\left[\frac{1}{c-3} - \frac{2}{c+1}\right]$$

$$+ \frac{(c-3)(c-2)x^{c+2}}{(c+1)^2(c+2)^2}\left[\frac{1}{c-3} + \frac{1}{c-2} - \frac{2}{c+1} - \frac{2}{c+2}\right]$$

$$+ \frac{(c-3)(c-2)(c-1)x^{c+3}}{(c+1)^2(c+2)^2(c+3)^2}\left[\frac{1}{c-3} + \frac{1}{c-2} + \frac{1}{c-1} - \frac{2}{c+1} - \frac{2}{c+2} - \frac{2}{c+3}\right]$$

$$+ \sum_{n=4}^{\infty} \frac{(c-3)(c-2)(c-1)cx^{c+n}}{[(c+1)\cdots(c+n)](c+n-3)(c+n-2)(c+n-1)(c+n)} \bullet [P(c) - Q(c)],$$

where $P(c) = \dfrac{1}{c-3} + \dfrac{1}{c-2} + \dfrac{1}{c-1} + \dfrac{1}{c}$ and $Q(c) = \dfrac{1}{c+1} + \cdots + \dfrac{1}{c+n} + \dfrac{1}{c+n-3} + \dfrac{1}{c+n-2} +$

$\dfrac{1}{c+n-1} + \dfrac{1}{c+n}$.

Substituting $c = 0$ gives the solutions

$$y_1 = 1 - 3x + \frac{3}{2}x^2 - \frac{1}{6}x^3,$$

$$y_2 = y_1 \ln x + 7x - \frac{23}{4}x^2 + \frac{11}{12}x^3 - 6\sum_{n=4}^{\infty} \frac{x^n}{n!\, n(n-1)(n-2)(n-3)}.$$

39. Set $L(y) = 2x(1-x)y'' + (1-2x)y' + 8y$. Then

$$L(y) = \sum_{n=0}^{\infty} \left[2(n+c)(n+c-1) + (n+c)\right]a_n x^{n+c-1}$$

$$- \sum_{n=0}^{\infty} \left[2(n+c)(n+c-1) + 2(n+c) - 8\right]a_n x^{n+c}$$

$$= \sum_{n=0}^{\infty} (n+c)(2n+2c-1)a_n x^{n+c-1} - \sum_{n=0}^{\infty} 2(n+c-2)(n+c+2)a_n x^{n+c}$$

$$= c(2c-1)a_0 x^{c-1} + \sum_{n=1}^{\infty} \left[(n+c)(2n+2c-1)a_n - 2(n+c-3)(n+c+1)a_{n-1}\right]x^{n+c-1}.$$

The indicial equation is $c(2c - 1) = 0$ and $a_n = \dfrac{2(n + c - 3)(n + c + 1)\, a_{n-1}}{(n + c)(2n + 2c - 1)}$ for $n \geq 1$.

Choosing $c = 0$ and $a_n = \dfrac{2(n - 3)(n + 1)\, a_{n-1}}{n(2n - 1)}$ we have $a_1 = -8a_0$, $a_2 = 8a_1$, and $a_n = 0$
for $n \geq 3$. Therefore one solution is $y_1 = 1 - 8x + 8x^2$. Choosing $c = 1/2$ and solving the recurrence relation gives us

$$a_n = \frac{3\big[5 \cdot 7 \cdots (2n + 3)\big]\, a_0}{2^n\, n!\, (2n - 3)(2n - 1)(2n + 1)},$$

$$y_2 = x^{1/2} + 3\sum_{n=1}^{\infty} \frac{5 \cdot 7 \cdots (2n + 3)x^{n+1/2}}{2^n\, n!\, (2n - 3)(2n - 1)(2n + 1)}.$$

41. Set $L(y) = x^2 y'' - 3x(1 + x)y' + 4(1 - x)y$. Then

$$L(y) = \sum_{n=0}^{\infty} \big[(n + c)(n + c - 1) - 3(n + c) + 4\big] a_n x^{n+c} - \sum_{n=0}^{\infty} \big[3(n + c) + 4\big] a_n x^{n+c+1}$$

$$= \sum_{n=0}^{\infty} (n + c - 2)^2 a_n x^{n+c} - \sum_{n=1}^{\infty} (3n + 3c + 1) a_{n-1} x^{n+c}$$

$$= (c - 2)^2 a_0 x^c + \sum_{n=1}^{\infty} \big[(n + c - 2)^2 a_n - (3n + 3c + 1)a_{n-1}\big] x^{n+c}.$$

The indicial equation is $(c - 2)^2 = 0$ and $a_n = \dfrac{(3n + 3c + 1)\, a_{n-1}}{(n + c - 2)^2}$ for $n \geq 1$. Solving this recurrence relation we get

$$a_n = \frac{(3c + 4) \cdots (3c + 3n + 1)\, a_0}{(c - 1)^2 \cdots (c + n - 2)^2},$$

$$y_c = x^c + \sum_{n=1}^{\infty} \frac{(3c + 4) \cdots (3c + 3n + 1)\, x^{n+c}}{(c - 1)^2 \cdots (c + n - 2)^2},$$

$$\frac{\partial y_c}{\partial c} = y_c \ln x + \sum_{n=1}^{\infty} \frac{(3c + 4) \cdots (3c + 3n + 1)\, x^{n+c}}{(c - 1)^2 \cdots (c + n - 2)^2} \bullet$$

$$\left[\frac{3}{3c + 4} + \cdots + \frac{3}{3c + 3n + 1} - \frac{2}{c - 1} - \cdots - \frac{2}{c + n - 2} \right].$$

Substituting $c = 2$ gives the solutions

$$y_1 = x^2 + \sum_{n=1}^{\infty} \frac{10 \cdots (3n + 7)\, x^{n+2}}{(n!)^2},$$

$$y_2 = y_1 \ln x + \sum_{n=1}^{\infty} \frac{10 \cdots (3n + 7)\, x^{n+2}}{(n!)^2} \left[\frac{3}{10} + \cdots + \frac{3}{3n + 7} - 2H_n \right].$$

43. Set $L(y) = 2xy'' + (3 - x)y' - 3y$. Then

$$L(y) = \sum_{n=0}^{\infty} \left[2(n+c)(n+c-1) + 3(n+c)\right]a_n x^{n+c-1} - \sum_{n=0}^{\infty}(n+c+3)a_n x^{n+c}$$

$$= \sum_{n=0}^{\infty}(n+c)(2n+2c+1)a_n x^{n+c-1} - \sum_{n=1}^{\infty}(n+c+2)a_{n-1} x^{n+c-1}$$

$$= c(2c+1)a_0 x^{c-1} + \sum_{n=1}^{\infty}\left[(n+c)(2n+2c+1)a_n - (n+c+2)a_{n-1}\right]x^{n+c-1}.$$

The indicial equation is $c(2c+1) = 0$ and $a_n = \dfrac{(n+c+2)\, a_{n-1}}{(n+c)(2n+2c+1)}$ for $n \geq 1$. Choosing $c = 0$ and solving the recurrence relation we have

$$a_n = \frac{(n+1)(n+2)\, a_0}{2\left[3 \cdot 5 \cdots (2n+1)\right]},$$

$$y_1 = 1 + \sum_{n=1}^{\infty} \frac{(n+1)(n+2)x^n}{2\left[3 \cdot 5 \cdots (2n+1)\right]}.$$

Choosing $c = -1/2$ and solving the recurrence relation gives us

$$a_n = \frac{5 \cdot 7 \cdots (2n+3)\, a_0}{2^n\, n!\, 1 \cdot 3 \cdots (2n-1)} = \frac{(2n+1)(2n+3)\, a_0}{3 \cdot 2^n n!},$$

$$y_2 = x^{-1/2} + \sum_{n=1}^{\infty} \frac{(2n+1)(2n+3)x^{n-1/2}}{3 \cdot 2^n\, n!}.$$

45. Set $L(y) = xy'' + 3y' - y$. Then

$$L(y) = \sum_{n=0}^{\infty} \left[(n+c)(n+c-1) + 3(n+c)\right]a_n x^{n+c-1} - \sum_{n=0}^{\infty} a_n x^{n+c}$$

$$= \sum_{n=0}^{\infty}(n+c)(n+c+2)a_n x^{n+c-1} - \sum_{n=1}^{\infty} a_{n-1} x^{n+c-1}$$

$$= c(c+2)a_0 x^{c-1} + \sum_{n=1}^{\infty}\left[(n+c)(n+c+2)a_n - a_{n-1}\right]x^{n+c-1}.$$

The indicial equation is $c(c+2) = 0$ and $a_n = \dfrac{a_{n-1}}{(n+c)(n+c+2)}$ for $n \geq 1$.

$$a_1 = \frac{a_0}{(c+1)(c+3)}$$

$$a_2 = \frac{a_0}{(c+1)(c+2)(c+3)(c+4)}$$

$$\vdots$$

$$a_n = \frac{a_0}{(c+1)(c+2)\big[(c+3)\cdots(c+n)\big]\big[(c+3)\cdots(c+n+2)\big]} \quad \text{for } n \geq 3.$$

We now take $a_0 = c+2$ and get

$$y_c = (c+2)x^c + \frac{(c+2)x^{c+1}}{(c+1)(c+3)} + \frac{x^{c+2}}{(c+1)(c+3)(c+4)}$$

$$+ \sum_{n=3}^{\infty} \frac{x^{n+c}}{(c+1)\big[(c+3)\cdots(c+n)\big]\big[(c+3)\cdots(c+n+2)\big]},$$

$$\frac{\partial y_c}{\partial c} = y_c \ln x + x^c + \frac{(c+2)x^{c+1}}{(c+1)(c+3)}\left[\frac{1}{c+2} - \frac{1}{c+1} - \frac{1}{c+3}\right]$$

$$+ \frac{x^{c+2}}{(c+1)(c+3)(c+4)}\left[-\frac{1}{c+1} - \frac{1}{c+3} - \frac{1}{c+4}\right]$$

$$+ \sum_{n=3}^{\infty} \frac{x^{n+c}}{(c+1)\big[(c+3)\cdots(c+n)\big]\big[(c+3)\cdots(c+n+2)\big]} \bullet$$

$$\left[-\frac{1}{c+1} - \left\{\frac{1}{c+3} + \cdots + \frac{1}{c+n}\right\} - \left\{\frac{1}{c+3} + \cdots + \frac{1}{c+n+2}\right\}\right].$$

Setting $c = -2$ gives the solutions

$$y_1 = -\frac{1}{2} - \sum_{n=3}^{\infty} \frac{x^{n-2}}{n!\,(n-2)!} = -\sum_{n=0}^{\infty} \frac{x^n}{n!\,(n+2)!},$$

$$y_2 = y_1 \ln x + x^{-2} - x^{-1} + \frac{1}{4} - \sum_{n=3}^{\infty} \frac{(1 - H_n - H_{n-2})x^{n-2}}{n!\,(n-2)!}$$

$$= y_1 \ln x + x^{-2} - x^{-1} + \frac{1}{4} - \sum_{n=1}^{\infty} \frac{(1 - H_n - H_{n+2})x^n}{n!\,(n+2)!}.$$

47. Set $L(y) = 3xy'' + 2(1 - x)y' - 2y$. Then

$$L(y) = \sum_{n=0}^{\infty} \left[3(n + c)(n + c - 1) + 2(n + c)\right]a_n x^{n+c-1} - \sum_{n=0}^{\infty}(2n + 2c + 2)a_n x^{n+c}$$

$$= \sum_{n=0}^{\infty}(n + c)(3n + 3c - 1)a_n x^{n+c-1} - \sum_{n=1}^{\infty} 2(n + c)a_{n-1}x^{n+c-1}$$

$$= c(3c - 1)a_0 x^{c-1} + \sum_{n=1}^{\infty}\left[(n + c)(3n + 3c - 1)a_n - 2(n + c)a_{n-1}\right]x^{n+c-1}.$$

The indicial equation is $c(3c - 1) = 0$ and $a_n = \dfrac{2(n + c)\, a_{n-1}}{(n + c)(3n + 3c - 1)}$ for $n \geq 1$. Choosing $c = 0$ and solving the recurrence relation we have

$$a_n = \frac{2^n\, a_0}{2 \cdot 5 \cdots (3n - 1)},$$

$$y_2 = 1 + \sum_{n=1}^{\infty} \frac{2^n x^n}{2 \cdot 5 \cdots (3n - 1)}.$$

Choosing $c = 1/3$ and solving the recurrence relation gives us

$$a_n = \frac{2^n\, a_0}{3^n\, n!},$$

$$y_1 = x^{1/3} + \sum_{n=1}^{\infty} \frac{2^n x^{n+1/3}}{3^n\, n!} = x^{1/3} \exp\left(\frac{2}{3}x\right).$$

49. Set $L(y) = xy'' + (3 - 2x)y' + 4y$. Then

$$L(y) = \sum_{n=0}^{\infty}\left[(n + c)(n + c - 1) + 3(n + c)\right]a_n x^{n+c-1} - \sum_{n=0}^{\infty}(2n + 2c - 4)a_n x^{n+c}$$

$$= \sum_{n=0}^{\infty}(n + c)(n + c + 2)a_n x^{n+c-1} - \sum_{n=0}^{\infty} 2(n + c - 2)a_n x^{n+c}$$

$$= c(c + 2)a_0 x^{c-1} + \sum_{n=1}^{\infty}\left[(n + c)(n + c + 2)a_n - 2(n + c - 3)a_{n-1}\right]x^{n+c-1}.$$

The indicial equation is $c(c+2) = 0$ and $a_n = \dfrac{2(c+n-3)a_{n-1}}{(n+c)(n+c+2)}$ for $n \geq 1$.

$$a_1 = \frac{2(c-2)\,a_0}{(c+1)(c+3)}$$

$$a_2 = \frac{2^2(c-2)(c-1)\,a_0}{(c+1)(c+2)(c+3)(c+4)}$$

$$a_3 = \frac{2^3(c-2)(c-1)c\,a_0}{(c+1)(c+2)(c+3)^2(c+4)(c+5)}$$

$$a_4 = \frac{2^4(c-2)(c-1)c\,a_0}{(c+2)(c+3)^2(c+4)^2(c+5)(c+6)}$$

$$\vdots$$

$$a_n = \frac{2^n(c-2)(c-1)c\,a_0}{(c+n-2)(c+n-1)(c+n)(c+3)(c+4)\cdots(c+n+2)} \quad \text{for } n \geq 5.$$

We now take $a_0 = c+2$ and get

$$y_c = (c+2)x^c + \frac{2(c-2)(c+2)x^{c+1}}{(c+1)(c+3)} + \frac{4(c-2)(c-1)x^{c+2}}{(c+1)(c+3)(c+4)}$$

$$+ \frac{8(c-2)(c-1)cx^{c+3}}{(c+1)(c+3)(c+3)(c+4)(c+5)} + \frac{16(c-2)(c-1)cx^{c+4}}{(c+3)(c+4)(c+3)(c+4)(c+5)(c+6)}$$

$$+ \sum_{n=5}^{\infty} \frac{2^n(c-2)(c-1)c(c+2)x^{n+c}}{(c+n-2)(c+n-1)(c+n)(c+3)(c+4)\cdots(c+n+2)},$$

$$\frac{\partial y_c}{\partial c} = y_c \ln x + x^c + \frac{2(c-2)(c+2)x^{c+1}}{(c+1)(c+3)}\left[\frac{1}{c-2} + \frac{1}{c+2} - \frac{1}{c+1} - \frac{1}{c+3}\right]$$

$$+ \frac{4(c-2)(c-1)x^{c+2}}{(c+1)(c+3)(c+4)}\left[\frac{1}{c-2} + \frac{1}{c-1} - \frac{1}{c+1} - \frac{1}{c+3} - \frac{1}{c+4}\right]$$

$$+ \frac{8(c-2)(c-1)cx^{c+3}}{(c+1)(c+3)^2(c+4)(c+5)}\left[\frac{1}{c-2} + \frac{1}{c-1} + \frac{1}{c} - \frac{1}{c+1} - \frac{2}{c+3} - \frac{1}{c+4} - \frac{1}{c+5}\right]$$

$$+ \frac{16(c-2)(c-1)cx^{c+4}}{(c+3)^2(c+4)^2(c+5)(c+6)}\left[\frac{1}{c-2} + \frac{1}{c-1} + \frac{1}{c} - \frac{2}{c+3} - \frac{2}{c+4} - \frac{1}{c+5} - \frac{1}{c+6}\right]$$

$$+ \sum_{n=5}^{\infty} \frac{2^n(c-2)(c-1)c(c+2)x^{n+c}}{(c+n-2)(c+n-1)(c+n)\left[(c+3)\cdots(c+n+2)\right]} \bullet$$

$$\left[\frac{1}{c-2} + \frac{1}{c-1} + \frac{1}{c} + \frac{1}{c+2} - \frac{1}{c+n-2} - \frac{1}{c+n-1} - \frac{1}{c+n} - \left\{\frac{1}{c+3}\cdots + \frac{1}{c+n+2}\right\}\right].$$

Setting $c = -2$ gives the solutions

$$y_1 = -24 + 32x - 8x^2,$$

$$y_2 = y_1 \ln x + x^{-2} + 8x^{-1} + 26 - \frac{70}{3}x + \frac{112}{3}x^2 - 24\sum_{n=5}^{\infty} \frac{2^n x^{n-2}}{n!\,(n-4)(n-3)(n-2)}.$$

51. Set $L(y) = 2x^2 y'' + 3xy' - (1 + x)y$. Then

$$L(y) = \sum_{n=0}^{\infty} \left[2(n + c)(n + c - 1) + 3(n + c) - 1 \right] a_n x^{n+c} - \sum_{n=0}^{\infty} a_n x^{n+c+1}$$

$$= \sum_{n=0}^{\infty} (n + c + 1)(2n + 2c - 1) a_n x^{n+c} - \sum_{n=1}^{\infty} a_{n-1} x^{n+c}$$

$$= (c + 1)(2c - 1) a_0 x^c + \sum_{n=1}^{\infty} \left[(n + c + 1)(2n + 2c - 1) a_n - a_{n-1} \right] x^{n+c}.$$

The indicial equation is $(c + 1)(2c - 1) = 0$ and $a_n = \dfrac{a_{n-1}}{(n + c + 1)(2n + 2c - 1)}$ for $n \geq 1$. Choosing $c = -1$ and solving the recurrence relation we have

$$a_n = \frac{a_0}{n! \, (-1) \cdot 1 \cdots (2n - 3)},$$

$$y_1 = x^{-1} + \sum_{n=1}^{\infty} \frac{x^{n-1}}{n! \, (-1) \cdot 1 \cdots (2n - 3)}.$$

Choosing $c = 1/2$ and solving the recurrence relation gives us

$$a_n = \frac{a_0}{n! \, 5 \cdot 7 \cdots (2n + 3)},$$

$$y_2 = x^{1/2} + \sum_{n=1}^{\infty} \frac{x^{n+1/2}}{n! \, 5 \cdot 7 \cdots (2n + 3)}.$$

Chapter 20

Partial Differential Equations

20.3 Method of Separation of Variables

1. $\dfrac{\partial^2 u}{\partial t^2} = a^2 \dfrac{\partial^2 u}{\partial x^2}$. We put $u(x, t) = X(x)T(t)$, separate the variables, and arrive at the equations

$\dfrac{T''}{a^2 T} = \dfrac{X''}{X} = c$. If $c = -\beta^2 < 0$ then $T'' + a^2\beta^2 T = 0$ and $X'' + a^2\beta^2 X = 0$, so that
$u(x, t) = (A_1 \cos a\beta t + A_2 \sin a\beta t)(B_1 \cos a\beta x + B_2 \sin a\beta x)$. If $c = 0$ then $T'' = 0$ and $X'' = 0$,
so that $u(x, t) = (A_3 + A_4 t)(B_3 + B_4 x)$. If $c = \beta^2 > 0$ then $T'' - a^2\beta^2 T = 0$ and $X'' - a^2\beta^2 X = 0$,
so that $u(x, t) = (A_5 \cosh a\beta t + A_6 \sinh a\beta t)(B_5 \cosh a\beta x + B_6 \sinh a\beta x)$.

3. $\dfrac{\partial^2 u}{\partial t^2} + 2b\dfrac{\partial u}{\partial t} = a^2 \dfrac{\partial^2 u}{\partial x^2}$. We set $u(x\,t) = X(x)T(t)$, separate the variables, and arrive at

$\dfrac{T'' + 2bT'}{a^2 T} = \dfrac{X''}{X} = c$. If $c = -k^2$ we have $T'' + 2bT' + k^2 a^2 T = 0$ and $X'' + k^2 X = 0$. The
auxiliary equation $m^2 + 2bm + k^2 a^2 = 0$ has roots $m = -b \pm \sqrt{b^2 - a^2 k^2}$. If $b^2 - a^2 k^2 = \gamma^2 > 0$
then $T = e^{-bt}(A_1 e^{\gamma t} + A_2 e^{-\gamma t})$ so that

$$u(x, t) = e^{-bt}(A_1 e^{\gamma t} + A_2 e^{-\gamma t})(B_1 \cos kx + B_2 \sin kx).$$

If $b^2 - a^2 k^2 = -\delta^2 < 0$ then $T = e^{-bt}(A_3 \cos \delta t + A_4 \sin \delta t)$, so that

$$u(x, t) = e^{-bt}(A_3 \cos \delta t + A_4 \sin \delta t)(B_1 \cos kx + B_2 \sin kx).$$

If $b^2 - a^2 k^2 = 0$ then $T = e^{-bt}(A_5 + A_6 t)$, so that

$$u(x, t) = e^{-bt}(A_5 + A_6 t)(B_1 \cos kx + B_2 \sin kx).$$

If $c = 0$ we have $T'' + 2bT' = 0$ and $X'' = 0$. The auxiliary equation $m^2 + 2bm = 0$ has roots
$m = 0, -2b$, so that

$$u(x, t) = (A_7 + A_8 e^{-2bt})(B_3 + B_4 x).$$

If $c = k^2 > 0$ we have $T'' + 2bT' - k^2 a^2 T = 0$ and $X'' - k^2 X = 0$. Here the auxiliary equation
$m^2 + 2bm - k^2 a^2 = 0$ has real roots $-b \pm \sqrt{b^2 + k^2 a^2} = -b \pm \mu$. It follows that

$$u(x, t) = e^{-bt}(A_9 e^{\mu t} + A_{10} e^{-\mu t})(B_5 e^{kx} + B_6 e^{-kx}).$$

5. $x\dfrac{\partial w}{\partial x} = w + y\dfrac{\partial w}{\partial y}$. We put $w = X(x)Y(y)$, separate the variables, and obtain the equations

$\dfrac{xX'}{X} = 1 + \dfrac{yY'}{Y} = k$. Thus $xX' - kX = 0$ or $yY' - (k-1)Y = 0$. These equations have solutions $X = Bx^k$ and $Y = Cy^{k-1}$, so that $w = Ax^k y^{k-1}$.

7. $\dfrac{\partial^2 u}{\partial x^2} + 4\dfrac{\partial^2 u}{\partial x \partial t} + 5\dfrac{\partial^2 u}{\partial t^2} = 0$. We make the substitution $u(x, t) = X(x)T(t)$ and get the equation $X''(x)T(t) + 4X'(x)T'(t) + 5X(x)T''(t) = 0$. Dividing by the product $X(x)T(t)$ yields the equation $\dfrac{X''}{X} + 4\dfrac{X'}{X} \cdot \dfrac{T'}{T} + 5\dfrac{T''}{T} = 0$, but the second term keeps us from separating the variables in this equation unless either $\dfrac{X'}{X}$ is independent of x or $\dfrac{T'}{T}$ is independent of t. One suggestion would be to set $T(t) = e^{kt}$ so that $\dfrac{T'}{T} = k$. Then the variables will separate. See Exercise 8.

9. $\dfrac{\partial^2 u}{\partial x^2} + 4x\dfrac{\partial u}{\partial x} + \dfrac{\partial^2 u}{\partial y^2} = 0$. We separate the variables by setting $u = f(x)g(y)$ and obtain $\dfrac{f'' + 4xf'}{f} = \dfrac{-g''}{g} = c$. If $c = -4k^2$ then $f'' + 4xf' + 4k^2 f = 0$ and $g'' - 4k^2 g = 0$. Thus

$$u(x, y) = \left[A_1 e^{2ky} + A_2 e^{-2ky}\right]\left[B_1 f_1(x) + B_2 f_2(x)\right],$$

where f_1 and f_2 are any linearly independent solutions of $f'' + 4xf' + 4k^2 f = 0$. These solutions may be found by setting $f(x) = \displaystyle\sum_{n=0}^{\infty} a_n x^n$ to obtain $\displaystyle\sum_{n=0}^{\infty}\left[(n+2)(n+1)a_{n+2} + 4(n+k^2)a_n\right]x^n = 0$. The recurrence relation can be solved to get

$$a_{2m} = \frac{(-4)^m k^2(k^2 + 2)\cdots(k^2 + 2m - 2)\, a_0}{(2m)!} \quad \text{and}$$

$$a_{2m+1} = \frac{(-4)^m (k^2 + 1)(k^2 + 3)\cdots(k^2 + 2m - 1)\, a_1}{(2m+1)!},$$

so that

$$f_1(x) = 1 + \sum_{m=1}^{\infty} \frac{(-4)^m k^2(k^2 + 2)\cdots(k^2 + 2m - 2)x^{2m}}{(2m)!} \quad \text{and}$$

$$f_2(x) = x + \sum_{m=1}^{\infty} \frac{(-4)^m (k^2 + 1)(k^2 + 3)\cdots(k^2 + 2m - 1)x^{2m+1}}{(2m+1)!}.$$

If $c = 4k^2$ then $f'' + 4xf' - 4k^2 f = 0$ and $g'' + 4k^2 g = 0$. Thus

$$u(x, y) = \left[A_3 \cos 2ky + A_4 \sin 2ky\right]\left[B_3 f_3(x) + B_4 f_4(x)\right],$$

where f_3 and f_4 are any linearly independent solutions of $f'' + 4xf' - 4k^2 f = 0$. These solutions may be found by replacing k^2 by $-k^2$ in f_1 and f_2. We get

$$f_3(x) = 1 + \sum_{m=1}^{\infty} \frac{(-4)^m (-k^2)(2 - k^2)(4 - k^2) \cdots (2m - 2 - k^2) x^{2m}}{(2m)!} \quad \text{and}$$

$$f_4(x) = x + \sum_{m=1}^{\infty} \frac{(-4)^m (1 - k^2)(3 - k^2) \cdots (2m - 1 - k^2) x^{2m+1}}{(2m + 1)!}.$$

If $c = 0$ then $f'' + 4xf' = 0$ and $g'' = 0$. These differential equations can be solved to obtain

$$u(x, y) = [A_5 + A_6 y]\left[B_5 + B_6 \int \exp\left(-2x^2\right) dx \right].$$

See Section 5.6 in the text for possible ways to treat the nonelementary integral in this set of solutions.

11. $u = f_1(x - at) + f_2(x + at)$. We let $y = x - at$ and $z = x + at$. Then

$$\frac{\partial u}{\partial x} = \frac{df_1}{dy}\frac{\partial y}{\partial x} + \frac{df_2}{dz}\frac{\partial z}{\partial x} = \frac{df_1}{dy} + \frac{df_2}{dz}, \qquad \frac{\partial^2 u}{\partial x^2} = \frac{d^2 f_1}{dy^2}\frac{\partial y}{\partial x} + \frac{d^2 f_2}{dz^2}\frac{\partial z}{\partial x} = \frac{d^2 f_1}{dy^2} + \frac{d^2 f_2}{dz^2},$$

$$\frac{\partial u}{\partial t} = \frac{df_1}{dy}\frac{\partial y}{\partial t} + \frac{df_2}{dz}\frac{\partial z}{\partial t} = a\frac{df_1}{dy} - a\frac{df_2}{dz}, \qquad \frac{\partial^2 u}{\partial t^2} = a\frac{d^2 f_1}{dy^2}\frac{\partial y}{\partial t} - a\frac{d^2 f_2}{dz^2}\frac{\partial z}{\partial t} = a^2\frac{d^2 f_1}{dy^2} + a^2\frac{d^2 f_2}{dz^2}.$$

We see that $\dfrac{\partial^2 u}{\partial t^2} = a^2 \dfrac{\partial^2 u}{\partial x^2}$.

13. $w = f(2x + y^2)$. We set $z = 2x + y^2$. Then $\dfrac{\partial w}{\partial x} = \dfrac{df}{dz}\dfrac{\partial z}{\partial x} = 2\dfrac{df}{dz}$ and $\dfrac{\partial w}{\partial y} = \dfrac{df}{dz}\dfrac{\partial z}{\partial y} = 2y\dfrac{df}{dz}$. It follows that $\dfrac{\partial w}{\partial y} = y\dfrac{\partial w}{\partial x}$.

Chapter 21

Orthogonal Sets of Functions

21.6   Other Orthogonal Sets

1. From equation (3) in the text we have

$$D\big[xe^{-x}L_n'(x)\big] + ne^{-x}L_n(x) = 0 \quad \text{and} \quad D\big[xe^{-x}L_m'(x)\big] + me^{-x}L_m(x) = 0.$$

Multiplying the first of these equations by $L_m(x)$, the second by $L_n(x)$, and subtracting we obtain

$$(m - n)e^{-x}L_m(x)L_n(x) = L_m(x)D\big[xe^{-x}L_n'(x)\big] - L_n(x)D\big[xe^{-x}L_m'(x)\big]. \tag{21.1}$$

We also note that

$$D\big[xe^{-x}L_n(x)L_m'(x)\big] = L_n(x)D\big[xe^{-x}L_m'(x)\big] + xe^{-x}L_n'(x)L_m'(x),$$
$$D\big[xe^{-x}L_m(x)L_n'(x)\big] = L_m(x)D\big[xe^{-x}L_n'(x)\big] + xe^{-x}L_n'(x)L_m'(x).$$

It follows that

$$L_m(x)D\big[xe^{-x}L_n'(x)\big] - L_n(x)D\big[xe^{-x}L_m'(x)\big] = D\big[xe^{-x}L_m(x)L_n'(x)\big] - D\big[xe^{-x}L_n(x)L_m'(x)\big].$$

Equation (21.1) can now be written

$$(m - n)e^{-x}L_m(x)L_n(x) = D\big[xe^{-x}\{L_m(x)L_n'(x) - L_n(x)L_m'(x)\}\big].$$

An integration gives us

$$(m - n)\int_0^\infty e^{-x}L_m(x)L_n(x)\ dx = \big[xe^{-x}\{L_m(x)L_n'(x) - L_n(x)L_m'(x)\}\big]_0^\infty.$$

As $x \to \infty$ the function on the right approaches zero because $L_m(x)$ and $L_n(x)$ are polynomials. Moreover xe^{-x} is zero for $x = 0$. Hence we have for $m \neq n$,

$$\int_0^\infty e^{-x}L_m(x)L_n(x)\ dx = 0.$$

The Laguerre polynomials are orthogonal with respect to the weight function e^{-x} on the interval $0 \leq x < \infty$.

3. From equation (6) in the text we have

$$D\big[\exp{(-x^2)}H_n'(x)\big] + 2n\exp{(-x^2)}H_n(x) = 0,$$
$$D\big[\exp{(-x^2)}H_m'(x)\big] + 2m\exp{(-x^2)}H_m(x) = 0.$$

Multiplying the first of these equations by $H_m(x)$, the second by $H_n(x)$, and subtracting we obtain

$$2(m-n)\exp{(-x^2)}H_m(x)H_n(x) = H_m(x)D\big[\exp{(-x^2)}H_n'(x)\big] - H_n(x)D\big[\exp{(-x^2)}H_m'(x)\big].$$
$$(21.2)$$

We also note that

$$D\big[\exp{(-x^2)}H_n(x)H_m'(x)\big] = H_n(x)D\big[\exp{(-x^2)}H_m'(x)\big] + \exp{(-x^2)}H_n'(x)H_m'(x),$$
$$D\big[\exp{(-x^2)}H_m(x)H_n'(x)\big] = H_m(x)D\big[\exp{(-x^2)}H_n'(x)\big] + \exp{(-x^2)}H_n'(x)H_m'(x).$$

It follows that

$$H_m(x)D\big[\exp{(-x^2)}H_n'(x)\big] - H_n(x)D\big[\exp{(-x^2)}H_m'(x)\big]$$
$$= D\big[\exp{(-x^2)}H_m(x)H_n'(x)\big] - D\big[\exp{(-x^2)}H_n(x)H_m'(x)\big].$$

Equation (21.2) can now be written

$$2(m-n)\exp{(-x^2)}H_m(x)H_n(x) = D\left[\exp{(-x^2)}\left\{H_m(x)H_n'(x) - H_n(x)H_m'(x)\right\}\right].$$

An integration gives us

$$2(m-n)\int_{-\infty}^{\infty}\exp{(-x^2)}H_m(x)H_n(x)\,dx = \left[\exp{(-x^2)}\left\{H_m(x)H_n'(x) - H_n(x)H_m'(x)\right\}\right]_{-\infty}^{\infty}.$$

As $x \to \infty$ and as $x \to -\infty$ the function on the right approaches zero because $H_m(x)$ and $H_n(x)$ are polynomials. Hence we have for $m \neq n$,

$$\int_{-\infty}^{\infty}\exp{(-x^2)}H_m(x)H_n(x)\,dx = 0.$$

The Hermite polynomials are orthogonal with respect to the weight function $\exp{(-x^2)}$ on the interval $-\infty < x < \infty$.

Chapter 22

Fourier Series

22.3   Numerical Examples of Fourier Series

1. Using equations (10), (11), and (12) from the text we have

$$a_0 = \frac{1}{c} \int_0^c (c - x)\, dx = \frac{c}{2}, \quad a_n = \frac{1}{c} \int_0^c (c - x) \cos \frac{n\pi x}{c}\, dx = \frac{c}{n^2\pi^2}\left[1 - (-1)^n\right],$$

$$b_n = \frac{1}{c} \int_0^c (c - x) \sin \frac{n\pi x}{c}\, dx = \frac{c}{n\pi}.$$

$$f(x) \sim \frac{c}{4} + \frac{c}{\pi^2} \sum_{n=1}^{\infty} \frac{1}{n^2} \left[\{1 - (-1)^n\} \cos \frac{n\pi x}{c} + n\pi \sin \frac{n\pi x}{c} \right].$$

3. Using equations (10), (11), and (12) from the text we have

$$a_0 = \frac{1}{c} \int_{-c}^c x^2\, dx = \frac{2c^2}{3}, \quad a_n = \frac{1}{c} \int_{-c}^c x^2 \cos \frac{n\pi x}{c}\, dx = \frac{4c^2(-1)^n}{n^2\pi^2},$$

$$b_n = \frac{1}{c} \int_{-c}^c x^2 \sin \frac{n\pi x}{c}\, dx = 0.$$

$$f(x) \sim \frac{c^2}{3} + \frac{4c^2}{\pi^2} \sum_{n=1}^{\infty} \frac{(-1)^n}{n^2} \cos \frac{n\pi x}{c}.$$

5. Using equations (10), (11), and (12) from the text we have

$$a_0 = \frac{1}{c} \int_0^c dx = 1, \quad a_n = \frac{1}{c} \int_0^c \cos \frac{n\pi x}{c}\, dx = 0,$$

$$b_n = \frac{1}{c} \int_0^c \sin \frac{n\pi x}{c}\, dx = \frac{1}{n\pi}\left[1 - (-1)^n\right].$$

$$f(x) \sim \frac{1}{2} + \sum_{n=1}^{\infty} \frac{[1 - (-1)^n]}{n\pi} \sin \frac{n\pi x}{c}.$$

7. Using equations (10), (11), and (12) from the text we have

$$a_0 = \frac{1}{\pi} \int_{-\pi}^{0} (3\pi + 2x) \, dx + \frac{1}{\pi} \int_{0}^{\pi} (\pi + 2x) \, dx = 4\pi,$$

$$a_n = \frac{1}{\pi} \int_{-\pi}^{0} (3\pi + 2x) \cos nx \, dx + \frac{1}{\pi} \int_{0}^{\pi} (\pi + 2x) \cos nx \, dx = 0,$$

$$b_n = \frac{1}{\pi} \int_{-\pi}^{0} (3\pi + 2x) \sin nx \, dx + \frac{1}{\pi} \int_{0}^{\pi} (\pi + 2x) \sin nx \, dx = \frac{-2}{n} \left[1 - (-1)^n \right].$$

$$b_{2k-1} = 0, \quad b_{2k} = \frac{-2}{k}, \quad f(x) \sim 2\pi - 2 \sum_{k=1}^{\infty} \frac{\sin 2kx}{k}.$$

9. Using equations (10), (11), and (12) from the text we have

$$a_0 = \frac{1}{2} \int_{-2}^{0} (x + 1) \, dx + \frac{1}{2} \int_{0}^{2} dx = 1,$$

$$a_n = \frac{1}{2} \int_{-2}^{0} (x + 1) \cos \frac{n\pi x}{2} \, dx + \frac{1}{2} \int_{0}^{2} \cos \frac{n\pi x}{2} \, dx = \frac{2 \left[1 - (-1)^n \right]}{n^2 \pi^2},$$

$$b_n = \frac{1}{2} \int_{-2}^{0} (x + 1) \sin \frac{n\pi x}{2} \, dx + \frac{1}{2} \int_{0}^{2} \sin \frac{n\pi x}{2} \, dx = \frac{2(-1)^n}{n\pi}.$$

$$f(x) \sim \frac{1}{2} + \frac{2}{\pi^2} \sum_{n=1}^{\infty} \frac{1}{n^2} \left[\{1 - (-1)^n\} \cos \frac{n\pi x}{2} + n\pi(-1)^{n+1} \sin \frac{n\pi x}{2} \right].$$

11. Using equations (10), (11), and (12) from the text we have

$$a_0 = \frac{1}{\pi} \int_{0}^{\pi} x^2 \, dx = \frac{\pi^2}{3}, \quad a_n = \frac{1}{\pi} \int_{0}^{\pi} x^2 \cos nx \, dx = \frac{2(-1)^n}{n^2},$$

$$b_n = \frac{1}{\pi} \int_{0}^{\pi} x^2 \sin nx \, dx = \frac{1}{n^3 \pi} \left[(-1)^{n+1} n^2 \pi^2 - 2 + 2(-1)^n \right].$$

$$f(x) \sim \frac{\pi^2}{6} + \sum_{n=1}^{\infty} \frac{2(-1)^n \cos nx}{n^2} + \frac{1}{\pi} \sum_{n=1}^{\infty} \frac{1}{n^3} \left[(-1)^{n+1} n^2 \pi^2 - 2 + 2(-1)^n \right] \sin nx.$$

13. Using equations (10), (11), and (12) from the text we have

$$a_0 = \frac{1}{\pi} \int_{-\pi}^{\pi} \cos \frac{x}{2} \, dx = \frac{4}{\pi}, \quad a_n = \frac{1}{\pi} \int_{-\pi}^{\pi} \cos \frac{x}{2} \cos nx \, dx = \frac{4(-1)^{n+1}}{\pi(4n^2 - 1)},$$

$$b_n = \frac{1}{\pi} \int_{-\pi}^{\pi} \cos \frac{x}{2} \sin nx \, dx = 0, \quad f(x) \sim \frac{2}{\pi} + \frac{4}{\pi} \sum_{n=1}^{\infty} \frac{(-1)^{n+1} \cos nx}{4n^2 - 1}.$$

15. Using equations (10), (11), and (12) from the text we have

$$a_0 = \frac{1}{c} \int_{-c}^{c} e^x \, dx = \frac{2 \sinh c}{c}, \quad a_n = \frac{1}{c} \int_{-c}^{c} e^x \cos \frac{n\pi x}{c} \, dx = \frac{2c(-1)^n \sinh c}{c^2 + n^2 \pi^2},$$

$$b_n = \frac{1}{c} \int_{-c}^{c} e^x \sin \frac{n\pi x}{c} \, dx = \frac{2(-1)^n \sinh c}{c^2 + n^2 \pi^2}(-n\pi).$$

$$f(x) \sim \frac{\sinh c}{c} + \sum_{n=1}^{\infty} \frac{2(-1)^n \sinh c \left[c \cos (n\pi x/c) - n\pi \sin (n\pi x/c) \right]}{c^2 + n^2 \pi^2}.$$

17. Using equations (10), (11), and (12) from the text we have

$$a_0 = \frac{1}{c} \int_{c/2}^{c} dx = \frac{1}{2}, \quad a_n = \frac{1}{c} \int_{c/2}^{c} \cos \frac{n\pi x}{c} \, dx = \frac{-1}{n\pi} \sin \frac{n\pi}{2},$$

$$b_n = \frac{1}{c} \int_{c/2}^{c} \sin \frac{n\pi x}{c} \, dx = \frac{-1}{n\pi} \left(\cos n\pi - \cos \frac{n\pi}{2} \right).$$

$$f(x) \sim \frac{1}{4} - \frac{1}{\pi} \sum_{n=1}^{\infty} \frac{1}{n} \left[\sin \frac{n\pi}{2} \cos \frac{n\pi x}{c} + \left(\cos n\pi - \cos \frac{n\pi}{2} \right) \sin \frac{n\pi x}{c} \right].$$

19. Let us put $c = 4$ in Exercise 17. Then if $f(x)$ is the function of Exercise 17, $1 - f(x)$ is the function of this exercise. Thus for the function of this exercise

$$f(x) \sim 1 - \left\{ \frac{1}{4} - \frac{1}{\pi} \sum_{n=1}^{\infty} \frac{1}{n} \left[\sin \frac{n\pi}{2} \cos \frac{n\pi x}{4} + \left(\cos n\pi - \cos \frac{n\pi}{2} \right) \sin \frac{n\pi x}{4} \right] \right\},$$

$$f(x) \sim \frac{3}{4} + \frac{1}{\pi} \sum_{n=1}^{\infty} \frac{1}{n} \left[\sin \frac{n\pi}{2} \cos \frac{n\pi x}{4} + \left(\cos n\pi - \cos \frac{n\pi}{2} \right) \sin \frac{n\pi x}{4} \right].$$

21. Using equations (10), (11), and (12) from the text we have

$$a_0 = \frac{1}{c} \int_{-c}^{0} (c + x) \, dx = \frac{c}{2}, \quad a_n = \frac{1}{c} \int_{-c}^{0} (c + x) \cos \frac{n\pi x}{c} \, dx = \frac{c}{n^2 \pi^2}(1 - \cos n\pi),$$

$$b_n = \frac{1}{c} \int_{-c}^{0} (c + x) \sin \frac{n\pi x}{c} \, dx = \frac{-c}{n\pi}.$$

$$f(x) \sim \frac{c}{4} + \sum_{n=1}^{\infty} \left[\frac{c}{n^2 \pi^2} \{ 1 - (-1)^n \} \cos \frac{n\pi x}{c} - \frac{c}{n\pi} \sin \frac{n\pi x}{c} \right].$$

23. $x^2 \sim \frac{c^2}{3} + \frac{4c^2}{\pi^2} \sum_{n=1}^{\infty} \frac{(-1)^n}{n^2} \cos \frac{n\pi x}{c}$. Since the function of the series is continuous at $x = 0$ we

have $0 = \frac{c^2}{3} + \frac{4c^2}{\pi^2} \sum_{n=1}^{\infty} \frac{(-1)^n}{n^2}$. It follows that $\sum_{n=1}^{\infty} \frac{(-1)^{n+1}}{n^2} = \frac{\pi^2}{12}$.

25. $x^4 \sim \dfrac{c^4}{5} + 8c^4 \displaystyle\sum_{n=1}^{\infty} \dfrac{(-1)^n(n^2\pi^2 - 6)}{n^4\pi^4} \cos \dfrac{n\pi x}{c}$. Since the function of the series is continuous at

$x = c$ we have $c^4 = \dfrac{c^4}{5} + 8c^4 \displaystyle\sum_{n=1}^{\infty} \dfrac{1}{n^2\pi^2} - 48c^4 \displaystyle\sum_{n=1}^{\infty} \dfrac{1}{n^4\pi^4}$. The second series can be summed with

the use of the answer to Exercise 24. We have $\dfrac{48}{\pi^4} \displaystyle\sum_{n=1}^{\infty} \dfrac{1}{n^4} = -1 + \dfrac{1}{5} + \dfrac{8}{\pi^2} \cdot \dfrac{\pi^2}{6} = \dfrac{8}{15}$. It follows

that $\displaystyle\sum_{n=1}^{\infty} \dfrac{1}{n^4} = \dfrac{\pi^4}{90}$.

27. $\displaystyle\lim_{c\to 0} \dfrac{c - \sinh c}{2c^2 \sinh c} = \lim_{c\to 0} \dfrac{1 - \cosh c}{2c^2 \cosh c + 4c \sinh c} = \lim_{c\to 0} \dfrac{-\sinh c}{2c^2 \sinh c + 8c \cosh c + 4 \sinh c}$

$= \displaystyle\lim_{c\to 0} \dfrac{-\cosh c}{2c^2 \cosh c + 12c \sinh c + 12 \cosh c} = \dfrac{-1}{12} = \displaystyle\sum_{n=1}^{\infty} \dfrac{(-1)^n}{n^2\pi^2}$. Hence $\displaystyle\sum_{n=1}^{\infty} \dfrac{(-1)^{n+1}}{n^2} = \dfrac{\pi^2}{12}$.

22.4 Fourier Sine Series

1. From equations (1) and (2) in the text we obtain

$$b_n = \dfrac{2}{c} \int_0^c \sin \dfrac{n\pi x}{c}\, dx = \dfrac{2\left[1 - (-1)^n\right]}{n\pi}, \quad b_{2k} = 0, \quad b_{2k+1} = \dfrac{4}{(2k+1)\pi}.$$

$$f(x) \sim \dfrac{4}{\pi} \sum_{k=0}^{\infty} \dfrac{1}{2k+1} \sin \dfrac{(2k+1)\pi x}{c}.$$

3. From equations (1) and (2) in the text we obtain

$$b_n = \dfrac{2}{c} \int_0^c x^2 \sin \dfrac{n\pi x}{c}\, dx = 2c^2 \left[\dfrac{(-1)^{n+1}}{n\pi} - \dfrac{2\left\{1 - (-1)^n\right\}}{n^3\pi^3}\right].$$

$$f(x) \sim 2c^2 \sum_{n=1}^{\infty} \left[\dfrac{(-1)^{n+1}}{n\pi} - \dfrac{2\left\{1 - (-1)^n\right\}}{n^3\pi^3}\right] \sin \dfrac{n\pi x}{c}.$$

5. The odd periodic extension of this function is the same as the odd periodic extension of the function in Exercise 4. Hence the two functions have ths same Fourier sine series. Thus $f(x) \sim \dfrac{2c}{\pi} \displaystyle\sum_{n=1}^{\infty} \dfrac{1}{n} \sin \dfrac{n\pi x}{c}$.

7. From equations (1) and (2) in the text we obtain

$$b_n = \dfrac{2}{c} \int_0^c x(c - x) \sin \dfrac{n\pi x}{c}\, dx = \dfrac{4c^2 \left[1 - (-1)^n\right]}{n^3\pi^3}, \quad b_{2k} = 0, \quad b_{2k+1} = \dfrac{8c^2}{(2k+1)^3\pi^3}.$$

$$f(x) \sim \dfrac{8c^2}{\pi^3} \sum_{k=0}^{\infty} \dfrac{1}{(2k+1)^3} \sin \dfrac{(2k+1)\pi x}{c}.$$

9. From equations (1) and (2) in the text we obtain

$$b_n = \frac{2}{t_1} \int_0^{t_0} \sin \frac{n\pi t}{t_1} \, dt = \frac{2}{n\pi} \left[1 - \cos \frac{n\pi t_0}{t_1} \right], \quad f(x) \sim \frac{2}{\pi} \sum_{n=1}^{\infty} \frac{1}{n} \left(1 - \cos \frac{n\pi t_0}{t_1} \right) \sin \frac{n\pi t}{t_1}.$$

11. From equations (1) and (2) in the text we obtain

$$b_n = 2 \int_{1/2}^1 (x - \tfrac{1}{2}) \sin n\pi x \, dx = -\frac{1}{n\pi} \cos n\pi - \frac{2}{n^2\pi^2} \sin \frac{n\pi}{2}.$$

$$f(x) \sim \sum_{n=1}^{\infty} \left[\frac{(-1)^{n+1}}{n\pi} - \frac{2}{n^2\pi^2} \sin \frac{n\pi}{2} \right] \sin n\pi x.$$

13. From equations (1) and (2) in the text we obtain

$$b_n = \frac{\mathbf{0}}{\pi} \int_0^\pi \cos 2x \sin nx \, dx, \quad b_2 = \frac{\mathbf{0}}{\pi} \int_0^\pi \frac{1}{2} \sin 4x \, dx = 0.$$

If $n \neq 2$ then $b_n = \frac{1}{\pi} \int_0^\pi [\sin (n+2)x + \sin (n-2)x] \, dx$. From this integral it follows that

$b_{2k} = 0$ and $b_{2k+1} = \frac{4}{\pi} \frac{2k+1}{(2k-1)(2k+3)}$. We get $f(x) \sim \frac{4}{\pi} \sum_{k=0}^{\infty} \frac{(2k+1) \sin [(2k+1)x]}{(2k-1)(2k+3)}$.

15. From equations (1) and (2) in the text we obtain

$$b_n = \frac{2}{c} \int_0^c e^{-x} \sin \frac{n\pi x}{c} \, dx = \frac{2n\pi [1 - (-1)^n e^{-c}]}{c^2 + n^2\pi^2}, \quad f(x) \sim \sum_{n=1}^{\infty} \frac{2n\pi [1 - (-1)^n e^{-c}]}{c^2 + n^2\pi^2} \sin \frac{n\pi x}{c}.$$

17. From equations (1) and (2) in the text we obtain

$$b_n = \frac{2}{c} \int_0^c \cosh kx \sin \frac{n\pi x}{c} \, dx = \frac{2n\pi [1 - (-1)^n \cosh kc]}{c^2 k^2 + n^2\pi^2},$$

$$f(x) \sim \sum_{n=1}^{\infty} \frac{2n\pi [1 - (-1)^n \cosh kc]}{c^2 k^2 + n^2\pi^2} \sin \frac{n\pi x}{c}.$$

19. From equations (1) and (2) in the text we obtain

$$b_n = \frac{2}{c} \int_0^c x^4 \sin \frac{n\pi x}{c} \, dx = \frac{2c^4}{\pi} \left[(-1)^{n+1} \left\{ \frac{1}{n} - \frac{12}{n^3\pi^2} - \frac{24}{n^5\pi^4} \right\} + \frac{24}{n^5\pi^4} \right],$$

$$f(x) \sim \frac{2c^4}{\pi} \sum_{n=1}^{\infty} \left[(-1)^{n+1} \left\{ \frac{1}{n} - \frac{12}{n^3\pi^2} - \frac{24}{n^5\pi^4} \right\} + \frac{24}{n^5\pi^4} \right] \sin \frac{n\pi x}{c}.$$

21. From equations (1) and (2) in the text we obtain

$$b_n = 2 \int_0^1 (x-1)^2 \sin n\pi x \, dx = \frac{2}{n^3 \pi^3} \left[n^2 \pi^2 - 2 + 2(-1)^n \right],$$

$$f(x) \sim \frac{2}{\pi^3} \sum_{n=1}^{\infty} \frac{n^2 \pi^2 - 2 + 2(-1)^n}{n^3} \sin nx.$$

22.5 Fourier Cosine Series

1. From equations (1) and (2) in the text we obtain

$$a_0 = \int_0^1 x \, dx + \int_1^2 (2-x) \, dx = 1,$$

$$a_n = \int_0^1 x \cos \frac{n\pi x}{2} \, dx + \int_1^2 (2-x) \cos \frac{n\pi x}{2} \, dx = \frac{4}{\pi^2 n^2} \left[2 \cos \frac{n\pi}{2} - 1 - (-1)^n \right],$$

$$f(x) \sim \frac{1}{2} + \frac{4}{\pi^2} \sum_{n=1}^{\infty} \frac{1}{n^2} \left[2 \cos \frac{n\pi}{2} - 1 - (-1)^n \right] \cos \frac{n\pi x}{2}.$$

3. From equations (1) and (2) in the text we obtain

$$a_0 = 2 \int_0^1 (x-1)^2 \, dx = \frac{2}{3}, \quad a_n = 2 \int_0^1 (x-1)^2 \cos n\pi x \, dx = \frac{4}{\pi^2 n^2},$$

$$f(x) \sim \frac{1}{3} + \frac{4}{\pi^2} \sum_{n=1}^{\infty} \frac{\cos n\pi x}{n^2}.$$

5. From equations (1) and (2) in the text we obtain

$$a_0 = \frac{2}{c} \int_0^c (c-x) \, dx = c, \quad a_n = \frac{2}{c} \int_0^c (c-x) \cos \frac{n\pi x}{c} \, dx = \frac{2c \left[1 - (-1)^n \right]}{n^2 \pi^2},$$

$$a_{2k} = 0, \quad a_{2k+1} = \frac{4c}{(2k+1)^2 \pi^2}, \quad f(x) \sim \frac{c}{2} + \frac{4c}{\pi^2} \sum_{k=0}^{\infty} \frac{1}{(2k+1)^2} \cos \frac{(2k+1)\pi x}{c}.$$

7. From equations (1) and (2) in the text we obtain

$$a_0 = 2 \int_{1/2}^1 (x - \tfrac{1}{2}) \, dx = \frac{1}{4}, \quad a_n = 2 \int_{1/2}^1 (x - \tfrac{1}{2}) \cos n\pi x \, dx = \frac{2}{\pi^2 n^2} \left(\cos n\pi - \cos \frac{n\pi}{2} \right),$$

$$f(x) \sim \frac{1}{8} + \frac{2}{\pi^2} \sum_{n=1}^{\infty} \frac{1}{n^2} \left(\cos n\pi - \cos \frac{n\pi}{2} \right) \cos n\pi x.$$

9. From equations (1) and (2) in the text we obtain

$$a_0 = \frac{2}{\pi} \int_0^\pi \cos 2x \, dx = 0, \quad a_n = \frac{2}{\pi} \int_0^\pi \cos 2x \cos nx \, dx = \frac{1}{\pi} \int_{-\pi}^\pi \cos 2x \cos nx \, dx.$$

But on the interval $-\pi < x < \pi$, $\cos 2x$ is orthogonal to $\cos nx$ for $n \neq 2$. Hence $a_n = 0$ for $n \neq 2$. Moreover $a_2 = \dfrac{1}{\pi} \displaystyle\int_{-\pi}^{\pi} \cos^2 2x \, dx = 1$, so that $f(x) \sim \cos 2x$.

11. From equations (1) and (2) in the text we obtain

$$a_0 = \frac{2}{c} \int_0^{c/2} x \, dx = \frac{c}{4}, \quad a_n = \frac{2}{c} \int_0^{c/2} x \cos \frac{n\pi x}{c} \, dx = \frac{c}{\pi^2 n^2} \left[n\pi \sin \frac{n\pi}{2} - 2 \left(1 - \cos \frac{n\pi}{2} \right) \right],$$

$$f(x) \sim \frac{c}{8} + \frac{c}{\pi^2} \sum_{n=1}^{\infty} \frac{1}{n^2} \left[n\pi \sin \frac{n\pi}{2} - 2 \left(1 - \cos \frac{n\pi}{2} \right) \right] \cos \frac{n\pi x}{c}.$$

13. From equations (1) and (2) in the text we obtain

$$a_0 = \frac{2}{c} \int_0^{c} \cosh kx \, dx = \frac{2 \sinh kc}{kc}, \quad a_n = \frac{2}{c} \int_0^{c} \cosh kx \cos \frac{n\pi x}{c} \, dx = \frac{\sinh kc \cdot 2kc(-1)^n}{k^2 c^2 + n^2 \pi^2},$$

$$f(x) \sim \frac{\sinh kc}{kc} + \sinh kc \sum_{n=1}^{\infty} \frac{2kc(-1)^n}{k^2 c^2 + n^2 \pi^2} \cos \frac{n\pi x}{c}.$$

15. From equations (1) and (2) in the text we obtain

$$a_0 = \frac{2}{c} \int_0^{c} x^3 \, dx = \frac{c^3}{2}, \quad a_n = \frac{2}{c} \int_0^{c} x^3 \cos \frac{n\pi x}{c} \, dx = \frac{6c^3}{\pi^2} \left[\frac{(-1)^n}{n^2} + \frac{2}{\pi^2} \cdot \frac{1 - (-1)^n}{n^4} \right],$$

$$f(x) \sim \frac{c^3}{4} + \frac{6c^3}{\pi^2} \sum_{n=1}^{\infty} \left[\frac{(-1)^n}{n^2} + \frac{2}{\pi^2} \cdot \frac{1 - (-1)^n}{n^4} \right] \cos \frac{n\pi x}{c}.$$

Chapter 23

Boundary Value Problems

23.1 The One-Dimensional Heat Equation

1. The boundary value problem to be solved is

$$\frac{\partial u}{\partial t} = h^2 \frac{\partial^2 u}{\partial x^2}, \quad \text{for } 0 < x < c,\ 0 < t;$$

$$\text{As } t \to 0^+,\ u \to u_0, \qquad \text{for } 0 < x < c,$$

$$\text{As } x \to 0^+,\ u \to 0, \qquad \text{for } 0 < t;$$

$$\text{As } x \to c^-,\ u \to 0, \qquad \text{for } 0 < t.$$

The standard procedure for separating the variables will be used. We seek solutions of the form $u = X(x)T(t)$. Thus we require that $\dfrac{T'}{h^2 T} = \dfrac{X''}{X} = k$. We now seek nontrivial (not identically zero) solutions of the system

$$X'' - kX = 0, \quad X(0) = X(c) = 0.$$

If $k = 0$ then $X = b_1 + b_2 x$.

$$X(0) = 0 \ \text{ implies that } b_1 = 0.$$
$$X(c) = 0 \ \text{ implies that } b_1 + b_2 c = 0.$$

Hence $b_1 = b_2 = 0$ and $X = 0$. That is there are no nontrivial solutions in the case where $k = 0$.

If $k = \beta^2 > 0$ then $X = c_1 \sinh \beta x + c_2 \cosh \beta x$.

$$X(0) = 0 \ \text{ implies that } c_2 = 0.$$
$$X(c) = 0 \ \text{ implies that } c_1 \sinh \beta c + c_2 \cosh \beta c = 0.$$

Hence $c_1 = c_2 = 0$ and there are no nontrivial solutions in the case where k is positive.

If $k = -\beta^2 < 0$ then $X = c_1 \sin \beta x + c_2 \cos \beta x$.

$$X(0) = 0 \ \text{ implies that } c_2 = 0.$$
$$X(c) = 0 \ \text{ implies that } c_1 \sin \beta c + c_2 \cos \beta c = 0.$$

We can now obtain nontrivial solutions by choosing $\beta c = n\pi$ or $\beta = n\pi/c$. These solutions are
$X(x) = c_n \sin \dfrac{n\pi x}{c}$.

The choice of the separation constant $k = -\beta^2 = \dfrac{-n^2\pi^2}{c^2}$ having been determined by the boundary conditions in x, we must now find functions that satisfy the differential equation $T' + \dfrac{h^2 n^2 \pi^2}{c^2} T = 0$. These functions are $T(t) = b_n \exp\left(-h^2 n^2 \pi^2 t/c^2\right)$.

We now have an infinite set of product functions

$$u(x,\ t) = A_n \sin \frac{n\pi x}{c} \exp\left(\frac{-h^2 n^2 \pi^2 t}{c^2}\right),$$

each of which satisfies the required boundary conditions in x, but none satisfying the initial condition in t. We consider the infinite series with undetermined coefficients A_n

$$u(x,\ t) = \sum_{n=1}^{\infty} A_n \sin \frac{n\pi x}{c} \exp\left(\frac{-h^2 n^2 \pi^2 t}{c^2}\right) \tag{23.1}$$

and attempt to choose the A_n in a manner that will satisfy the initial value condition. Setting $t = 0$, we see that we want

$$u_0 = \sum_{n=1}^{\infty} A_n \sin \frac{n\pi x}{c}, \quad \text{for } 0 < x < c.$$

That is we want the A_n to be the Fourier sine coefficients for the constant function u_0. Thus

$$A_n = \frac{2}{c} \int_0^c u_0 \sin \frac{n\pi x}{c}\ dx = \frac{2u_0}{n\pi}(1 - \cos n\pi).$$

It follows that $A_{2k} = 0$ and $A_{2k+1} = \dfrac{4u_0}{\pi(2k+1)}$. Using these coefficients we obtain the final solution

$$u(x,\ t) = \frac{4u_0}{\pi} \sum_{k=0}^{\infty} \frac{1}{2k+1} \sin \frac{(2k+1)\pi x}{c} \cdot \exp\left[\frac{-h^2(2k+1)^2 \pi^2 t}{c^2}\right].$$

3. If u is independent of t the heat equation reduces to $u'' = 0$. Thus $u(x) = a_1 + a_2 x$. The conditions $u(0) = A$ and $u(c) = 0$ force us to choose a_1 and a_2 so that $a_1 = A$ and $a_1 + a_2 c = 0$. Thus $a_1 = A$ and $a_2 = -A/c$. It follows that $u(x) = A(c - x)/c$.

5. The solution proceeds as in Exercise 4. We want a solution of the form

$$u(x,\ t) = A(c - x)/c + \sum_{n=1}^{\infty} b_n \exp\left[-\left(\frac{hn\pi}{c}\right)^2 t\right] \sin \frac{n\pi x}{c}.$$

Each term of the right hand side is a solution of the heat equation. Moreover the two boundary conditions are satisfied by each term. We need to choose b_n so that

$$f(x) = \frac{A(c-x)}{c} + \sum_{n=1}^{\infty} b_n \sin \frac{n\pi x}{c}.$$

That is the constants b_n must be the coefficients of the Fourier sine series expansion of the function $f(x) - A(c-x)/c$ on the interval $0 < x < c$.

7. The problem to be solved is

$$\frac{\partial u}{\partial t} = \frac{.4}{\pi^2} \frac{\partial^2 u}{\partial x^2}, \quad \text{for } 0 < x < 20, \ 0 < t;$$

$$\text{As } t \to 0^+, \ u \to 120, \qquad\qquad \text{for } 0 < x < 15,$$

$$\text{As } t \to 0^+, \ u \to 30, \qquad\qquad \text{for } 15 < x < 20;$$

$$\text{As } x \to 0^+, \ u \to 30, \qquad\qquad \text{for } 0 < t;$$

$$\text{As } x \to 20^-, \ u \to 30, \qquad\qquad \text{for } 0 < t.$$

We choose a new variable $v = u - 30$. The new problem becomes

$$\frac{\partial v}{\partial t} = \frac{.4}{\pi^2} \frac{\partial^2 v}{\partial x^2}, \quad \text{for } 0 < x < 20, \ 0 < t;$$

$$\text{As } t \to 0^+, \ v \to 90, \qquad\qquad \text{for } 0 < x < 15,$$

$$\text{As } t \to 0^+, \ v \to 0, \qquad\qquad \text{for } 15 < x < 20;$$

$$\text{As } x \to 0^+, \ v \to 0, \qquad\qquad \text{for } 0 < t;$$

$$\text{As } x \to 20^-, \ v \to 0, \qquad\qquad \text{for } 0 < t.$$

This problem is much like the problem in Exercise 1. We must replace h^2 with $\frac{.4}{\pi^2}$, c with 20, and the constant initial temperature function with a step function. Equation (23.1) in Exercise 1 becomes

$$v(x, \ t) = \sum_{n=1}^{\infty} A_n \exp\left(\frac{-n^2 t}{1000}\right) \sin \frac{n\pi x}{20}.$$

The initial condition now requires that $f(x) \sim \sum_{n=1}^{\infty} b_n \sin \frac{n\pi x}{20}$, where $f(x)$ is the step function of this problem. Thus

$$b_n = \frac{2}{20} \int_0^{20} f(x) \sin \frac{n\pi x}{20} \, dx = \frac{1}{10} \int_0^{15} 90 \sin \frac{n\pi x}{20} \, dx = \frac{180}{n\pi} \left[1 - \cos \frac{3n\pi}{4}\right].$$

It follows that

$$u(x, \ t) = 30 + v(x, \ t) = 30 + \frac{180}{\pi} \sum_{n=1}^{\infty} \frac{[1 - \cos(3n\pi/4)]}{n} \cdot \exp\left(\frac{-n^2 t}{1000}\right) \sin \frac{n\pi x}{20}.$$

9. We have for $t = 0$, $\dfrac{4A}{\pi} \displaystyle\sum_{n=1}^{\infty} \dfrac{1}{n} \sin^3 \dfrac{n\pi}{3} = A$. But $\sin \dfrac{3k\pi}{3} = 0$ and $\sin \dfrac{(3k+1)\pi}{3} = \dfrac{(-1)^k\sqrt{3}}{2} =$ $\sin \dfrac{(3k+2)\pi}{3}$. Thus

$$\sum_{n=1}^{\infty} \frac{1}{n} \sin^3 \frac{n\pi}{3} = \sum_{k=0}^{\infty} \left[\frac{(-1)^k 3\sqrt{3}}{(3k+1)8} + \frac{(-1)^k 3\sqrt{3}}{(3k+2)8} \right] = \sum_{k=0}^{\infty} \frac{(-1)^k 9\sqrt{3}}{8} \cdot \frac{2k+1}{(3k+1)(3k+2)}.$$

It follows that $\displaystyle\sum_{k=0}^{\infty} \frac{(-1)^k(2k+1)}{(3k+1)(3k+2)} = \frac{2\pi}{9\sqrt{3}}$.

11. The problem can be interpreted as that of heat conduction in a slab of width c. The face at $x = c$ is held at temperature zero while the face at $x = 0$ is insulated. The initial temperature in the slab is given by $f(x)$. We let $u(x, t) = X(x)T(t)$ in the heat equation and obtain $\dfrac{T'}{h^2 T} = \dfrac{X''}{X} = a$. The differential equation in X is accompanied by conditions $X'(0) = 0$ and $X(c) = 0$. As in Exercise 1 the cases where $a = 0$ and $a > 0$ have only the trivial solutions $X = 0$. When $a = -\alpha^2$ the differential equation has the solutions $X = A\cos \alpha x + B \sin \alpha x$, so that $X' = -A\alpha \sin \alpha x + B\alpha \cos \alpha x$. The condition $X'(0) = 0$ implies that $B = 0$. The condition that $X(c) = 0$ forces us to choose α so that $\cos \alpha c = 0$. That is $\alpha = \dfrac{(2k+1)\pi}{2c}$ for $k = 0, 1, 2, \cdots$, and $X = A_k \cos \dfrac{(2k+1)\pi x}{2c}$. Now the differential equation in T becomes $T' + \dfrac{h^2(2k+1)^2\pi^2}{4c^2}T = 0$, so that $T = B_k \exp\left[\dfrac{-h^2(2k+1)^2\pi^2 t}{4c^2} \right]$. We therefore have a solution of the form

$$u(x, t) = \sum_{k=0}^{\infty} b_k \exp\left[\frac{-h^2(2k+1)^2\pi^2 t}{4c^2} \right] \cos \frac{(2k+1)\pi x}{2c}.$$

The initial condition forces us to choose b_k so that $f(x) \sim \displaystyle\sum_{k=0}^{\infty} b_k \cos \dfrac{(2k+1)\pi x}{2c}$.

23.4   Heat Conduction in a Sphere

1. We first solve the problem of equations (8) through (11). Separation of the variables can be accomplished by substituting $v(\rho, t) = Q(\rho)T(t)$ to obtain $\dfrac{T'}{h^2 T} = \dfrac{Q''}{Q} = k$. The boundary value problem in Q becomes

$$Q'' - kQ = 0, \quad \text{as } \rho \to R^-, \ Q \to 0, \text{ and as } \rho \to 0^+, \ \frac{Q}{\rho} \to \text{a limit.}$$

If $k = 0$ then $Q = c_1 + c_2\rho$ and the boundary conditions force us to satisfy $c_1 + c_2 R = 0$ and as $\rho \to 0$, $\dfrac{c_1}{\rho} + c_2$ has a limit. These equations can be satisfied only if $c_1 = c_2 = 0$.

If $k = \alpha^2 > 0$ we then obtain $Q = c_3 \cosh \alpha\rho + c_4 \sinh \alpha\rho$ and the boundary conditions require that $c_3 \cosh \alpha R + c_4 \sinh \alpha R = 0$ and that $\dfrac{c_3 \cosh \alpha\rho + c_4 \sinh \alpha\rho}{\rho}$ have a limit as $\rho \to 0$. This second condition forces us to choose $c_3 = 0$ and now the first equation forces $c_4 = 0$.

If $k^2 = -\beta^2$ the solution of the differential equation takes the form $Q = c_5 \cos \beta\rho + c_6 \sin \beta\rho$. The condition as $\rho \to 0$ again forces the choice $c_5 = 0$. Finally $Q(R) = 0$ requires that $\beta = \dfrac{n\pi}{R}$ and we get $Q = c_n \sin \dfrac{n\pi\rho}{R}$. The equation in T is now $T' + \dfrac{h^2 n^2 \pi^2}{R^2} T = 0$ with solutions $T = d_n \exp\left[\dfrac{-h^2 n^2 \pi^2 t}{R^2}\right]$. This gives us

$$v(\rho, t) = \sum_{n=1}^{\infty} b_n \exp\left[\frac{-h^2 n^2 \pi^2 t}{R^2}\right] \sin \frac{n\pi\rho}{R}.$$

We now force the condition in equation (11) and need $\rho f(\rho) = \sum_{n=1}^{\infty} b_n \sin \dfrac{n\pi\rho}{R}$. Thus b_n must be the coefficients of the Fourier sine series for $\rho f(\rho)$. That is

$$b_n = \frac{2}{R} \int_0^R \rho f(\rho) \sin \frac{n\pi\rho}{R} \, d\rho.$$

FInally with these values for b_n

$$u(\rho, t) = \frac{v(\rho, t)}{\rho} = \frac{1}{\rho} \sum_{n=1}^{\infty} b_n \exp\left[\frac{-h^2 n^2 \pi^2 t}{R^2}\right] \sin \frac{n\pi\rho}{R}.$$

23.5 The Simple Wave Equation

1. The problem to be solved is

$$\frac{\partial^2 y}{\partial t^2} = a^2 \frac{\partial^2 y}{\partial x^2}, \quad \text{for } 0 < x < c, \ 0 < t;$$

$$\text{As } t \to 0^+, \ y \to f(x), \qquad \text{for } 0 < x < c;$$

$$\text{As } t \to 0^+, \ \frac{\partial y}{\partial t} \to 0, \qquad \text{for } 0 < x < c;$$

$$\text{As } x \to 0^+, \ y \to 0, \qquad \text{for } 0 < t;$$

$$\text{As } x \to c^-, \ y \to 0, \qquad \text{for } 0 < t.$$

If we seek product solutions of the differential equation of the form $y(x, t) = X(x)T(t)$, we have $\dfrac{T''}{a^2 T} = \dfrac{X''}{X} = k$. The two boundary conditions now require that we find nontrivial solutions of the system

$$X'' - kX = 0, \quad X(0) = X(c) = 0.$$

The only nontrivial solutions occur when $k = -n^2\pi^2/c^2$ and are of the form $\sin n\pi x/c$. That particular choice of the separation constant k requires us to find solutions of the system

$$T'' + \frac{n^2\pi^2 a^2}{c^2}T = 0, \quad T'(0) = 0.$$

These solutions are of the form $\cos n\pi a t/c$. Thus the desired product solutions are of the form $A_n \sin \dfrac{n\pi x}{c} \cos \dfrac{n\pi a t}{c}$. We are therefore led to consider the series

$$y(x,\,t) = \sum_{n=1}^{\infty} A_n \sin \frac{n\pi x}{c} \cos \frac{n\pi a t}{c}.$$

In order to satisfy the remaining initial condition we need to choose A_n in such a way that

$$f(x) = \sum_{n=1}^{\infty} A_n \sin \frac{n\pi x}{c}, \quad \text{for } 0 < x < c. \tag{23.2}$$

That is, we require that

$$A_n = \frac{2}{c} \int_0^c f(x) \sin \frac{n\pi x}{c}\, dx. \tag{23.3}$$

For the particular $f(x)$ of this problem we have

$$A_n = \frac{2}{c} \int_0^{c/2} x \sin \frac{n\pi x}{c}\, dx + \frac{2}{c} \int_{c/2}^{c} (c - x) \sin \frac{n\pi x}{c}\, dx = \frac{4c}{n^2\pi^2} \sin \frac{n\pi}{2}.$$

It follows that $A_{2k} = 0$ and $A_{2k+1} = \dfrac{4c(-1)^k}{(2k+1)^2\pi^2}$. We finally have

$$y(x,\,t) = \frac{4c}{\pi^2} \sum_{k=0}^{\infty} \frac{(-1)^k}{(2k+1)^2} \sin \frac{(2k+1)\pi x}{c} \cos \frac{(2k+1)\pi a t}{c}.$$

3. From equations (22.1) and (22.2) of Exercise 1 we have

$$A_n = \frac{2}{c} \int_0^{c/4} x \sin \frac{n\pi x}{c}\, dx + \frac{2}{c} \int_{c/4}^{3c/4} \frac{c}{4} \sin \frac{n\pi x}{c}\, dx + \frac{2}{c} \int_{3c/4}^{c} (c - x) \sin \frac{n\pi x}{c}\, dx$$

$$= \frac{2c}{n^2\pi^2} \left(\sin \frac{n\pi}{4} + \sin \frac{3n\pi}{4} \right),$$

$$y = \frac{2c}{\pi^2} \sum_{n=1}^{\infty} \frac{1}{n^2} \left(\sin \frac{n\pi}{4} + \sin \frac{3n\pi}{4} \right) \cos \frac{n a\pi t}{c} \sin \frac{n\pi x}{c}.$$

5. From equations (22.1) and (22.2) of Exercise 1 we have

$$A_n = \frac{2}{c} \int_0^{c/4} x \sin \frac{n\pi x}{c}\, dx + \frac{2}{c} \int_{c/4}^{c/2} (\tfrac{1}{2}c - x) \sin \frac{n\pi x}{c}\, dx = \frac{2c}{n^2\pi^2} \left(2\sin \frac{n\pi}{4} - \sin \frac{n\pi}{2} \right),$$

$$y = \frac{2c}{\pi^2} \sum_{n=1}^{\infty} \frac{1}{n^2} \left(2\sin \frac{n\pi}{4} - \sin \frac{n\pi}{2} \right) \cos \frac{n a\pi t}{c} \sin \frac{n\pi x}{c}.$$

7. This is like Exercise 1 except that the system in T becomes

$$T'' + \frac{n^2\pi^2 a^2}{c^2} T = 0, \quad T(0) = 0,$$

which has solutions $\sin \dfrac{n\pi a t}{c}$ so that $y(x,\,t) = \displaystyle\sum_{n=1}^{\infty} A_n \sin \dfrac{n\pi x}{c} \sin \dfrac{n\pi a t}{c}$. Now we need to satisfy

the remaining condition at $t = 0$, namely $\dfrac{\partial y}{\partial t} = \displaystyle\sum_{n=1}^{\infty} \dfrac{n\pi a A_n}{c} \sin \dfrac{n\pi x}{c} \cos \dfrac{n\pi a t}{c}$ must approach

$\phi(x)$ as $t \to 0^+$. Thus we need $\phi(x) = \displaystyle\sum_{n=1}^{\infty} \dfrac{n\pi a A_n}{c} \sin \dfrac{n\pi x}{c}$. This can be accomplished by

choosing

$$A_n = \frac{c}{n\pi a} \cdot \frac{2}{c} \int_{c/3}^{2c/3} v_0 \sin \frac{n\pi x}{c}\, dx = \frac{2v_0 c}{an^2\pi^2}\left[\cos\frac{n\pi}{3} - \cos\frac{2n\pi}{3}\right].$$

Thus

$$y(x,\,t) = \frac{2v_0 c}{\pi^2 a} \sum_{n=1}^{\infty} \frac{1}{n^2}\left[\cos\frac{n\pi}{3} - \cos\frac{2n\pi}{3}\right]\sin\frac{n\pi a t}{c}\sin\frac{n\pi x}{c}.$$

9. This problem is like Exercise 7 in that

$$y(x,\,t) = \sum_{n=1}^{\infty} B_n \sin\frac{n\pi x}{c}\sin\frac{n\pi a t}{c},$$

$$\frac{\partial y}{\partial t} = \sum_{n=1}^{\infty} B_n \frac{n\pi a}{c}\sin\frac{n\pi x}{c}\cos\frac{n\pi a t}{c}.$$

We want $\phi(x) = \displaystyle\sum_{n=1}^{\infty} B_n \dfrac{n\pi a}{c}\sin\dfrac{n\pi x}{c}$; that is we must choose

$$\frac{n\pi a}{c} B_n = \frac{2}{c} \int_0^c \phi(x)\sin\frac{n\pi x}{c}\, dx.$$

23.6 Laplace's Equation in Two Dimensions

1. This is the problem of equations (2) to (6) in the text. We seek product solutions of the form $u(x,\,y) = X(x)Y(y)$. Separation of the variables leads to $\dfrac{X''}{X} = -\dfrac{Y''}{Y} = k$. Equations (3) and (4) together with this differential equation in X require that

$$X'' - kX = 0, \quad X(0) = X(a) = 0.$$

The only nontrivial solutions of this system come from taking $k = -n^2\pi^2/a^2$. These solutions are of the form $\sin(n\pi x/a)$.

If we use these same valus for the separation constant k, we must solve the following system in the variable Y:

$$Y'' - \frac{n^2\pi^2}{a^2}Y = 0, \quad Y(0) = 0.$$

The solutions of this system are $\sinh(n\pi y/a)$. We therefore have an infinite set of product solutions, $\sin(n\pi x/a)\sinh(n\pi y/a)$, each of which satisfies equations (2) to (5), but none of which satisfies equation (6).

We now consider the infinite series

$$u(x, y) = \sum_{n=1}^{\infty} c_n \sin\frac{n\pi x}{a}\sinh\frac{n\pi y}{a},$$

and attempt to determine the constants c_n in order to satisfy the remaining condition (6). We therefore require that

$$f(x) = \sum_{n=1}^{\infty} c_n \sinh\frac{n\pi b}{a}\sin\frac{n\pi x}{a}.$$

This may be accomplished by choosing

$$c_n \sinh\frac{n\pi b}{a} = \frac{2}{a}\int_0^a f(x)\sin\frac{n\pi x}{a}\,dx.$$

3. This is the same problem as Exercise 1 with $f(x) = 1$ for $0 < x < \frac{1}{2}a$, $f(x) = 0$ for $\frac{1}{2}a < x < a$. We need

$$c_n = \frac{2}{a\sinh(n\pi b/a)}\int_0^{a/2}\sin\frac{n\pi x}{a}\,dx = \frac{2[1-\cos(n\pi/2)]}{n\pi\sinh(n\pi b/a)},$$
$$u = \frac{2}{\pi}\sum_{n=1}^{\infty}\frac{[1-\cos(n\pi/2)]}{n\sinh(n\pi b/a)}\sinh\frac{n\pi y}{a}\sin\frac{n\pi x}{a}.$$

5. Separation of the variables as in Exercise 1 gives us $\dfrac{X''}{X} = \dfrac{-Y''}{Y} = c$. This time however the boundary value problem in X is

$$X'' - cX = 0, \quad X'(0) = X(a) = 0.$$

Nontrivial solutions can occur only if $c = -\beta^2 < 0$, $X = c_1\cos\beta x + c_2\sin\beta x$, and also $X' = -c_1\beta\sin\beta x + c_2\beta\cos\beta x$. The condition $X'(0) = 0$ causes us to choose $c_2 = 0$ and then $X(a) = 0$ requires that $\cos\beta a = 0$. This can happen only if $\beta = \dfrac{(2k+1)\pi}{2a}$ and for these choices of β we have $X = A_n\cos\dfrac{(2k+1)\pi x}{2a}$. We now have for the equation in Y

$$Y'' - \frac{(2k+1)^2\pi^2}{4a^2}Y = 0, \quad Y(0) = 0.$$

Thus $Y = B_n \sinh \dfrac{(2k+1)\pi y}{2a}$. This leads to the series of products

$$u(x, y) = \sum_{k=0}^{\infty} C_n \sinh \frac{(2k+1)\pi y}{2a} \cos \frac{(2k+1)\pi x}{2a}.$$

Now we require that $u(x, b) = 1$, so that

$$1 = \sum_{k=0}^{\infty} C_n \sin \frac{(2k+1)\pi b}{2a} \cos \frac{(2k+1)\pi x}{2a}.$$

It follows that we must take

$$C_n = \frac{2}{a \sinh\left[(2k+1)\pi b/(2a)\right]} \int_0^a \cos \frac{(2k+1)\pi x}{2a}\, dx = \frac{4(-1)^k}{\pi(2k+1) \sinh\left[(2k+1)\pi b/(2a)\right]}.$$

Finally we have

$$u(x, y) = \frac{4}{\pi} \sum_{k=0}^{\infty} \frac{(-1)^k}{(2k+1) \sinh\left[(2k+1)\pi b/(2a)\right]} \sinh \frac{(2k+1)\pi y}{2a} \cos \frac{(2k+1)\pi x}{2a}.$$

7. The solution of this problem is the same as that of Exercise 5 except that here we require $Y'(0) = 0$ instead of $Y(0) = 0$. That change will result in each sinh function being replaced by a cosh function. All other details remain the same. We get

$$u(x, y) = \frac{4}{\pi} \sum_{k=0}^{\infty} \frac{(-1)^k}{(2k+1) \cosh\left[(2k+1)\pi b/(2a)\right]} \cosh \frac{(2k+1)\pi y}{2a} \cos \frac{(2k+1)\pi x}{2a}.$$

9. Let the temperature at the center of the given square plate be the constant u_c. Now consider four separate plates, each having a different face held at temperature unity while the other three faces are held at temperature zero. For each such plate the temperature at the center will be u_c. Now if the four plates are superimposed we obtain a plate with all four boundaries held at temperature unity and the temperature at the center will be $4u_c$. But such a plate will have in the steady state a temperature of one at its center. Thus $u_c = \frac{1}{4}$. Now from Exercise 8, with $b = a$, we get

$$\frac{2}{\pi} \sum_{k=0}^{\infty} \frac{(-1)^k \operatorname{sech}\left[(2k+1)\pi/2\right]}{2k+1} = \frac{1}{4}.$$

It follows that

$$\sum_{k=0}^{\infty} \frac{(-1)^k \operatorname{sech}\left[(2k+1)\pi/2\right]}{2k+1} = \frac{\pi}{8}.$$

11. Because of the symmetry of the boundary conditions the temperature in the plate will be a function of r only. Hence the problem to be solved is

$$\frac{d^2u}{dr^2} + \frac{1}{r}\frac{du}{dr} = 0 \quad \text{for } a < r < b.$$

$$\text{As } r \to b^-, \ u \to B,$$

$$\text{As } r \to a^+, \ u \to A.$$

The general solution of the differential equation is $u = c_1 \ln r + c_2$. The boundary conditions require that $B = c_1 \ln b + c_2$ and $A = c_1 \ln a + c_2$, so that

$$u = \frac{(B-A)\ln r}{\ln(b/a)} + \frac{A\ln b - B\ln a}{\ln(b/a)} = \frac{B\ln(r/a) - A\ln(r/b)}{\ln(b/a)}.$$

13. The problem to be solved is

$$\frac{\partial^2 u}{\partial r^2} + \frac{1}{r}\frac{\partial u}{\partial r} + \frac{1}{r^2}\frac{\partial^2 u}{\partial \theta^2} = 0 \quad \text{for } 0 < r < R, \ 0 < \theta < \beta.$$

$$\text{As } \theta \to 0^+, \ u \to 0,$$

$$\text{As } \theta \to \beta^-, \ u \to 0,$$

$$\text{As } r \to R^-, \ u \to 1,$$

$$\text{As } r \to 0^+, \ u \text{ has a limit.}$$

We separate the variables by letting $u(r, \theta) = F(r)G(\theta)$. We get $\dfrac{r^2 F'' + r F'}{F} = \dfrac{-G''}{G} = c$. G must now satisfy

$$G'' + cG = 0, \quad G(0) = 0, \ G(\beta) = 0.$$

Nontrivial solutions occur only if $c = \dfrac{n^2\pi^2}{\beta^2}$ and they are of the form $G = A_n \sin \dfrac{n\pi\theta}{\beta}$. F must now satisfy

$$r^2 F'' + r F' - \frac{n^2\pi^2}{\beta^2} F = 0, \quad \text{and } F(0) \text{ exists.}$$

The general solution of this Cauchy type equation is $F(r) = c_1 r^{n\pi/\beta} + c_2 r^{-n\pi/\beta}$. To satisfy the condition at $r = 0$, we must take $c_2 = 0$. Thus $F(r) = B_n r^{n\pi/\beta}$. It follows that

$$u(r, \theta) = \sum_{n=1}^{\infty} C_n r^{n\pi/\beta} \sin \frac{n\pi\theta}{\beta}.$$

Now we require that $1 = \displaystyle\sum_{n=1}^{\infty} C_n R^{n\pi/\beta} \sin \dfrac{n\pi\theta}{\beta}$, so that

$$C_n = \frac{2}{\beta R^{n\pi/\beta}} \int_0^\beta \sin \frac{n\pi\theta}{\beta} \, d\theta = \frac{2(1 - \cos n\pi)}{n\pi R^{n\pi/\beta}}.$$

It follows that $C_{2k} = 0$ and $C_{2k+1} = \dfrac{4}{(2k+1)\pi R^{(2k+1)\pi/\beta}}$. Finally we have

$$u(r,\,\theta) = \frac{4}{\pi} \sum_{k=0}^{\infty} \left(\frac{r}{R}\right)^{(2k+1)\pi/\beta} \frac{\sin\left[(2k+1)\pi\theta/\beta\right]}{2k+1}.$$

Chapter 24

Additional Properties of the Laplace Transform

24.1   Power Series and Inverse Transforms

1. We use equations (4) and (7) and obtain

$$L\left\{\frac{\sin kt}{t}\right\} = L\left\{\sum_{n=0}^{\infty}\frac{(-1)^n k^{2n+1}t^{2n}}{(2n+1)!}\right\} = \sum_{n=0}^{\infty}\frac{(-1)^n k^{2n+1}(2n)!}{(2n+1)!\ s^{2n+1}}$$

$$= \sum_{n=0}^{\infty}\frac{(-1)^n k^{2n+1}}{(2n+1)s^{2n+1}} = \arctan\frac{k}{s}, \quad s > 0.$$

3. We use equations (6) and (10) and obtain

$$L\left\{\frac{\sinh kt}{t}\right\} = L\left\{\sum_{n=0}^{\infty}\frac{k^{2n+1}t^{2n}}{(2n+1)!}\right\} = \sum_{n=0}^{\infty}\frac{k^{2n+1}(2n)!}{(2n+1)!\ s^{2n+1}}$$

$$= \sum_{n=0}^{\infty}\frac{k^{2n+1}}{(2n+1)s^{2n+1}} = \frac{1}{2}\ln\frac{1+(k/s)}{1-(k/s)} = \frac{1}{2}\ln\frac{s+k}{s-k}, \quad s > k > 0.$$

5. We use equation (1) and obtain

$$F(t) = L^{-1}\left\{\frac{1}{s^3(1-e^{-2s})}\right\} = L^{-1}\left\{\sum_{n=0}^{\infty}\frac{e^{-2ns}}{s^3}\right\} = \frac{1}{2}\sum_{n=0}^{\infty}(t-2n)^2\alpha(t-2n).$$

$$F(5) = \frac{1}{2}\left[5^2\alpha(5) + 3^2\alpha(3) + 1^2\alpha(1)\right] = 17.5.$$

7. We use equation (1) and obtain

$$\phi(t) = L^{-1}\left\{\frac{3}{s^4\sinh(3s)}\right\} = L^{-1}\left\{\frac{6}{s^4(e^{3s}-e^{-3s})}\right\} = L^{-1}\left\{\frac{6e^{-3s}}{s^4(1-e^{-6s})}\right\}$$

$$= L^{-1}\left\{\sum_{n=0}^{\infty}\frac{6e^{-3s}e^{-6sn}}{s^4}\right\} = L^{-1}\left\{6\sum_{n=0}^{\infty}\frac{e^{-(6n+3)s}}{s^4}\right\} = \sum_{n=0}^{\infty}(t-6n-3)^3\alpha(t-6n-3).$$

$$\phi(10) = 7^3\alpha(7) + 1^3\alpha(1) = 344.$$

9. We use equation (1) and obtain

$$L^{-1}\left\{f(s)\tanh(cs)\right\} = L^{-1}\left\{f(s)\frac{e^{cs}-e^{-cs}}{e^{cs}+e^{-cs}}\right\} = L^{-1}\left\{f(s)\left[1-\frac{2e^{-2cs}}{1+e^{-2cs}}\right]\right\}$$

$$= L^{-1}\left\{f(s)\left[1-2\sum_{n=0}^{\infty}(-1)^n\exp\left(-2ncs-2cs\right)\right]\right\}$$

$$= L^{-1}\left\{f(s)\left[1+2\sum_{n=0}^{\infty}(-1)^{n+1}\exp\left[-2(n+1)cs\right]\right]\right\}$$

$$= L^{-1}\left\{f(s)\left[1+2\sum_{n=1}^{\infty}(-1)^n e^{-2cns}\right]\right\}$$

$$= F(t)+2\sum_{n=1}^{\infty}(-1)^n F(t-2nc)\alpha(t-nc).$$

11. There is a notational problem here. We use $Q(t)$ for the charge on the capacitor in the circuit and $Q(t, c)$ for the square-wave function. Note also that C is the capacitance and c is half the period of the square-wave function. If we let $i(t) = Q'(t)$ then the problem to be solved is

$$RQ'(t)+\frac{1}{C}Q(t) = EQ(t, c), \quad Q(0) = 0.$$

Using the Laplace transform on this equation together with equation (16) from Section 14.10 we obtain

$$Rsq(s)+\frac{1}{C}q(s) = \frac{E}{s}\tanh\left(\frac{cs}{2}\right),$$

$$q(s) = \frac{E/R}{s\left[s+(1/RC)\right]}\tanh\left(\frac{cs}{2}\right) = CE\left[\frac{1}{s}-\frac{1}{s+(1/RC)}\right]\tanh\left(\frac{cs}{2}\right).$$

To obtain the inverse transform we make use of the result in Exercise 9. We get

$$Q(t) = CE\left[1+2\sum_{n=1}^{\infty}(-1)^n\alpha(t-nc)\right]$$

$$-CE\left[\exp\left(\frac{-t}{RC}\right)+2\sum_{n=1}^{\infty}(-1)^n\exp\left\{\frac{-(t-nc)}{RC}\right\}\alpha(t-nc)\right].$$

The current in the circuit can now be found by a differentiation. We have

$$I(t) = \frac{E}{R}\left[\exp\left(\frac{-t}{RC}\right)+2\sum_{n=1}^{\infty}(-1)^n\exp\left\{\frac{-(t-nc)}{RC}\right\}\alpha(t-nc)\right].$$

24.2   The Error Function

1. Consider the graph of $y = \text{erf } x$. From equation (4) in the text we see that the slope of the graph is always positive. A second derivative of erf x shows that for $x > 0$ the graph is concave up. There is an inflection point at $x = 0$. From equation (7) we also conclude for $x > 0$ that the graph of erf x is asymptotic to the line $y = 1$. A similar argument for $x < 0$ yields the result that the graph is asymptotic to the line $y = -1$. Thus the entire graph lies between the lines $y = 1$ and $y = -1$.

3. Using l'Hopital's rule we have

$$\lim_{x \to 0} \frac{\text{erf } x}{x} = \lim_{x \to 0} \frac{2}{\sqrt{\pi}} \exp\left(-x^2\right) = \frac{2}{\sqrt{\pi}}.$$

5. Since $|\text{erfc } x| < 2$ it follows that for $m \leq 0$, $\lim_{x \to \infty} x^m \text{ erfc } x = 0$. The indeterminate form occurs if $m > 0$. In this case

$$\lim_{x \to \infty} x^m \text{ erfc } x = \lim_{x \to \infty} \frac{\text{erfc } x}{x^{-m}} = \lim_{x \to \infty} \frac{1 - \text{erf } x}{x^{-m}}$$

$$= \lim_{x \to \infty} \frac{(-2/\sqrt{\pi}) \exp\left(-x^2\right)}{-mx^{-m-1}} = \lim_{x \to \infty} \frac{2}{m\sqrt{\pi}} x^{m+1} \exp\left(-x^2\right).$$

But we know from Chapter 14 that polynomials are of exponential order so that this last limit is zero.

7. Using the suggestions made in the text we have

$$L^{-1}\left\{\frac{1}{1 + \sqrt{1+s}}\right\} = L^{-1}\left\{-\frac{1}{s} + \frac{1+s}{s\sqrt{1+s}}\right\} = L^{-1}\left\{-\frac{1}{s} + \frac{1}{s\sqrt{1+s}} + \frac{1}{\sqrt{1+s}}\right\}.$$

Now from equation (3) and the equation preceding equation (2) we have

$$L^{-1}\left\{\frac{1}{1 + \sqrt{1+s}}\right\} = -1 + \text{erf}(\sqrt{t}) + \frac{e^{-t}}{\sqrt{xt}} = \frac{e^{-t}}{\sqrt{xt}} - \text{erfc}(\sqrt{t}).$$

9. From Exercise 8 we have

$$L^{-1}\left\{\frac{1}{\sqrt{s}(\sqrt{s}+1)}\right\} = L^{-1}\left\{\frac{1}{\sqrt{s}} - \frac{1}{\sqrt{s}+1}\right\} = L^{-1}\left\{\frac{1}{\sqrt{s}}\right\} - L^{-1}\left\{\frac{1}{\sqrt{s}+1}\right\} = e^t \text{erfc}(\sqrt{t}).$$

Now we use equation (1) from Section 14.6 to obtain $\dfrac{1}{\sqrt{\pi t}} - L^{-1}\left\{\dfrac{1}{\sqrt{s}+1}\right\} = e^t \text{ erfc}(\sqrt{t})$. It follows that $L^{-1}\left\{\dfrac{1}{\sqrt{s}+1}\right\} = \dfrac{1}{\sqrt{\pi t}} - e^t \text{ erfc}(\sqrt{t})$.

11. We are given that $L\{\phi(t)\} = \text{erf}\left(\dfrac{1}{s}\right)$. From equation (5) in the text we get the series

$$L\{\phi(t)\} = \frac{2}{\sqrt{\pi}} \sum_{n=0}^{\infty} \frac{(-1)^n}{(2n+1)n!\, s^{2n+1}}.$$ Applying the inverse transform to both sides gives us

$$\phi(t) = \frac{2}{\sqrt{\pi}} \sum_{n=0}^{\infty} \frac{(-1)^n t^{2n}}{(2n+1)\, n!\, (2n)!}.$$ It follows that $\phi(\sqrt{t}) = \dfrac{2}{\sqrt{\pi}} \displaystyle\sum_{n=0}^{\infty} \frac{(-1)^n t^n}{(2n+1)!\, n!}$. Therefore

$$L\{\phi(\sqrt{t})\} = \frac{2}{\sqrt{\pi}} \sum_{n=0}^{\infty} \frac{(-1)^n}{(2n+1)!\, s^{n+1}} = \frac{2}{\sqrt{\pi}} \sum_{n=0}^{\infty} \frac{(-1)^n}{(2n+1)!\, (\sqrt{s})^{2n+2}} = \frac{2}{\sqrt{\pi s}} \sin\left(\frac{1}{\sqrt{s}}\right).$$

13. The derivation of equation (29) that is given in the text gives us a clue as to how to proceed here. We have

$$\text{csch}\left(x\sqrt{s}\right) = \frac{2}{\exp\left(x\sqrt{s}\right) - \exp\left(-x\sqrt{s}\right)} = \frac{2\exp\left(-x\sqrt{s}\right)}{1 - \exp\left(-2x\sqrt{s}\right)}$$

$$= 2\exp\left(-x\sqrt{s}\right) \sum_{n=0}^{\infty} \exp\left(-2nx\sqrt{s}\right) = 2 \sum_{n=0}^{\infty} \exp\left[-(2n+1)x\sqrt{s}\right].$$

It follows that

$$L^{-1}\left\{\frac{\text{csch}\left(x\sqrt{s}\right)}{s}\right\} = 2 \sum_{n=0}^{\infty} L^{-1}\left\{\frac{1}{s} \exp\left[-(2n+1)x\sqrt{s}\right]\right\}.$$

Now we use equation (24) from the text to obtain

$$L^{-1}\left\{\frac{\text{csch}\left(x\sqrt{s}\right)}{s}\right\} = 2 \sum_{n=0}^{\infty} \text{erfc}\left[\frac{(2n+1)x}{2\sqrt{t}}\right].$$

24.3   Bessel Functions

1. We expand the exponential function into a power series to get

$$L^{-1}\left\{\frac{1}{s^{n+1}} \exp\left(\frac{x}{s}\right)\right\} = L^{-1}\left\{\sum_{k=0}^{\infty} \frac{x^k}{k!\, s^{k+n+1}}\right\} = \sum_{k=0}^{\infty} \frac{x^k}{k!} L^{-1}\left\{\frac{1}{s^{k+n+1}}\right\}$$

$$= \sum_{k=0}^{\infty} \frac{x^k t^{k+n}}{k!\, (k+n)!} = \left(\frac{t}{x}\right)^{n/2} \sum_{k=0}^{\infty} \frac{(\sqrt{xt})^{2k+n}}{k!\, (k+n)!} = \left(\frac{t}{x}\right)^{n/2} I_n\left(2\sqrt{xt}\right).$$

Chapter 25

Partial Differential Equations: Transform Methods

25.1   Boundary Value Problems

1. We apply the Laplace transform in the variable t, letting $L\{y(x, t)\} = u(x, s)$. We get

$$\frac{du}{dx} + 4su = \frac{-8}{s^2}, \quad x \to 0, \quad u \to \frac{4}{s^3}.$$

This linear differential equation has general solution $u = \frac{-2}{s^3} + c_1(s)e^{-4sx}$. As $x \to 0$ we need to take $\frac{4}{s^3} = \frac{-2}{s^3} + c_1(s)$. That is $c_1(s) = \frac{6}{s^3}$. We therefore have $u = \frac{-2}{s^3} + \frac{6}{s^3}e^{-4xs}$. An inverse transform now yields the solution $y(x, t) = -t^2 + 3(t - 4x)^2\alpha(t - 4x)$.

3. We apply the Laplace transform in the variable t, letting $L\{y(x, t)\} = u(x, s)$. We get

$$\frac{du}{dx} + 4su - 4x = \frac{-8}{s^2}, \quad x \to 0, \quad u \to \frac{4}{s^3},$$

$$\frac{du}{dx} + 4su = 4x - \frac{8}{s^2}, \quad x \to 0, \quad u \to \frac{4}{s^3}.$$

This linear differential equation has general solution $u = \frac{x}{s} - \frac{1}{4s^2} - \frac{2}{s^3} + c_1(s)e^{-4sx}$. As $x \to 0$ we need to take $\frac{4}{s^3} = -\frac{1}{4s^2} - \frac{2}{s^3} + c_1(s)$. That is $c_1(s) = \frac{6}{s^3} + \frac{1}{4s^2}$. We therefore have $u = \frac{x}{s} - \frac{1}{4s^2} - \frac{2}{s^3} + \left[\frac{6}{s^3} + \frac{1}{4s^2}\right]e^{-4xs}$. An inverse transform now yields the solution $y(x, t) = x - \frac{1}{4}t - t^2 + \left[3(t - 4x)^2 + \frac{1}{4}(t - 4x)\right]\alpha(t - 4x)$.

5. We apply the Laplace transform in the variable t, letting $L\{y(x, t)\} = u(x, s)$. We get

$$\frac{d^2u}{dx^2} = 16(s^2u + 2), \quad x \to 0, \quad u \to \frac{1}{s^2}, \quad \lim_{x \to \infty} \text{ exists},$$

$$\frac{d^2u}{dx^2} - 16s^2u = 32, \quad x \to 0, \quad u \to \frac{1}{s^2}, \quad \lim_{x \to \infty} \text{ exists}.$$

This linear differential equation has general solution $u(x, s) = c_1(s)e^{4sx} + c_2(s)e^{-4sx} - \frac{2}{s^2}$. In order for $\lim_{x\to\infty}$ to exist we must take $c_1(s) = 0$. As $x \to 0$ we need to take $\frac{1}{s^2} = c_2(s) - \frac{2}{s^2}$. That is $c_2(s) = \frac{3}{s^2}$. We therefore have $u(x, s) = \frac{3}{s^2}e^{-4sx} - \frac{2}{s^2}$. An inverse transform now yields the solution $y(x, t) = 3(t - 4x)\alpha(t - 4x) - 2t$.

25.2 The Wave Equation

1. Direct application of the Laplace transform gives us the transformed system

$$s^2 u - s(x - x^2) = \frac{d^2 u}{dx^2}, \quad x \to 0^+, \ u \to 0, \ x \to 1^-, \ u \to 0.$$

The linear equation has as its general solution $u = \frac{x}{s} - \frac{x^2}{s} - \frac{2}{s^3} + c_1 e^{-sx} + c_2 e^{sx}$. The condition $x \to 0^+$, $u \to 0$ implies that $c_1 + c_2 = 2/s^3$. The condition $x \to 1^-$, $u \to 0$ implies that $e^{-s}c_1 + e^s c_2 = 2/s^3$. From these two equations we obtain

$$c_1 = \frac{2}{s^3(1 + e^{-s})} = \frac{2}{s^3}\sum_{n=0}^{\infty}(-1)^n e^{-ns} \quad \text{and} \quad c_2 = \frac{2e^{-s}}{s^3(1 + e^{-s})} = \frac{2}{s^3}\sum_{n=0}^{\infty}(-1)^n e^{-(n+1)s}.$$

Therefore

$$u(x, s) = \frac{x}{s} - \frac{x^2}{s} - \frac{2}{s^3} + \frac{2}{s^3}\sum_{n=0}^{\infty}(-1)^n e^{-(n+x)s} + \frac{2}{s^3}\sum_{n=0}^{\infty}(-1)^n e^{-(n+1-x)s}.$$

The inverse transform now yields

$$y(x, t) = x - x^2 - t^2 + \sum_{n=0}^{\infty}(-1)^n\left[(t - n - x)^2\alpha(t - n - x) + (t - n - 1 + x)^2\alpha(t - n - 1 + x)\right].$$

25.5 Diffusion in a Slab of Finite Width

1. Direct application of the Laplace transform gives us the transformed system

$$sw - 1 = \frac{d^2 w}{dx^2}, \quad x \to 0^+, \ w \to 0, \ x \to 1^-, \ \frac{dw}{dx} \to 0.$$

The linear equation has as its general solution $w = \frac{1}{s} + c_1 \sinh(x\sqrt{s}) + c_2 \cosh(x\sqrt{s})$. The condition $x \to 0^+$, $w \to 0$ implies that $c_2 = -1/s$. Thus

$$\frac{dw}{dx} = \sqrt{s}c_1 \cosh(x\sqrt{s}) - \frac{1}{\sqrt{s}}\sinh(x\sqrt{s}).$$

dition $x \to 1^-$, $dw/dx \to 0$ implies that $c_1 = (\tanh \sqrt{s})/s$. Thus

$$w(x, s) = \frac{1}{s} + \frac{\tanh \sqrt{s} \sinh (x\sqrt{s})}{s} - \frac{\cosh (x\sqrt{s})}{s} = \frac{1}{s} - \frac{\cosh [(x-1)\sqrt{s}]}{s \cosh \sqrt{s}}.$$

... argument on pages 475 and 476 in the text that produced equation (29) can be slightly altered to give the result

$$L^{-1}\left\{ \frac{\cosh (x\sqrt{s})}{s \cosh \sqrt{s}} \right\} = \sum_{n=0}^{\infty} (-1)^n \left[\operatorname{erfc}\left(\frac{1 - x + 2n}{2\sqrt{t}} \right) + \operatorname{erfc}\left(\frac{1 + x + 2n}{2\sqrt{t}} \right) \right].$$

Using this inverse transform with x replaced by $x - 1$ allows us to write

$$u(x, t) = 1 - \sum_{n=0}^{\infty} (-1)^n \left[\operatorname{erfc}\left(\frac{2n + 2 - x}{2\sqrt{t}} \right) + \operatorname{erfc}\left(\frac{2n + x}{2\sqrt{t}} \right) \right].$$

25.6 Diffusion in a Quarter-Infinite Solid

1. $u = \operatorname{erf}\left(\dfrac{x}{2\sqrt{t}} \right) \operatorname{erf}\left(\dfrac{y}{2\sqrt{t}} \right) = \dfrac{4}{\pi} \left[\displaystyle\int_0^{\frac{x}{2\sqrt{t}}} \exp\left(-\beta^2\right) d\beta \right] \left[\displaystyle\int_0^{\frac{y}{2\sqrt{t}}} \exp\left(-\alpha^2\right) d\alpha \right]$. Since the inte-

grand of each of these integrals is a positive decreasing function, the integrals themselves are positive monotone increasing functions of x and y respectively for any fixed $t > 0$. Moreover the number $\displaystyle\int_0^{\infty} \exp\left(-\beta^2\right) d\beta = \dfrac{\sqrt{\pi}}{2}$ is an upper bound for each integral. Hence for any fixed $x > 0$, $t > 0$ each of these integrals has value less than $\sqrt{\pi}/2$. It follows that $0 < u < 1$.

To Rhonda, Rachel, Alex, and Meghan, thank you for your loving support.
—MEW

To my mother, Frances Perkins Godwin; it is a wonderful life.
—HJM

Brief Table of Contents